AF581688

Leaching of Rare Earth Elements from Various Sources

Leaching of Rare Earth Elements from Various Sources

Editor

Kenneth N. Han

MDPI • Basel • Beijing • Wuhan • Barcelona • Belgrade • Manchester • Tokyo • Cluj • Tianjin

Editor
Kenneth N. Han
South Dakota School of Mines and Technology
USA

Editorial Office
MDPI
St. Alban-Anlage 66
4052 Basel, Switzerland

This is a reprint of articles from the Special Issue published online in the open access journal *Minerals* (ISSN 2075-163X) (available at: https://www.mdpi.com/journal/minerals/special_issues/LREEVS).

For citation purposes, cite each article independently as indicated on the article page online and as indicated below:

LastName, A.A.; LastName, B.B.; LastName, C.C. Article Title. *Journal Name* **Year**, *Volume Number*, Page Range.

ISBN 978-3-0365-2566-2 (Hbk)
ISBN 978-3-0365-2567-9 (PDF)

Contents

About the Editor

Kenneth N. Han obtained his B.S. and M.S. degrees from Seoul National University, an M.S. from the University of Illinois, and a Ph.D. from the University of California, Berkeley. He was with the Department of Chemical Engineering, Monash University, Melbourne, Australia from 1971 to 1980. In 1981, he joined South Dakota School of Mines and Technology (SDSMT). He was head of the Department of Metallurgical Engineering, Dean of the College of Materials Science, Engineering and Dean of Graduate Studies at SDSMT. He is currently a Distinguished Professor Emeritus at SDSMT. He has published more than 200 papers in national and international journals in the areas of hydrometallurgy, leaching kinetics, interfacial phenomena, corrosion, and manufacturing of nano-metal particles. He is a member of the National Academy of Engineering.

MDPI

Editorial

Editorial for Special Issue "Leaching of Rare Earth Elements from Various Sources"

Kenneth N. Han

Materials and Metallurgical Engineering, South Dakota School of Mines & Technology, Rapid City, SD 57701-3995, USA; kennethhydro@gmail.com

Rare earth elements (REEs) have become an important group of metals used in many high-tech industries, including high-strength magnets, plasma TVs, various military applications, and clean and efficient green energy industries. Unlike many commercially available metals, REEs are rarely concentrated in mineable ore deposits. The principal concentrations of REEs are associated with uncommon varieties of igneous rocks, alkaline rocks, and carbonatites [1,2]. As the urgent need and demand for REEs to fuel the high-tech industries are growing, and relatively REE-rich ore deposits are getting depleted, there have been considerable efforts in identifying low-grade ores, and also the possible recovering of REEs from recycling resources. These sources include clay minerals, and coal and coal byproducts, especially in the eastern United States [3,4]. There have been attempts made to recover REEs from wasted used magnets and other recycling streams [5].

Most of the traditional deposits containing REE-bearing minerals are refractory in nature due to the unusually strong binding energies of these metals with the surrounding media. As a result, the strategy involving the extraction of these elements requires careful consideration of the chemical and physical nature of the energy associated between these metals and the surrounding media [6,7].

Citation: Han, K.N. Editorial for Special Issue "Leaching of Rare Earth Elements from Various Sources". *Minerals* **2021**, *11*, 164. https://doi.org/10.3390/min11020164

Received: 27 January 2021
Accepted: 3 February 2021
Published: 4 February 2021

This Special Issue carries some important findings from the most current investigations into the leaching of REEs from various sources. There are 10 papers grouped into four categories: 1. Leaching from low-grade ores including coal, clay, soil, and red mud; 2. Pre-treatment and leaching; 3. Recycling; and 4. Precipitation and nano-hydrometallurgy. Four papers are dedicated to low-grade ores, such as coal and its derivates, clay minerals, soils, and red mud. The grade of the REEs of some of these ores ranges from 200 to 1000 ppm, with some remarkable advantages, including the easy leaching ability due to the simple ion-exchange mechanism [8,9]. Zhang et al. [8] described a comprehensive literature survey with relevant scientific and technological discussions. They covered a vast amount of literature on the recovery of REEs from coal-based sources, including coal refuse, coal combustion ash, and acid coal mine drainage. Some process flowsheets integrating various technologies, such as physical beneficiation, acid leaching, pre-leach roasting, and hydrothermal pretreatment, have been designed, and some of these have been tested at bench scale. High-purity rare earth concentrates have been successfully produced from coal refuse and acid coal mine drainage. A rare-earth pilot plant was also constructed and tested, enabling the continuous production of REEs from coal refuse.

Montrose et al. [10] investigated the leaching behavior of REEs from a coal seam underclay using an organic acid and in combination with inorganic acids. Some of the underclay associated with coal seams in the Lower Freeport, Middle Kittanning, and Pittsburgh formations contain REEs at 250–457 ppm. Clay minerals such as illite, halloysite, and kaolinite make up >55% of the total bulk mineralogy of the rock. Organic acid and ammonium sulfate are often used for clay-type minerals, whereon the leaching is carried out through the ion-exchange mechanism. However, the current coal seam clay materials seem to contain REEs mainly as refractory REE-phosphates, and therefore, only the partial recovery of REEs was possible. Chaikin et al. [11] made an attempt to improve the extraction of REEs from an intermediate product of the bauxite sintering process, especially from the electrostatic precipitator dust, and have concluded that the best results were obtained by a double leaching process of leaching first at 95 °C, and then again at 240 °C.

Most of the REE-bearing minerals are refractory in nature, and therefore, in order to extract REEs from them it is necessary to increase the acidity, or they have to be subjected to pre-treatment with lixiviants such as sulfuric acid or sodium hydroxide before subjecting them to leaching. Kim et al. [12] attempted to control the amounts of additives in the heat-treatment to optimize the effectiveness of the pre-heating process. Others have also attempted pre-heating before leaching [8,13]. Xiao et al. [13] attempted alkaline fusion at 900 °C, resulting in a better than 95% recovery of scandium in the subsequent leaching.

A future trend in recovering REEs from valuable resources involves recovering them from wasted products. Sarfo et al. [14] attempted to recover Nd, Pr, and Dy from rare earth magnets by sulfuric acid leaching, followed by the precipitation of these REEs by ammonium hydroxide and ammonium bifluoride. Preferential precipitation of the dissolved REEs from the leach liquor can be a key to the success of the overall leaching process. Researchers have demonstrated the effective precipitation of REEs using various precipitants [15,16]. The effect of the anions present in the leach liquor has been found to be significant in the precipitation process. For example, anions such as Cl^-, NO_3^-, and SO_4^{2-} derived from acids used in the leaching process display a significant effect on the precipitation [15]. Watts and Leong have hypothesized that the distribution coefficient, λ, may be approximated from thermodynamic terms, including the solubility product (K_{Sp}) of each rare earth oxalate and the stability constant (β_1) for the mono oxalato complex of each rare earth. The proposed model was used to calculate λ between pairs of rare earths [16]. Melo et al. [17] introduced magnetic nanohydrometallurgy to separate REE-ions using nano-sized iron oxide particles functionalized with ethylenediaminepropylsilane and diethylenediaminepentatcetic acid under a super magnetic field.

Acknowledgments: The guest editor wishes to sincerely acknowledge all authors, reviewers, the editor Anker He, and other editorial staff at *Minerals* for their kind and timely efforts to successfully complete this Special Issue.

Conflicts of Interest: The Guest Editor declares no conflict of interest.

References

1. Haskin, L.A. On rare-earth element behavior in igneous rocks. *Phys. Chem. Earth* **1979**, *11*, 175–189. [CrossRef]
2. Dostal, J. Rare Earth Element Deposits of Alkaline Igneous Rocks. *Resources* **2017**, *6*, 34. [CrossRef]
3. Haxel, G.B.; Hedrick, J.B.; Hendley, J.W. *Rare Earth Elements, Critical Resources for High Technology: U.S. Geological Survey Fact Sheet*; U.S. Dept. of the Interior, U.S. Geological Survey: Reston, VA, USA, 2002.
4. Hower, J.C.; Granite, E.J.; Mayfield, D.B.; Lewis, A.S.; Finkelman, R.B. Notes on contributions to the science of rare earth element enrichment in coal and coal combustion byproducts. *Minerals* **2016**, *6*, 32. [CrossRef]
5. Önal, M.A.R.; Borra, C.R.; Guo, M.; Blanpain, B.; Van Gerven, T. Hydrometallurgical recycling of NdFeB magnets: Complete leaching, iron removal and electrolysis. *J. Rare Earths* **2017**, *35*, 574–584. [CrossRef]
6. Gupta, C.K.; Krishnamurthy, N. Extractive metallurgy of rare earths. *Int. Mater. Rev.* **1992**, *37*, 197–248. [CrossRef]
7. Abhilash, S.S.; Sinha, M.K.; Padney, B.D. Metallurgical processes for the recovery and recycling of lanthanum from various resources—A review. *Hydrometallurgy* **2016**, *160*, 47–59.
8. Zhang, W.; Noble, A.; Yang, X.; Honaker, R. A Comprehensive review of rare earth elements recovery from coal-related materials. *Minerals* **2020**, *10*, 451. [CrossRef]
9. Chen, L.; Jin, X.; Chen, H.; He, Z.; Qiu, L.; Duan, H. Grain size distribution and clay minerals distinction of rare earth ores implication for studying of regional soil surveys, in situ conditions and simulated leaching. *Minerals* **2020**, *10*, 353. [CrossRef]
10. Montross, S.N.; Yang, J.; Britton, J.; McKoy, M.; Verba, C. Leaching of rare earth elements from Central Appalachian coal seam underclays. *Minerals* **2020**, *10*, 577. [CrossRef]
11. Chaikin, L.; Shoppert, A.; Valeev, D.; Loginova, I.; Napol'Skikh, J. Concentration of rare earth elements (Sc, Y, La, Ce, Nd, Sm) in red mud by leaching dust from electrostatic precipitators in bauxite sintering process. *Minerals* **2020**, *10*, 500. [CrossRef]
12. Kim, R.; Cho, H.; Jeong, J.; Kim, J.; Lee, S.; Chung, K.W.; Yoon, H.-S.; Kim, C.-J. Effect of sulfuric acid baking and caustic digestion enhancing rare earth elements recovery from a refractory ore. *Minerals* **2020**, *10*, 532. [CrossRef]
13. Xiao, J.X.; Peng, Y.; Ding, W.; Chen, T.; Zou, K.; Wang, Z. Recovering scandium from scandium rough concentrate using roasting-hydrolysis-leaching process. *Processes* **2020**, *8*, 365. [CrossRef]

14. Sarfo, P.; Frasz, T.; Das, A.; Young, C. Hydrometallurgical production of rare earth fluorides from recycled magnets and process optimization. *Minerals* **2020**, *10*, 340. [CrossRef]
15. Han, K.N. Characteristics of precipitation of rare earth elements with various precipitants. *Minerals* **2020**, *10*, 178. [CrossRef]
16. Watts, H.; Leong, Y.K. Predicting the logarithmic distribution factors for coprecipitation into an organometallic compound: Selective precipitation of rare earths into a mixed oxalate. *Minerals* **2020**, *10*, 712. [CrossRef]
17. de Melo, F.M.; Almeida, S.N.; Toma, H.E. Magnetic nanohydrometallurgy applied to lanthanide separation. *Minerals* **2020**, *10*, 530. [CrossRef]

Article

Predicting the Logarithmic Distribution Factors for Coprecipitation into an Organic Salt: Selection of Rare Earths into a Mixed Oxalate

Harry Watts [1,2,*] and Yee-Kwong Leong [1]

[1] Department of Chemical Engineering, University of Western Australia, Crawley, WA 6009, Australia; yeekwong.leong@uwa.edu.au
[2] Watts & Fisher Pty Ltd., Crawley, WA 6009, Australia
* Correspondence: harry@wattsandfisher.com

Received: 30 June 2020; Accepted: 6 August 2020; Published: 12 August 2020

Abstract: Thermodynamic modelling of a leaching system that involves concurrent precipitation depends on an understanding of how the metals distribute into the precipitate before an assessment of solubility can be made. It has been suggested in the past that a pair of rare earths (A and B) in solution will separate from each other by oxalate precipitation according to a logarithmic distribution coefficient (λ), determined by the kinetics of the precipitation. By contrast, the present study hypothesises that λ may be approximated from thermodynamic terms, including the solubility product (K_{Sp}) of each rare earth oxalate and the stability constant (β_1) for the mono-oxalato complex of each rare earth. The proposed model was used to calculate λ between pairs of rare earths. An experimental study was conducted to determine λ between selected pairs using homogenous precipitation through the hydrolysis of an oxalic acid ester, with fairly close agreement to the values under the proposed model. Though this model requires more thorough testing, as well as application to other organic salts, it may provide insight into distribution factors of a precipitate formed by a sequence of organic complexes.

Keywords: rare earths; distribution factor; selective precipitation; oxalates; organic complexes

1. Introduction

In rare earth extraction, a common technique for overcoming the poor solubility of rare earth minerals is to transfer the rare earth ions into a more stable solid. In sulfuric acid cracking, for example, the rare earths are transferred into solid sulphates. In caustic conversion, the rare earths are put into hydroxides. In more recent studies, minerals of rare earth phosphates have been dissolved in oxalic acid, while the rare earths concurrently precipitate as rare earth oxalates [1,2]. In each of these techniques, an effective transfer of the rare earths depends on each rare earth having a lower solubility or a higher stability in the destination compound.

The way that rare earths distribute into an oxalate precipitate has not been modelled, beyond describing the results of specific experimental conditions. This is despite oxalate coprecipitation being a common technique in a broad range of areas. Oxalate coprecipitation is a standard method for producing precursor powders that are calcined into mixed metal oxides. Examples in the literature abound, such as precursors for magnetic materials [3], piezoelectric oxides [4], superconducting materials [5,6] and alloys [7]. It is also used to separate rare earths and actinides from other ions in solution by precipitation. Breakdown of rare earth minerals such as monazite are often done in extreme conditions [8] that dissolve impurity ions along with the rare earths. The very low solubility of rare earth and actinide oxalates enables an isolation of rare earths from mixed solutions [9].

The most advanced modelling of oxalate co-precipitation appears to have come from experiments to separate rare earths from each other in controlled precipitation. Oxalic acid was generated homogeneously throughout the solution by hydrolysis of dimethyl oxalate, in order to minimize concentration gradients within the solution and allow separations closer to those predicted by solubility [10]. Feibush, Rowley and Gordon [11] found, however, that the separations were much lower than predicted, according to solubility for the rare earths tested in their study. There remained no method to predict the separation from fundamental terms, and the techniques for oxalate coprecipitation more generally appear to be based on trial and error.

Feibush, Rowley and Gordon [11] did find that the precipitation between pairs of rare earth oxalates follows a Doerner–Hoskins or logarithmic distribution coefficient. Such a distribution assumes that the precipitating solid does not re-order its whole self to be in equilibrium with the solution; rather, the precipitated surface is in equilibrium with the solution and once precipitated it forms an unchanging substrate for the next surface layer [12]. Doerner and Hoskins were the first to describe this running equilibrium mathematically [13]. It can be described between concentrations of rare earth ions $[A]$ and $[B]$ by Equation (1).

$$\frac{d[A]}{d[B]} = \lambda \frac{[A]}{[B]} \quad (1)$$

Here, $\frac{d[A]}{d[B]}$ is the ratio of the concentration of A to B lost from the solution in an infinitesimal increment, or, in other words, precipitated onto the surface in an infinitesimal increment. The letter λ signifies the logarithmic distribution coefficient. This expression can be integrated to give Equation (2).

$$ln \frac{[A] final}{[A]\ initial} = \lambda ln \frac{[B]\ final}{[B]\ initial} \quad (2)$$

Doerner and Hoskins [13] did not attempt to correlate the coefficient to fundamental terms, probably in the knowledge that different salts may be governed by different terms. Feibush, Rowley and Gordon [11] suggested that with rare earth oxalates, the coefficient should be the ratio between the square roots of the solubility products of the pure rare earth oxalates (square root because each unit of rare earth oxalate has two rare earth ions), as shown in Equation (3).

$$\lambda \cong \frac{\sqrt{K_{Sp,\ A}}}{\sqrt{K_{Sp,\ B}}} \quad (3)$$

This equation is based on $K_{Sp,\ A}$, as the solubility product for the equation shown in Equation (4).

$$A_2(C_2O_4)_3 \rightarrow 2A^{3+} + 3C_2O_4{}^{2-} \quad (4)$$

Their experiments, however, showed that the experimental values of the coefficients, while being logarithmic in nature, were far from what was predicted by solubility in the form of Equation (3). Equation (3) does not form a basis in the present study for further development of a model.

Some more recent work on the solubility of rare earth oxalates has determined that the saturated concentration of rare earth A in a solution of concentrated oxalic acid is determined by Equation (5) [14].

$$[A_{total}] = \frac{\sqrt{K_{Sp,\ A}}}{\gamma_3} \left(a_{C_2O_4{}^{2-}}\right)^{-3/2} + \frac{\sqrt{K_{Sp,\ A}}\,\beta_{1,A}}{\gamma_1} \left(a_{C_2O_4{}^{2-}}\right)^{-1/2} \quad (5)$$

In Equation (5), $\beta_{1,A}$ is the equilibrium constant for the reaction in Equation (6), with the subscript number denoting the number of oxalates in the complex formed. The expression for $\beta_{1,A}$ is shown in Equation (7), where a is the activity of the species denoted as subscript. The symbol γ_1 is the activity co-efficient for the rare earth mono-oxalato complex, and γ_3 is the activity co-efficient for the free rare earth ion, the subscript based on the magnitude of the charge.

$$A^{3+} + C_2O_4{}^{2-} \rightarrow AC_2O_4{}^{1+} \quad (6)$$

$$\beta_{1,A} = \frac{a_{AC_2O_4{}^{1+}}}{a_{A^{3+}} a_{C_2O_4{}^{2-}}} \tag{7}$$

It can be seen in Equation (5) that the total concentration of rare earth A is made up of two main species, the free or weakly-complexed rare earth ion in the first term, and the mono-oxalato complex in the second term. In one sense, Equation (5) represents the saturated concentrations of each of these species in the particular system. The challenge from a theoretical point of view is how a model of precipitation can be constructed if there are complexes present. For example, the precipitate may be formed from a sequence of complexation reactions, where the mono-oxalato complex is the precursor to the precipitate, rather than the free rare earth ion.

2. Materials and Methods

Rare earth oxides at a minimum of 99.9% purity were obtained from Treibacher Industries in Austria, along with certificates of assay. Dimethyl oxalate (99%) was obtained from Alfa Aesar, and diethyl oxalate (99%) from Sigma Aldrich. Analytical grade hydrochloric acid was used. Brand-new Duran 100 mL conical flasks were used for each experimental flask.

The method underwent considerable development. In the original method, each conical flask was initially weighed. To each conical flask was added at least two, but up to six, rare earth oxides. The amount of each rare earth oxide added was determined by a random number generator in Microsoft Excel, to be an amount between 0.00 and 0.10 g. As close to this number as possible was weighed and added to the flask. Hydrochloric acid was then added to dissolve the rare earth oxides into chlorides until a clear solution was obtained. The flask was then heated on a hotplate to evaporate the remaining hydrochloric acid and water, and leave a precipitate of rare earth chlorides. To the flask was added about 100 g of deionized water. The mass of dimethyl oxalate (DMO) required for a complete precipitation was calculated, and a fraction of this was chosen at random, weighed and added to the flask. The solution was swirled until all the dimethyl oxalate was dissolved. This also had the effect of dissolving any residual rare earth particles (either difficult crystals of chlorides or hydroxides). A stopper was placed in the opening of the flask. The flask was submerged in a water bath at 25 °C and left for a week. One week was chosen based on informal testing—after about four days, the solution could be decanted into another vessel without further precipitation. A sample of the solution was taken by pipette and diluted with hydrochloric acid (0.1 M), for analysis by ICP-AES. The density of the solution was checked to be 1 g/mL.

This method produced internally consistent results between yttrium, erbium and thulium. However, in the case of lanthanum or neodymium, the initial stages of the precipitation proceeded without the involvement of one or more of the elements. To minimize the impact of this initial precipitation, a higher initial concentration of rare earths was used so that the final concentration covered a greater range of concentration. Also, to minimize errors of measurement of the initial solution, a stock solution was prepared for use in every conical flask. To allow a slower, more gradual precipitation, diethyl oxalate (DEO) was chosen as the oxalic acid ester. This appears to have a slower rate of hydrolysis than dimethyl oxalate.

About 5 g of each of lanthanum, neodymium and yttrium oxides were dissolved together in the one flask in hydrochloric acid, and evaporated to rare earth chlorides. Deionized water and 1 mL of hydrochloric acid (10 M) was added to a total of 1000 g. The 1 mL of hydrochloric acid was to dissolve any rare earth hydroxides that may be present. The pH was not measured, since the acidity during the experiment increases as more of the oxalic acid ester hydrolyses, so that a constant pH could not be maintained without the use of a buffer that may complex the rare earth ions. To each 100 mL conical flask was added to about 100 g of rare earth solution and a weighed amount of diethyl oxalate solution. The diethyl oxalate was added by a glass dropper, as diethyl oxalate can be a solvent of plastic. A sample of the solution was taken by pipette and diluted with hydrochloric acid (0.1 M) for analysis by ICP-AES. The density of the solution was checked to be 1 g/mL.

It was decided to plot the results between two rare earths from every experimental vessel on the one graph. The graph was based on Equation (2), so that the logarithm of the quotient of the starting concentration and final concentration was plotted on an axis for one element, while the same was plotted on the other axis for the other element. In this way, the slope of the graph should give the logarithmic distribution coefficient. The reason why this graphical method was chosen rather than tabulating the coefficient for each experiment was because the graph allows checks of internal consistency. Firstly, the line of best fit should be passing through the origin. This is because any constants of integration would be cancelled out in the integration of Equation (1) to Equation (2). Secondly, the ratios between a pair should be expressible in terms of the others; for example, pairs in a triplet should multiply to one, as illustrated by Equation (8):

$$K_{\frac{A}{B}} = K_{\frac{C}{B}} \times K_{\frac{A}{C}} \quad (8)$$

This is because the logarithmic distribution coefficient should be consistent between a pair regardless of how many rare earths are in the system.

3. Results

Table 1 shows which rare earths and ester were in each experimental vessel or conical flask, as well as the solution masses used to calculate the initial concentration. The first five experiments used the original experimental method. In this procedure, the total solution mass was weighed after the rare earth chlorides and dimethyl oxalate were dissolved. The last four experiments used the stock solution. In this method, the assayed stock solution was weighed into the conical flask and then the addition of diethyl oxalate was weighed.

Table 1. Identity of solution components and masses for each experiment.

Experiment Number	Solution Components	Solution Mass, g	Ester	Ester Mass, g	Total Mass, g
1	La, Pr, Nd, Y, Er, Tm	100.00	DMO	-	100.00
2	Y, Er, Tm	100.16	DMO	-	100.16
3	La, Pr, Nd, Y, Er, Tm	100.01	DMO	-	100.01
4	Y, Er, Tm	100.04	DMO	-	100.04
5	Y, Er	100.84	DMO	-	100.84
6	La, Nd, Y	102.62	DEO	2.04	104.66
7	La, Nd, Y	102.24	DEO	4.20	106.45
8	La, Nd, Y	102.65	DEO	1.05	103.70
9	La, Nd, Y	101.50	DEO	3.15	104.65

Table 2 shows the calculated initial concentrations and the measured final concentrations in each experimental vessel. In Experiment 7, too much diethyl oxalate was added, and complete precipitation occurred, so that the rare earths were below the detection limit (BDL).

Figures 1–3 show the results of the experiments with diethyl oxalate and concentrations of lanthanum, neodymium and yttrium. In addition, Figure 4 shows the result of the experiments with dimethyl oxalate and yttrium and erbium, in which the intercepts were also small, so that the use of diethyl oxalate was not required. The graphical results involving thulium also had a small intercept but are not shown here, as no thermodynamic data for thulium could be found for correlating the slopes of the graphs.

It can be seen in Figures 1–4 that the results have an internal consistency, with the lines of best fit having an intercept close to the origin. Also, when the slopes of Figures 1–3 are multiplied together as a triplet, the product is very close to one, indicating another internal consistency. The data points also fit fairly closely to a straight line, although deviations will appear to be diminished somewhat by the logarithmic axes.

Table 2. Initial and final solution concentrations for each experiment.

Experiment Number		La (ppm)	Pr (ppm)	Nd (ppm)	Y (ppm)	Er (ppm)	Tm (ppm)
1	Initial	486	880	183	603	450	883
	Final	289	493	90	508	340	621
2	Initial	-	-	-	802	661	600
	Final	-	-	-	323	170	168
3	Initial	549	115	920	430	191	514
	Final	199	15	66	198	57	158
4	Initial	-	-	-	589	493	385
	Final	-	-	-	330	202	163
5	Initial	-	-	-	962	274	-
	Final	-	-	-	667	150	-
6	Initial	5603	-	4831	3681	-	-
	Final	1300	-	40	910	-	-
7	Initial	5488	-	4732	3606	-	-
	Final	*BDL*	-	*BDL*	*BDL*	-	-
8	Initial	5656	-	4877	3717	-	-
	Final	3885	-	1410	2600	-	-
9	Initial	5542	-	4779	3642	-	-
	Final	680	-	4	445	-	-

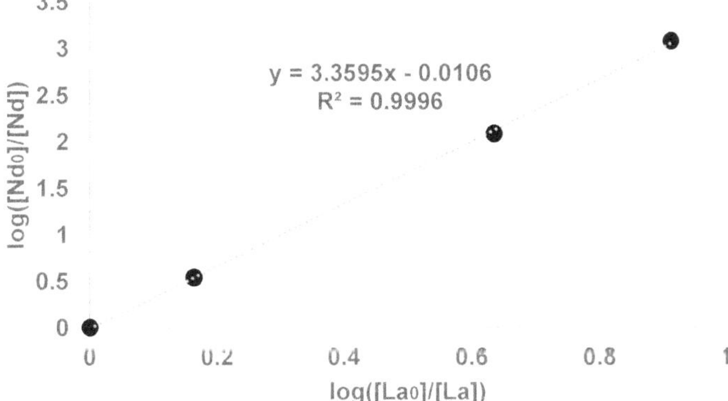

Figure 1. Enrichment of Nd against La: logarithm of the quotient of the starting concentration and final concentration for each rare earth.

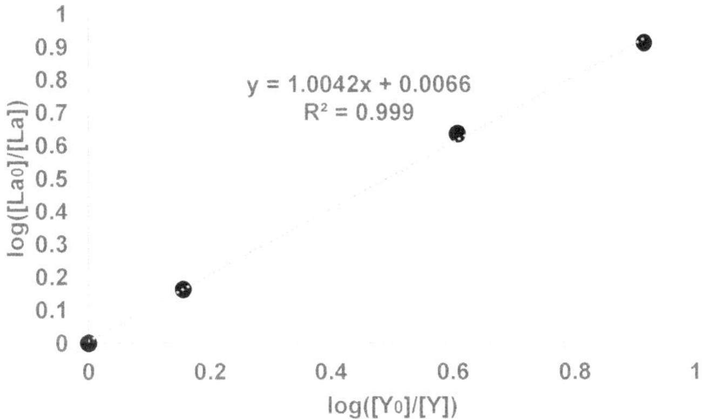

Figure 2. Enrichment of La against Y: logarithm of the quotient of the starting concentration and final concentration for each rare earth.

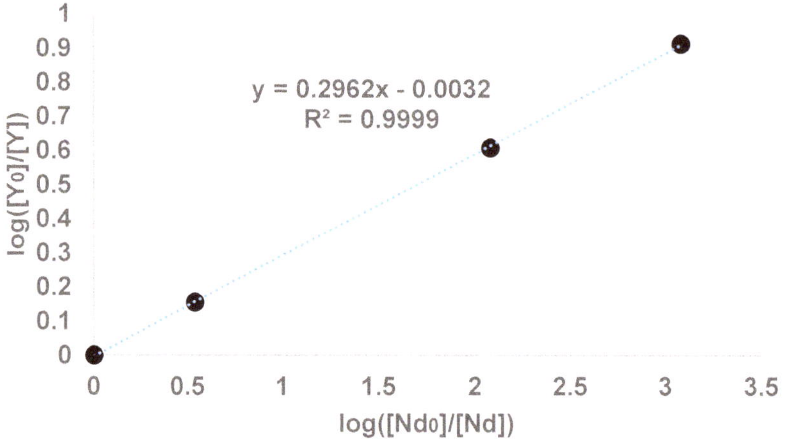

Figure 3. Enrichment of Y against Nd: logarithm of the quotient of the starting concentration and final concentration for each rare earth.

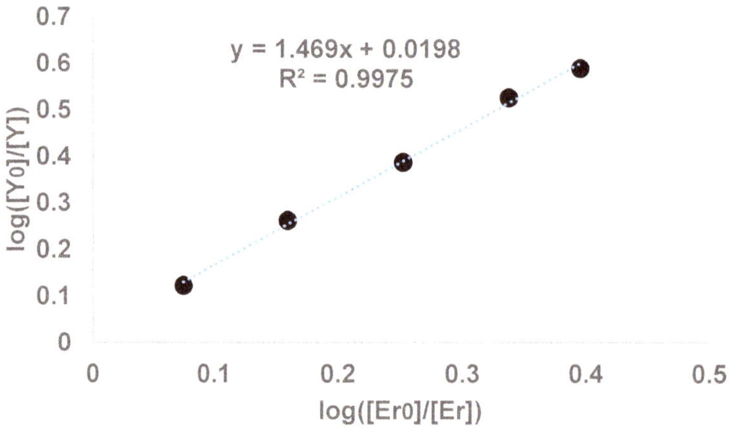

Figure 4. Enrichment of Y against Er: logarithm of the quotient of the starting concentration and final concentration for each rare earth.

It would have been preferable to have had a greater number of data points on each graph. However, each data point comes from a separate experimental vessel, and there was no exclusion of outliers, apart from Experiment 7, in which the precipitation was complete. In terms of repetition, in one sense each data point on the one graph is a repetition of the same separation, although repetition of the same specific conditions would also provide greater confidence. Overall, it seems that the slope values in Figures 1–4 can be described as the logarithmic distribution coefficient (λ), found experimentally.

4. Discussion

Given that Equation (3) did not describe the distribution coefficient λ [11], and that it is based on the solubility of the free or weakly-complexed rare earth ion, an alternative basis for describing λ was needed. It was hypothesized that the precipitation was governed by the energy difference between the solid rare earth oxalate and the mono-oxalato complex in Equation (9).

$$AC_2O_4{}^{1+} \leftrightarrow 0.5\,A_2(C_2O_4)_3 \tag{9}$$

Equation (10) was posed as a way to describe λ, because it represents the ratio of the equilibrium constants for Equation (9), which in turn is based on Equations (4) and (6). It also approximates the ratio of the saturated concentrations of the mono-oxalato complexes of each rare earth in question on the basis of Equation (5), assuming that the activity coefficient γ_1 is similar between rare earths. The term in Equation (5) for the free or weakly-complexed rare earth ion was neglected, as it seemed unlikely that solid rare earth oxalates would form directly from this simple species.

Equation (10) was used to calculate theoretical values in Table 3. The experimentally found values match these theoretical values fairly closely.

$$\lambda_{\frac{B}{A}} \cong \frac{\sqrt{K_{Sp,\,A}}\,\beta_{1,A}}{\sqrt{K_{Sp,\,B}}\,\beta_{1,B}} \tag{10}$$

Table 3. Comparison between theoretical and experimentally obtained values for the logarithmic distribution coefficient.

Pair	Theoretical	Experimental
Nd/La	3.25	3.36
La/Y	0.96	1.00
Y/Nd	0.32	0.30
Y/Er	1.47	1.47

The theoretical values for λ were calculated using Equation (10) on the results obtained by Chung, Kim, Lee and Yoo [14], shown in Table 4. The theoretical values for λ for the range of rare earths for which data is available are shown in Table 5.

Table 4. Solubility products [14], equilibrium constants [14] and their products $\sqrt{K_{Sp}}\,\beta_1$.

Element	Solubility Product, K_{Sp}	Equilibrium Constant, β_1	$\sqrt{K_{Sp}}{\cdot}\beta_1$
Y	5.1×10^{-30}	2.3×10^{7}	5.2×10^{-8}
La	6.0×10^{-30}	2.2×10^{7}	5.4×10^{-8}
Nd	1.3×10^{-31}	4.6×10^{7}	1.7×10^{-8}
Sm	4.5×10^{-32}	3.2×10^{7}	6.8×10^{-9}
Eu	4.2×10^{-32}	3.3×10^{7}	6.8×10^{-9}
Gd	4.25×10^{-32}	3.5×10^{7}	7.2×10^{-9}
Dy	2.0×10^{-31}	4.9×10^{7}	2.2×10^{-8}
Er	9.0×10^{-31}	8.0×10^{7}	7.6×10^{-8}

The reason why Equation (10) should approximate λ is a conundrum. Firstly, the most abundant form of each rare earth will be the mono-oxalato complex only in a concentrated solution of oxalic acid according to Equation (5). In the present case, a hydrolyzing ester will only produce a very dilute concentration of oxalic acid before precipitation occurs, meaning that the most abundant form of rare earth in solution will be the free or weakly-complexed ion according to Equation (5). This implies that the measured final concentrations are mainly made up of free or weakly-complexed rare earth ions, and that the model is still a relation between the concentration of these ions and the fractions on the surface of the precipitate. Yet, according to Equation (9), the free rare earth ion is not involved in that equilibrium directly.

Table 5. Calculated logarithmic distribution coefficients (precipitation of the element in the horizontal row for every one of the elements in the vertical column).

Element	La	Nd	Sm	Eu	Gd	Dy	Er
Y	0.96	3.1	7.7	7.7	7.2	2.4	0.68
La		3.25	7.9	8.0	7.5	2.5	0.71
Nd			2.4	2.5	2.3	0.76	0.22
Sm				1.0	0.94	0.31	0.09
Eu					0.94	0.31	0.09
Gd						0.33	0.10
Dy							0.29

Nor is there an obvious kinetic argument, as a higher stability of mono-oxalato complex will cause less precipitation into the solid oxalate compound under this particular model. It would perhaps be expected that a higher concentration of mono-oxalato complexes would lead to a higher rate of collisions leading to precipitation.

The notion of a stepwise formation of organic complexes into an organic salt also does not yield an adequate explanation. The concentration of the mono-oxalato complex of each rare earth would be determined by the stability constant in Equation (7), as well as the activity fraction of the free rare earth ion. Adding the chemical equations for the formation of the mono-oxalato complex (Equation (6)), and the formation of the precipitate from the mono-oxalato complex (Equation (9)) merely yields Equation (4), which represents the solubility product.

The system is behaving as though the ratio of the concentrations of the mono-oxalato complex of each rare earth is the same as that for the free rare earth ions. It may be conjectured that Equation (6) for the formation of the mono-oxalato complex is not attaining equilibrium before precipitation is occurring. It is possible that this may be connected to the very low concentration of oxalic acid and the very low solubility of the oxalate precipitate. Overall, it remains difficult to explain the model from kinetic and thermodynamic principles.

5. Conclusions

It was found that the logarithmic distribution co-efficient λ for precipitation between a pair of rare earths was fairly close to the ratio of the equilibrium constants for the step from mono-oxalato complex to the rare earth oxalate salt. This model relates the total concentration of rare earths (mostly free rare earth ions) to the fractions precipitating into the solid, so it is unclear why only the last step of the precipitation counts for λ. There appears to be no simple kinetic explanation either, with λ being derived from thermodynamic terms, and the curious result that a higher stability of mono-oxalato complex results in less precipitation.

Author Contributions: H.W. was responsible for designing and conducting the experiments and drafting. Y.-K.L. provided academic supervision and feedback. All authors have read and agreed to the published version of the manuscript.

Funding: This research received no external funding.

Acknowledgments: Acknowledged is the Minerals Research Institute of Western Australia for its funding of this project. Also acknowledged is the Australian Government Research Training Program for its funding.

Conflicts of Interest: The authors declare no conflict of interest.

References

1. Lapidus, G.T.; Doyle, F.M. Selective thorium and uranium extraction from monazite: I. Single-stage oxalate leaching. *Hydrometallurgy* **2015**, *154*, 102–110. [CrossRef]
2. Lazo, D.E.; Dyer, L.G.; Alorro, R.D.; Browner, R. Treatment of monazite by organic acids 1: Solution coversion of rare earths. *Hydrometallurgy* **2017**, *174*, 202–209. [CrossRef]

3. Gadkari, A.B.; Shinde, T.J.; Vasambekar, P.N. Magnetic properties of rare earth ion (Sm^{3+}) added nanocrystalline Mg–Cd ferrites, prepared by oxalate co-precipitation method. *J. Magn. Magn. Mater.* **2010**, *322*, 3823–3827. [CrossRef]
4. Choy, J.H.; Han, Y.H.; Kim, S.J. Oxalate coprecipitation route to the piezoelectric $Pb(Zr,Ti)O_3$ oxide. *J. Mater. Chem.* **1997**, *7*, 1807–1813. [CrossRef]
5. Cui, L.J.; Zhang, P.X.; Li, J.S.; Yan, G.; Wang, D.Y.; Pan, X.F.; Liu, G.Q.; Hao, Q.B.; Liu, X.H.; Feng, Y. Preparation and Characterization of Large-Quantity Bi-2223 Precursor Powder by Oxalate Coprecipitation. *IEEE Trans. Appl. Supercond.* **2016**, *26*, 1–4. [CrossRef]
6. Shter, G.E.; Grader, G.S. YBCO Oxalate Coprecipitation in Alcoholic Solutions. *J. Am. Ceram. Soc.* **1994**, *77*, 1436–1440. [CrossRef]
7. Drozd, V.A.; Gabovich, A.M.; Kała, M.P.; Nedilko, S.A.; Gierłowski, P. Oxalate coprecipitation synthesis and transport properties of polycrystalline $Sr_{1-x}La_xPbO_{3-\delta}$ solid solutions. *J. Alloy. Compd.* **2004**, *367*, 246–250. [CrossRef]
8. Krishnamurthy, N.; Gupta, C.K. *Extractive Metallurgy of Rare Earths*, 2nd ed.; CRC Press: Boca Raton, FL, USA, 2015.
9. Rodríguez-Ruiz, I.; Teychené, S.; Vitry, Y.; Biscans, B.; Charton, S. Thermodynamic modeling of neodymium and cerium oxalates reactive precipitation in concentrated nitric acid media. *Chem. Eng. Sci.* **2018**, *183*, 20–25. [CrossRef]
10. Gordon, L.; Shaver, K.J. Fractionation of some rare earth pairs: By precipitation from homogeneous solution. *Anal. Chem.* **1953**, *25*, 784–787. [CrossRef]
11. Feibush, A.M.; Rowley, K.; Gordon, L. Coprecipitation in Some Binary Systems of Rare Earth Oxalates. *Anal. Chem.* **1958**, *30*, 1605–1609. [CrossRef]
12. Matsui, M. The Coprecipitation Behavior of Rare Earth Elements with Calcium Oxalate upon Precipitation from a Homogeneous System. *Bull. Chem. Soc. Jpn.* **1966**, *39*, 1114. [CrossRef]
13. Doerner, H.A.; Hoskins, W.M. Co-precipitation of radium and barium sulfates 1. *J. Am. Chem. Soc.* **1925**, *47*, 662–675. [CrossRef]
14. Chung, D.Y.; Kim, E.H.; Lee, E.H.; Yoo, J.H. Solubility of rare earth oxalate in oxalic and nitric acid media. *J. Ind. Eng. Chem.* **1998**, *4*, 277–284.

Article

Leaching of Rare Earth Elements from Central Appalachian Coal Seam Underclays

Scott N. Montross [1,2,*], Jonathan Yang [1,3], James Britton [4], Mark McKoy [5] and Circe Verba [1]

1 National Energy Technology Laboratory, Albany, OR 97321, USA; Jonathan.yang@netl.doe.gov (J.Y.); Circe.Verba@netl.doe.gov (C.V.)
2 Leidos Research Support Team, Albany, OR 97321, USA
3 Oak Ridge Institute for Science and Education, Oak Ridge, TN 37830, USA
4 West Virginia Geological and Economic Survey, Morgantown, WV 26507, USA; britton@geosrv.wvnet.edu
5 National Energy Technology Laboratory, Morgantown, WV 26507, USA; Mark.McKoy@netl.doe.gov
* Correspondence: Scott.Montross@netl.doe.gov

Received: 1 May 2020; Accepted: 23 June 2020; Published: 26 June 2020

Abstract: Rare earth elements (REE) are necessary for advanced technological and energy applications. To support the emerging need, it is necessary to identify new domestic sources of REE and technologies to separate and recover saleable REE product in a safe and economical manner. Underclay rock associated with Central Appalachian coal seams and prevalent in coal utilization waste products is an alternative source of REE to hard rock ores that are mainly composed of highly refractory REE-bearing minerals. This study utilizes a suite of analytical techniques and benchtop leaching tests to characterize the properties and leachability of the coal seam underclays sampled. Laboratory bench-top and flow-through reactor leaching experiments were conducted on underclay rock powders to produce a pregnant leach solution (PLS) that has relatively low concentrations of gangue elements Al, Si, Fe, and Th and is amenable to further processing steps to recover and produce purified REE product. The leaching method described here uses a chelating agent, the citrate anion, to solubilize elements that are adsorbed, or weakly bonded to the surface of clay minerals or other mineral solid phases in the rock. The citrate PLS produced from leaching specific underclay powders contains relatively higher concentrations of REE and lower concentrations of gangue elements compared to PLS produced from sequential digestion using ammonium sulfate and mineral acids. Citrate solution leaching of underclay produces a PLS with lower concentrations of gangue elements and higher concentrations of REE than achieved with hydrochloric acid or sulfuric acid. The results provide a preliminary assessment of the types of REE-bearing minerals and potential leachability of coal seam underclays from the Central Appalachian basin.

Keywords: rare earth elements; coal utilization byproducts; pregnant leach solution; underclay; organic acid

1. Introduction

Rare earth elements (REE) are essential for the development of low-carbon, renewable energy technologies. In the United States (U.S.), a lack of domestic REE production is forcing end-users in energy, high-end technology, and manufacturing sectors to seek overseas sources. Exploration and production of new domestic sources of REE and critical minerals (CM) is essential to meet future demands. The U.S. Department of Energy report—2017 Report to Congress on Rare Earth Elements from Coal and Coal Utilization Byproducts—on rare earth elements from coal and coal byproducts outlines the strategic plan for expanding the U.S. REE reserve base [1]. The plan calls for identification of coal and coal byproducts with the highest known concentration of REE and the development of cost-effective separation technologies to recover the resource. A diversified REE product slate that includes recovery of REE from domestic coal byproducts and various types of sedimentary geologic materials can contribute to supply security and help to limit risks to market disruptions [1,2].

Production of REE in the United States, primarily sourced from bastnaesite and other accessory minerals, has increased between 2018–2019 from 18,000 to 26,000 tons, but is low compared to Chinese production (132,000 tons) [2]. The United States' domestic coal and coal utilization byproducts (CUB) are nevertheless promising sources of recoverable REE [1–6]. The term CUB includes a range of materials that are produced during coal utilization [1,3–6]. Coal mining waste rock and coal preparation plant refuse are two types of byproduct that contain underclay, a clay-rich sedimentary rock that is found adjacent to a coal seam. Underclay is commonly categorized as roof or floor rock and it is exposed and sometimes excavated during the mining of a coal seam.

Clay-rich horizons in Central Appalachian (CentApp) coal seams, for example, commonly contain higher concentrations of REE than the coal or other non-clay bearing rock adjacent to the coal seam [3,4]. In the CentApp region, there are approximately 840 coal refuse piles that overlie nearly 40 square kilometers of abandoned mines and coal fields. The amount of coal refuse in Pennsylvania alone is estimated to be 1.5×10^9 cubic meters [7]. Waste refuse piles, which plausibly contain a high percentage of clay minerals, may be heap leached or processed with limited beneficiation techniques (e.g., crushing and grinding, calcining, roasting, and floatation), compared to mineral bound ore. REE and CM can be leached from produced and stockpiled waste materials. With the availability of potential resource material, numerous studies have investigated REE recovery from coal-related materials, including coal fly ash [8,9], coal middlings [5,10], and underclays [3,4,10]. Underclays have an increased resource potential [4] as the rock is often subjected to previous diagenetic events and natural processes that transport and concentrate REE and CM in forms that may be easier to extract, compared to minerals bound in crystalline rock. Ease of extraction makes this type of material a more promising geologic source of REE and CM.

Organic acids and their degradation products provide ligands and chelating agents for heavy metals [11]. The citric acid-citrate system forms a relatively stable complex with alkaline earth metals [12] as well as heavy metals and lanthanides. The citric acid molecule is composed of one alpha position hydroxyl and carboxyl group and two beta position carboxylic acid groups, together the molecule contains at least seven potential O-donor sites that are capable of coordinating metal ions [12,13]. Carboxyl groups of citric acid have been shown to complex with both bivalent and trivalent metal ions in biological systems [13,14] and during interaction with alkaline earth metal ions [12].

Effects of organic acid on the leaching process of REE from ion-adsorbed clays was investigated by Wang et al. [15]. The leaching experiments by Wang et al. [15] showed REE recovery using citric acid was highest (10.4 mmol/kg) at pH range 3.5–4.0. The experiments were conducted at varying pH with the same carboxylic group concentration of 10^{-4} mol/L. Rare earth element concentrations decreased in the solutions with increasing pH (from 2–6). Increased pH should lead to greater acid dissociation because of pKa shifts increasing the number of complexation sites for REE on the organic ligands. The results confirm that organic anions, including anions of citric acid, can act as assistant leaching agents both through the complexation of REE in solution and the interaction with the clay surface to promote changes in the zeta potential of the clay. This process can lead to greater leaching of sorbed cations from the clay surface or for better dispersion of individual clay grains [15].

Citric acid anion recovery of REE from coal seam underclay is a promising method that may liberate higher concentrations of ion exchangeable REE from the clay compared to traditional lixiviants such as ammonium sulfate or sodium chloride. We chose to investigate the influence of organic acids on the recovery of exchangeable, or weakly bonded, REE ions from the surfaces of clay grains, as well as liberation of the nonexchangeable (stronger bonded) REE ions. Our selective approach of using pH buffered organic acid-based solutions amended with ionic constituents (e.g., $(NH_4)_2SO_4$ or NaCl) is designed to isolate and recover the exchangeable REE fraction type through partial dissolution of the clay matrix. The approach is based on previous studies which have shown that optimum recovery of REE adsorbed on clay minerals using ionic lixiviants occurs at the 3–4 pH range [15–17].

A mildly acidic organic acid-based ionic recovery solution with the presence of a monovalent salt likely liberates clay surface adsorbed REE and some inorganic REE mineral phases embedded in the clay matrix. The leaching solution may be pH buffered for optimum recovery from different clay mineral assemblages. Clay grains have a high surface area typically with a negative net charge. Exchangeable and nonexchangeable ions are present on the surface of clay grains. The amount of cation exchange capacity (CEC) or anion exchange capacity (AEC) is dependent on the clay mineral type or presence of organic matter. Exchangeable ions are weakly held in contact with the clay solution and are readily replaced by ions in solution. Positively charged ions, such as Al^{3+}, Ca^{2+}, K^{+}, and REE^{3+}, may be present on clay surfaces as exchangeable ions. Nonexchangeable ions are typically adsorbed by strong bonds or held in inaccessible places within the clay matrix (e.g., K^{+} between layers of illite). The use of organic acids can potentially increase the recovery of REE from clay-rich rocks by: (a) Maintaining a balanced charge on clay surfaces and increasing cation exchange capacity (CEC); (b) selectively dissolving matrix rock and increasing pore space connectivity and transmissivity of fluids; and (c) solubilizing phosphate bound REE [18,19].

Rozelle et al. [3] identified clay-rich ore deposits that contain up to 90% of the total REE in the rock, bound as ion-adsorbed REE, with the balance existing as colloids (e.g., Fe, Mn-oxides) and crystalline minerals (e.g., REE-phosphates). In China, about 10,000 tons of rare earth oxide (REO) concentrate are produced annually from weathered elution-deposits derived from lateritic weathering of granitic rock and in situ aqueous mining yields ~200 tons of REE annually [17]. This study characterizes and tests the leaching behavior of underclay rock from geologic formations associated with coal production in West Virginia, Pennsylvania, and Ohio.

2. Materials and Methods

2.1. Sample Preparation

The National Energy Technology Laboratory Research and Innovation Center (NETL-RIC) obtained underclay core samples from the West Virginia Geological and Economic Survey (WVGES) (Figure 1). The core samples were taken from strata associated with production coal seams—Lower Freeport, Middle Kittanning, and Pittsburgh No.5—in West Virginia.

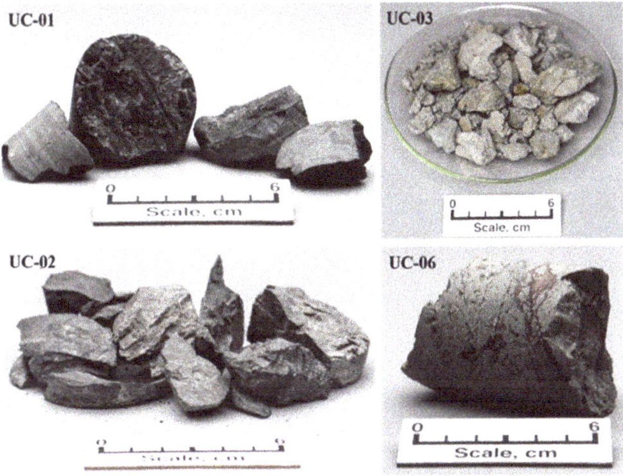

Figure 1. Photographs of as-received underclay core samples examined in this study. Information and descriptions of rock from the core samples are shown in Table 1.

A sample of approximately 200 g was either cut from the core or collected as chips and pieces. Each sample was pulverized first using a small jaw crusher then Zr-Ti lined shatter box to reduce the grain size to 149 μm or less. A mortar and pestle was used to hand-grind the material until it all passed through a 100 mesh (<149 μm) sieve. After grinding and sieving the powdered sample was homogenized using the cone and quarter technique [20]. The homogenized sample was split into specific subsamples for X-ray diffraction (30 g), elemental analysis (3 g), sequential digest (5 g), and leaching tests (0.5 g and ~40 g). Sample preparation for X-ray diffraction analysis included an additional step. The homogenized sample was ground for 60 s in a micronizing mill to achieve a grain size of <65 μm. The subsamples for the different analyses and tests were stored in chemical-free paper envelopes in a nitrogen purged desiccator until needed. The goal for sampling and sample preparation was to create a bulk homogenized underclay powder that could be used to test the recovery of REE from clay-rich material using different leaching solutions composed of organic acid anions. Bulk analysis and characterization of each sample was conducted to determine basic mineralogical and physical properties of the material. Rock chips or slices of core (3 cm × 2 cm) were used for electron microscopy imaging and X-ray microanalysis. Contextual information about the samples and core descriptions of the material are shown in Table 1.

Table 1. Sample information and core descriptions for underclay samples.

Sample ID	Material Type Formation	Depth (ft)	Core Description
UC-01	Underclay, shale Lower Freeport	1352.0–1352.5	Dark black to gray; fine grain matrix with visible pyrite and calcite cement; irregular sharp wavy contact. Directly underlies base of lower Freeport coal.
UC-02	Flint-clay, underclay Middle Kittanning	1463.0–1463.5	Medium to light gray/olive green rock fragments. Sub angular fine to medium clasts; few fine root traces and plant debris; few fine to medium black shale and coal streaks; clear lower contact.
UC-03	Underclay Pittsburgh	740.5–741.0	Medium gray and olive yellow-brown; common fine distinct olive mottles; few fine faint black mottles; few fine faint red mottles; clear lower contact.
UC-06	Paleosol, seat earth Lower Freeport	712.6–713.0	Pistachio green; extremely brecciated; paleosol; spiderweb calcite cement; siderite banding; soft sediment deformation structures present.

2.2. *Scanning Electron Microscopy and X-Ray Microanalysis*

Thin section and epoxy-mounted samples were evaporatively coated with carbon and imaged with a field emission scanning electron microscope (FE-SEM, FEI Inspect F) equipped with an energy-dispersive X-ray spectrometer (EDS, Oxford Instruments, Abingdon, UK). SEM imaging and EDS analysis was done at 20 kV, ~100 nA; a working distance of 10.4 mm, beam aperture 3, and spot size 5.0–5.5 nm. The entire area of the sample—thin section or epoxy mounted rock slice—was viewed frame by frame in x and y directions at low magnification (300×) in backscattered electron mode (BSE). Electron microscope images and EDS data were collected from single spots and full fields of view at multiple locations within the sample. Large area images (4 mm^2) with corresponding EDS maps were collected and constructed using the automate function in the Oxford INCA SEM-EDS software package (Version 5.05, Oxford Instruments, Abingdon, UK). Standards-based quantitative EDS was accomplished using REE-phosphate standards REEP25-15+ FC (Astimex Standards Ltd., Toronto, ON, Canada) and REE-oxide standard #489 (Gellar Analytical, Topsfield, MA, USA) for all analyses. Standard block #489 is certified to ISO 9001 and 17025 standards. Putative mineral phase identifications were made using images and elemental data from SEM-EDS analysis.

2.3. X-Ray Diffraction

Bulk mineralogy of rock samples was determined by X-ray diffraction (XRD) of randomly oriented powder mounts. Each sample was powdered to <63 µm using a micronizing mill. The powdered samples were spiked with 10 wt. % ZnO and mounted on an automatic 6-position sample changer equipped with a sample spinner. XRD patterns were collected using a Rigaku III Ultima diffractometer with Cu K-alpha radiation at 40 kV and 44 mA from 3.0–65.0 degrees-two-theta with a step size of 0.02° at 2.4 s. Initial peak alignments and identifications, and mineral IDs were made via comparison of the diffraction peaks against the ICDD-4 database using HighScore Plus XRD software (Version 3.0, Malvern PANalytical Ltd., Malvern, UK). Basic Rietveld fitting was performed using the software to quantify mineral percentages and estimate amorphous content (wt. %). Semi-quantitative analysis of crystalline components and mineral phase identifications were done by diffraction pattern analysis using the RockJock 7.0 computer program (U.S. Geological Survey, Boulder, CO, USA) [21].

Oriented mounts were prepared for clay identification by XRD analysis following the methods outlined in the U.S. Geological Survey (USGS) Open file report 01-041 [22]. The prepared mounts of each sample were scanned with a Rigaku III Ultima diffractometer (Rigaku, Tokyo, Japan) with Cu K-alpha radiation at 40 kV and 44 mA from 2.0–30.0 degrees-two-theta with a step size of 0.02° at 0.5° per minute. After the initial scan, the samples were treated sequentially with ethylene glycol and two separate heat treatments (400 and 550 °C). The samples were scanned after each treatment. Phase IDs, peak alignments, and mineral identifications of clay mineral peaks were made via comparison of the diffraction peaks against the ICDD-4 database using MDI Jade 6.0 XRD software (MDI Jade, Livermore, CA, USA). Diffraction patterns for untreated and treated samples were compared using Jade and basic Rietveld fitting was performed using the software to quantify mineral percentages. Presence of specific clay mineral phases was determined by changes in diffraction peak patterns across treatments following the identification flow chart in the USGS report [22].

2.4. Particle Size

Particle size analyses were completed on unreacted and reacted solid samples using a Malvern Mastersizer2000 (Malvern PANalytical Ltd., Malvern, UK) following procedures outlined in Sperazza et al. [23]. Briefly, ~5.5 g/L sodium hexametaphosphate was added to the solid samples as a dispersant and vortexed. The resulting slurry was then added to the sample introduction unit and the laser obscuration value adjusted to fall between 10–20% by adding tap water or additional sample. For unreacted samples, 60 s of ultrasonication was applied in the pre-measurement routine. For reacted samples, ultrasonication was turned off to preserve the particle size distribution from the reaction. Standardization and accuracy of measurements was monitored with QA standard QAS3002 (15–150 µm) from Malvern. Analytical procedure and results are shown in the Supplementary Materials.

2.5. Sequential Acid Digestion

Underclay powders were reacted sequentially using ammonium sulfate $(NH_4)_2SO_4$, hydrochloric acid (HCl), sulfuric acid (H_2SO_4) and the residual solids were subjected to lithium borate ($LiBO_2$) fusion and digestion. The procedure was designed to operationally evaluate common lixiviants used in commercial leaching and extraction of REE from geologic materials [24] and to provide a first-order comparison to leaching with organic acid-based reagents. The reagents and conditions for each step of the sequential digest are shown in Table 2. Dry, powdered sample was combined with different reagents and mixed in polypropylene tubes on a rotator at 25 rpm or stirred and heated in 100-mL Teflon beakers on a magnetic hot plate. At the end of each step the extraction solutions were separated by centrifugation 3500× *g* for 20 min. The extraction solution was collected from the centrifuge tube with a syringe and the liquid passed through a 0.45 µm nylon filter and collected for analysis. Major, trace, and rare earth element concentrations in the liquids collected were determined by inductively coupled plasma optical emission spectroscopy (ICP-OES) and mass spectroscopy (ICP-MS) following the methods in Bank et al. [25]. Post extraction solids were collected by centrifugation and washed by rinsing the solids with ~30 mL of MilliQ, mixing solids and water on the rotator for 5 min and then separated by centrifugation (3500× *g*, 20 min). The wash step was repeated three times. The washed solids were dried at 60 °C overnight and weighed for dry weight. Weight loss for each step was always less than 5%. The steps were repeated for each solution. The remaining solid material was collected for $LiBO_2$ fusion and digestion [25] and reported as residual.

Table 2. Extraction reagents and conditions used in sequential extraction. Concentration of reagents shown in mol/L (M). Mass of starting.

Step	Reagent/Target Fraction	Solids (%)	Temperature (°C)	Time (h)	pH
1	0.5 M $(NH_4)_2SO_4$ Exchangeable	1	22	4	5.0
2	1 M HCl Colloid	1	22	24	1.0
3	1.2 M H_2SO_4 Colloid + Mineral	1	70	1	0.86
4	$LiBO_2$-Digestion Mineral + Residual	200 mg	-	-	-

2.6. Citrate Leaching of REE

Benchtop leaching experiments were conducted on subsamples of underclay powders using various formulations of a water based leaching solution. The composition of each solution tested and a list of samples and conditions for each leach are shown in Table 3. The solutions were composed of combination of citric acid, sodium chloride, and conjugate buffer salt sodium citrate tribasic dihydrate ($HOC(COONa)(CH_2COONa)_2{\cdot}2H_2O$). Initial benchtop tests were conducted using rock powders from sample UC-01, 02, 03, and 06 and liquid–solid ratio of 1% solids (e.g., 50 mL of leaching solution—0.5 g of powder). The powders and solution were combinedin 50 mL polypropylene tubes and mixed using a tube mixer. Subsequent testing was performed on sample UC-02 using 20 g of powder and 200 mL of citrate leaching solution (10% solids)The test was conducted to (a) to verify the 10% slurry was properly mixed during the leaching steps, and (b) produce a PLS from a larger sample of material. Element concentrations in pregnant leach solutions (PLS) were determined by ICP-OES and ICP-MS. All leaching solutions contained a citrate concentration [citric acid + Na-citrate] of 0.1 mol/L (M) (See Table 3). In some cases, $(NH_4)_2SO_4$, or NaCl was added to the organic acid solution to provide an additional source of ions for ion exchange. Solutions were buffered to pH 3, 4, 5, or 6 using a mixture of citric acid and sodium citrate at a final concentration of 0.1 M citrate. The leaching solutions were mixed with a 0.5 g subsample of rock powder that was originally collected from the bulk homogenized sample (see above section) dried and powdered (<150 μm, 100 mesh) underclay sample in a 50 mL polypropylene centrifuge tube. The tubes were mixed on a rotational mixer for a range of times from 4–24 h at room temperature. The solution was separated from the slurry at the end of the leaching time by centrifugation (3500 rpm for 25 min.). The liquid was recovered from the tube with a syringe then passed through a 0.45-micron nylon filter. The solution pH was measured, and the remaining liquid sample was analyzed by ICP-MS at the NETL Pittsburgh Analytical Laboratory following the methods in Bank et al. [25].

Table 3. Leaching solutions and conditions used in benchtop experiments.

Solution ID	Composition	Samples Tested	Solid (%)	Temp (°C)/Time (h)	pH
RS-1	0.1 M Citrate * ($C_6H_5O_7^{3-}$ + 0.5 M Ammonium Sulfate ($(NH_4)_2SO_4$)	UC-02, UC-06	1	22/4	5.1
RS-2	0.1 M Citrate * + 0.5 M Sodium Chloride (NaCl)	UC-01, UC-02, UC-06	1, 10	22/24	3.5
RS-3	0.1 M Citrate *	UC-02, UC-06	1	22/24	2.0
RS-4	0.5 M Sodium Chloride (NaCl)	UC-01, UC-02, UC-06	1	22/24	5.0
RS-7	0.1 M Citric acid + Na-Citrate ($NaC_6H_5O_7^{3-}$) + 0.5 M NaCl	UC-01, UC-02,UC-03, UC-06	1	21/24	3.0–6.0

* Citrate from citric acid.

2.7. Flow Through

Powdered samples and fractured core samples were flooded with leaching solution RS-2 (see Table 3) a citrate buffered solution amended with NaCl. Fluid flow was established for time periods of 1–24 h. Hold-in times, referring to the length of time the solution is in contact with the sample (either powder or core) without flow, were varied from 20 min to 5 days. For the powdered clay samples, saturated flow was initially established and maintained for 5–6 h, after which flow was discontinued and sampling proceeded in a stepwise function at discrete time points of 6 h, 7 h, 24 h, and 5 days from the initial contact of fluid with the sample. Detailed descriptions of equipment, experimental setup, and the experimental parameters are found in the Supplementary Materials section on flow-through experiments.

3. Results and Discussion

3.1. Characterization of Underclay

The concentration of elements in the underclay samples, expressed as oxides, are shown in Table 4. Silicon (as SiO_2) and Al (as Al_2O_3) are the dominant cations in all samples. The range of Al_2O_3 concentrations (see Table 4) indicate the material comes from highly weathered crustal materials [26]. Additionally, low Ca, Na, and K values are also indicative of highly weathered horizons or zones of intense leaching. The exception is the Ca concentration in UC-06, withcalcite and siderite cement present throughout the rock matrix and visible in hand specimens (See Figure 1).

Table 4. Concentration of major cations as oxides (wt. %) in underclay samples.

Sample	Al	Ca	Fe	K	Mg	Na	P	Si	Ti	Zr
UC-01	18.5	0.4	9.0	3.1	1.6	0.3	0.1	62.8	0.9	<0.04
UC-02	28.0	0.8	1.7	3.2	0.5	0.4	0.1	63.3	1.4	<0.04
UC-03	18.9	0.7	4.5	4.0	1.2	0.6	0.0	68.4	0.9	<0.04
UC-06	24.9	5.4	2.4	3.9	0.4	0.4	0.4	65.3	1.2	<0.04

The extensive network of carbonate cement throughout the clay matrix is due to late diagenesis. The presence of diagenetic calcite in sample UC-06 results in a high concentration of Ca (5.4%) compared to the other underclay samples analyzed <1.0%. Rare earth element concentrations in each sample are shown in Table 5. The four samples have REE concentrations ranging from 262–353 mg/kg (See Table 5). The results reported in Table 5 are from the powdered and homogenized samples used for characterization, sequential digest, and 1% solids leaching tests. The results from a subsequent analysis of powder samples used in the leaching tests under different pH conditions are shown in Table S1 of the Supplementary Materials. Sample UC-02, a flint clay/underclay from the Middle Kittanning formation, had the highest REE concentration.

Table 5. Rare earth element concentrations (mg/kg) in West Virginia coal underclay samples. Values are reported on a whole sample basis.

Sample	La	Ce	Pr	Nd	Sm	Eu	Gd	Tb	Dy	Y	Ho	Er	Tm	Yb	Lu	ΣREE
UC-01	51.0	103.1	12.6	49.5	10.0	2.1	10.9	1.2	8.9	43.8	1.7	4.6	0.6	4.1	0.6	305
UC-02	82.9	119.3	13.2	45.4	9.4	2.8	9.7	1.0	8.0	50.3	1.5	4.3	0.6	4.2	0.6	353
UC-03	58.3	74.5	8.7	32.9	7.3	2.1	7.2	0.7	6.7	54.0	1.2	3.7	0.6	3.8	0.5	262
UC-06	71.3	102.1	9.7	30.9	7.1	1.9	6.0	0.4	5.1	35.4	0.9	2.9	0.4	3.0	0.5	278

Scanning electron microscope images of polished samples (see Figure 2) and various powdered samples (data not shown) were used to evaluate the composition and texture of the rock samples. SEM images and EDS results were used to make putative identifications of primary and secondary minerals in the samples, including the REE-bearing phases present. Rare earth element phosphate minerals were observed in all samples using SEM backscatter mode. The predominant REE-bearing mineral phases observed were rhabdophane, apatite, churchite, monazite, xenotime, and crandallite sp. Examples of REE-bearing minerals observed in the underclay samples analyzed are shown in Figure 2. Yang et al. [27] provides a comprehensive characterization of all REE-bearing minerals in the samples discussed here, as well as other underclay samples from West Virginia coal seam strata.

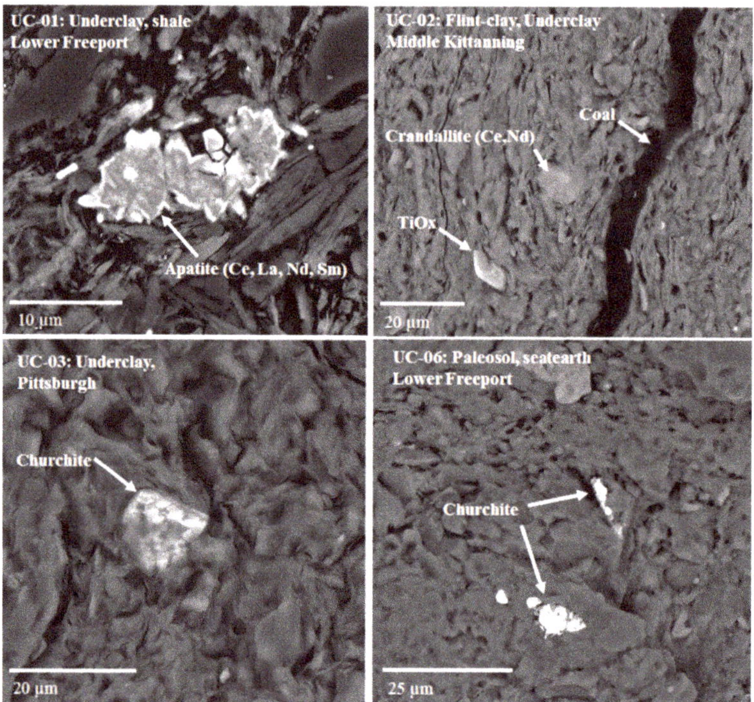

Figure 2. SEM micrographs of rock matrix and examples of rare earth elements (REE)-bearing minerals present in West Virginia underclay samples.

In sample UC-04 trace concentrations (1–4 wt. %) of Ce, La, and Nd were detected in grain-coating clay associated with pore filling framboidal pyrite and pyrite cement. Sample UC-02 contained Ce, La, Nd associated with aluminum phosphate ($AlPO_4$) mineral grains that were dispersed throughout the rock matrix and have a similar chemical composition and morphology to the mineral crandallite, a hydrous aluminum phosphate. Crandallite contains LREE, Ba, and Sr and is present throughout the clay matrix as 5–50-µm sized crystals that oval to round in grain mount samples and appear as spongy, porous in thin section (Figure 2, upper right). A summary of the REE-bearing minerals identified in the samples and general observations of texture and other mineral phases present in the samples are shown in Table 6.

Table 6. General characterization results from SEM-EDS analysis.

Sample ID	REE-Bearing Minerals Identified	General Observations
UC-01	Apatite in pore space contains Ce, La, Nd REE phosphate grains with Ce, La, Nd, Sm, and Eu up to 20 µm long, also contain U/Th. Ce, La, and Nd detected in clay coating on framboidal pyrite. Ytterbium detected in pore filling pyrite cement.	Abundant pyrite in bands and isolated matrix grains Large euhedral pyrite grains (up to 50 µm) in matrix and as pore filling cements. Apatite grains contain ~1–3 wt. % U and Th.
UC-02	Ce, La, Nd phosphate (rhabdophane) and monazite) in clay pore space. Y, Gd, Dy, Er, Yb phosphate (xenotime) grains present in clay pore space. Range of size 1–10 µm long. Xenotime grains bound in massive iron oxide. Ce, La, Nd associated with aluminum phosphate ($AlPO_4$) dispersed throughout the rock matrix. Aluminum phosphate grains with similar chemical composition and morphology of the mineral crandallite. Crandallite present throughout the matrix as 5–50 µm size grains.	Abundant Ti Oxide with Hf (0.5 wt. %) and Sc (0.25 to 0.5 wt. %). Aluminum phosphate grains present. $AlPO_4$ contains equimolar concentrations of S, Sr, Ba, and REE (Ce, La, Nd, Sm). Stoichiometrically constant with the mineral crandallite an illite conversion product. Massive iron oxide. Mixed Cu, Se, Pb sulfides Plant root fossils
UC-03	Ce, La, Nd phosphate (rhabdophane or monazite) mineral grains in clay. Y phosphate (churchite or xenotime) grains in matrix.	Sample contains abundant Fe-oxide and Ti oxide. Fe oxide band collocated within coal layer. Zircon present. Clay matrix composed of illite and smectite. Abundant quartz
UC-06	Monazite present as large crystals 10–25 microns. Monazite contains up to 5.0 wt. % U/Th. Xenotime and monazite present as embedded grains within siderite or calcite. Cementation by Ca/P mineral phase with LREE detectable using EDS.	Abundant pore filling and pore lining clay. Matrix clay composed of fibrous, tubular morphology typical of Halloysite and platy particles resembling kaolinite. Diagenetic spider web calcite, banded siderite, and pore filling clay, Calcium phosphate mineralization Pyrite, barite, zircon, rutile, and galena present in the matrix. Light grey zones have high quartz content, lack extensive carbonate cementation. Dark grey zones contain massive siderite and calcite.

3.2. XRD Results

Analysis of diffraction patterns collected from randomly oriented and oriented powders showed that predominant crystalline non-clay components are quartz, calcite, and ilmenite. Illite and smectite are the most abundant clay minerals in the samples. Halloysite is present in minor (5–7%) abundance in all samples except UC-03. Clay minerals make up more than 55% of the bulk material in each sample. Semi-quantitative results for all non-clay and clay minerals identified are shown in the Table S2 in the Supplementary Materials. The samples are all composed of two-component mixed clays from the groups Kaolin (kaolinite and Halloysite) and Mica (illite). Kaolinite and halloysite were identified by evaluating changes to the 7 Å XRD peak present in the scans. In all samples the peak remained unchanged or there was a small increase in d-spacing following glycol treatment, both of which are attributed to the presence of kaolinite or halloysite. The distinguishing treatment was the destruction of the 7 Å XRD peak after heating to 550 °C. Further confirmation of the presence and classification of kaolinite and halloysite were made using electron micrographs. Kaolinite displays a hexagonal morphology whereas halloysite has a tubular morphology. Analysis of SEM images confirms the presence of grains with hexagonal and tubular morphology. Illite was confirmed through the presence of a 10 Å XRD peak that remains unchanged by glycol and heat treatments.

3.3. *Leaching of Rare Earth Elements from Underclay*

The results of the sequential extraction of underclay powders and other leaching tests conducted on the samples in this study show that REEs are distributed across different mineral phases in the underclay. The results provided a basic screening of the distribution of REE in the bulk samples, specifically we were interested in the fraction of REE that was in the exchangeable phase. Other authors have presented results on ion-exchangeable clay from CentApp coal seam strata [2,18] and the results are not consistent across units in the basin [2,18,25]. Heterogeneity likely exists between individual clay units at the formation scale and plausibly within core samples collected from specific units. The samples may not be representative of an entire formation or basin, however, the results presented here are meaningful to evaluating different leaching solutions to recover REE from the material. The results of basic characterization efforts and leaching tests provide valuable information on not only the mineralogy and nature of REE-bearing minerals in the clay, but how the material responds to leaching with citrate solutions. In order to mature this technology and raise to a higher readiness level (TRL), currently at TRL 4-5, both the type and scale of sampling must be reconsidered, and the amount of material tested (e.g., scale) will have to be increased several fold. This is necessary due to the prevalence of heterogeneity and spatial variability of elements in geologic materials, as well as the presence of hot spots in such materials and within specific areas [28].

The samples discussed here and in Yang et al. [27] were taken from existing core samples. The characterization and analysis results reported here are from subsamples taken from the bulk, homogenized material prepared for the leaching tests. The results of basic sequential digest of the material, which can be considered a step-wise leaching test, are compared to the results of leaching using citrate. The results of the leaching tests are significant to evaluating alternative lixiviants (e.g., citric acid-citrate) and to compare the results to standard conventional mineral acid or salt leaches. The characteristics of the new solution, at minimum, should leach the ion-exchangeable fraction of REE from the samples, not produce excess liquid or solid hazardous waste, and be amenable to downstream processing to separate and recover REE in its pure form. Our leaching tests and demonstration of sorbent capture of REE from the citrate leachate demonstrate the potential of this technology.

We trialed a suite of different organic acids (e.g., acetic, indole-3-acetic, citrate) and formulations during our initial testing and development of an organic acid based lixiviant for recovering REE from clay [29]. Based on initial results we chose to pursue further testing and optimization of a citrate-based solution. A majority of the REE in the underclay samples analyzed here are bound in the residual phase (Table 7) and not extractable by exchange, or dissolution using 1 M HCl, or warm (70 °C) 1.2 M H_2SO_4. The complete set of analytical results for the sequential digests are shown in the Table S3 in the Supplementary Materials.

Table 7. Fraction of REE in pregnant leach solution (PLS) from sequential digestion of underclay powders. Values reported as percent of total REE leached from the solid sample.

Sample	$(NH_4)_2SO_4$ Exchangeable	HCl Extractable	H_2SO_4 Extractable	Residual
UC-01	0.3	12.1	8.2	79
UC-02	7.5	19.0	6.8	67
UC-03	4.9	31.1	20.7	43
UC-06	1.3	10.1	3.5	85

The $(NH_4)_2SO_4$ exchangeable fraction of REE in the samples ranged from 0.3–7.5% and are low but generally fall within a range of other published values for ion-exchangeable REE in CentApp coal and coal byproducts [6,29,30], with the exception of the values reported by Rozelle et al. [2].

Underclay from the Middle Kittanning formation (MKT)—sample ID UC-02—and from the Pittsburgh coal seam—sample ID UC-03—contained the highest concentrations of $(NH_4)_2SO_4$ exchangeable and HCl extractable REE of the samples tested. Both samples contain ~25–35% of the total REE in the exchangeable and HCl leached fraction (Figure 3). This fraction is likely comprised of REE bound to the surface of clay minerals, carbonates, and Fe-oxides. While the remaining REE is bound to phosphatic minerals or within the structure of more recalcitrant phases and may only be recovered using hot sulfuric acid or more destructive dissolution techniques such as microwave digestion or treatment with hydrofluoric acid. The concentration of REE in the HCl extracts may be higher in these samples due to the abundance of metal oxides that may bind colloidal ions or carbonate that may have ion adsorbed/exchangeable REE (See Table 8). The concentration of REE and gangue elements in leachates recovered from sequential extraction were compared to bench-top leaching experiments with citrate, and the results are shown in Figure 3 and Table 8. Ammonium sulfate used for recovery of ion exchangeable leached ~10 µg of REE or approximately 7% of the total REE present in UC-02 (Figure 3 and Table 8), other samples tested had low recovery (<10% of the total REE) when using ammonium sulfate. Citrate leaching using a cycle of citrate solution amended with NaCl (pH 5) leached ~45 µg of REE into the PLS. This amount correlates to nearly 33% of the total REE in the clay sample. Two different slurry concentrations using sample UC-02, 1 and 10%, were leached with citrate + NaCl. The results of the comparison are shown in Table S5.

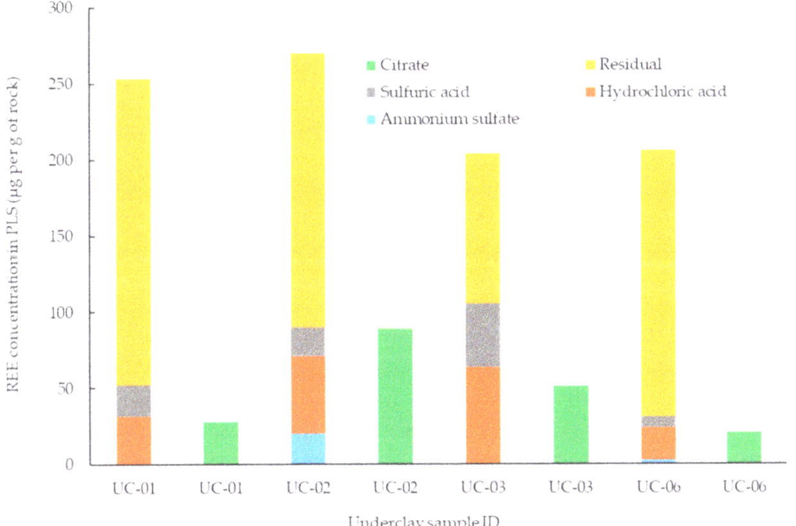

Figure 3. Concentration of REE in PLS after leaching with citrate solution and sequential digestion solutions. Concentrations leached into PLS for each step are denoted by the different bar colors. Citrate recovery values are shown as green bars. Values are reported as concentration of element in leached into PLS per gram of material (µg/g).

The results from the benchtop leaching experiments indicates that the citrate anion solution is more effective at recovering potentially ion exchangeable, or easily liberated REE that is present in the clay-rich sample. Additional REE, excess of ion adsorbed, is likely leached via chelation or complexation of the element from the clay or other mineral surface and solubilized into the leachate, presumably as an REE-citrate complex. In samples with less than < 1% ion exchangeable REE (e.g., UC-01, 03, and 06), citrate leached 10× more REE from the underclay than ammonium sulfate or NaCl, both of which are used commonly to recover exchangeable ions from soil and rock (Table S6, Supplementary Materials). One plausible explanation for the greater recovery of REE from citrate + sodium chloride treatment compared to ammonium sulfate is the effect of solution electrolyte concentration on dispersion of clay grains [31,32].

Table 8. Concentration (in µg/L) of Al, Si, Fe, Th, and REE in mineral acid and organic acid PLS generated from leaching of Middle Kittanning underclay (sample UC-02). Values are adjusted to mass of material leached.

PLS Composition	Al	Si	Fe	Th	REE
Mineral acids					
$(NH_4)_2(SO_4)$	5.0	1320	<DL	0.1	202
HCl	27,610	35,006	56,860	16	508
H_2SO_4	21,696	19,766	39,230	6	184
Σ Mineral acids + $(NH_4)_2(SO_4)$	76,907	56,092	57,252	24	894
Organic acids					
Citrate + NaCl (step 1 of 2)	21,230	13,491	3246	7.0	976
NaCl only (step 2 of 2)	1060	bd	843	0.5	44
Σ Citrate + NaCl	24,520	13,491	4099	7.5	1020
Citrate + $(NH_4)_2(SO_4)$	22,860	16,740	3380	18	146

Sodium chloride amended citrate may balance surface charge and increase total dissolved solids (TDS) conductivity in the leachate that may enhance colloid and particle dispersion [32]. We hypothesize that an increase in clay particle dispersion would lead to increased solution-clay grain interaction and higher recovery. However, the citrate anion leaches a greater fraction of REE than simply the exchangeable fraction, determined by $(NH_4)_2SO_4$. It is plausible that exchangeable RE concentrations in these samples are low and do not exceed more than 10% and that citrate liberates an additional fraction of the RE that is not exchangeable and not liberated by ammonium sulfate. Future work will be aimed at optimizing the solution in order to determine if additional mechanisms are at work where clay grains or RE mineral bearing phases are more susceptible to solubilization in the presence of citrate, compared to mineral acids or inorganic salts such as ammonium sulfate.

The concentration of REE and gangue elements in the different leachates from tests conducted on the Middle Kittanning underclay (UC-02) are shown in Table 8. The highest concentration of REE was released (during leaching using one single solution) by leaching with 0.1 M citrate solution amended with NaCl (solution RS-2). The high concentration of Al, Si, and Fe released from samples UC-01 and 06 during treatment with RS-2 leach solution indicates some dissolution of the mineral phases in these samples, as Si is not typically present as an exchangeable ion. The presence of weakly crystalline, amorphous phases and water-soluble species may also contribute to the higher release of these elements. Whether or not the solution can leach REE from refractory or phosphatic minerals such as monazite was not clear. Notably, the citrate PLS from leaching of UC-02 contained measurable P. Further testing is required to evaluate the leachability of these phases using a citrate or other organic acid-based solution. The recovery results for the different solutions tested indicate that the organic acid solution Na-citrate buffered citric acid (solution RS-2) leaches a greater mass of REE compared to ammonium sulfate (Table 8 and Figure 4), in some cases citrate leaching exceeds leaching with inorganic mineral acids (see Figure 4 and Tables S4 and S5).

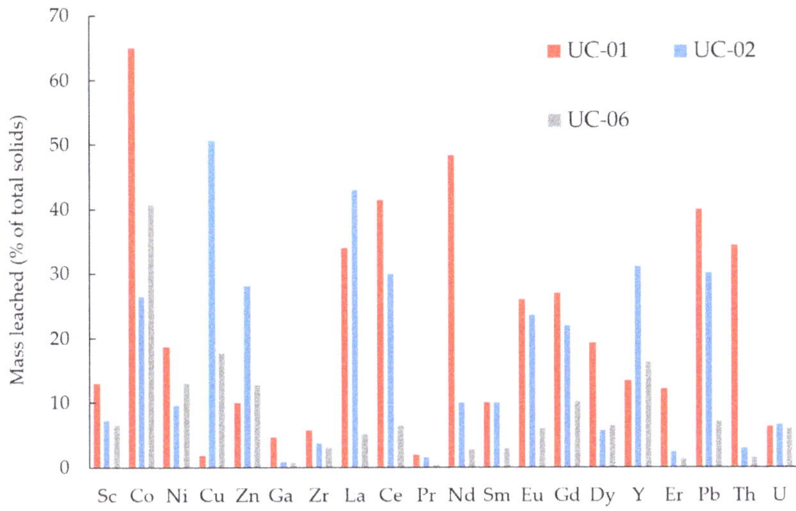

Figure 4. Concentration of REE and other trace metals leached from underclay with 0.1 M citrate + NaCl. Values are reported as the percent of total solids leached from the sample.

Citrate + sodium chloride leaching solution (RS-2) and the other citrate-based solutions tested released a higher concentration of base cations—Al, Si, and Fe—than ammonium sulfate (SEQ-1) but far less than the concentrations of cations released by low-pH inorganic mineral acids (Table 8, Table S3). The ratio of REE to base cations in the leachate accounts for the proportion of REE recovered to the proportion of "contaminants" such as cations Si, Al, and Fe. In recovering 45.9 μg of REE, RS-2, a citrate solution at pH 5, had a REE to base cation ratio of 2.5, compared to a significantly lower value of 0.3–0.2 for the acid and heat-based recovery solutions tested during the sequential extraction. The citrate solution (RS-2) performed best, based on the quantity of REE recovered and the reduction in the concentration of base cations and radioactive Th in the leachate. The citrate + sodium chloride solution, had a significantly higher recovery of REE than $(NH_4)_2SO_4$ from all three different underclay materials tested (Table 8, Tables S2 and S3).

The citrate solution leached greater than 30% of the total Ce, La, Nd, Eu, Gd, and Dy from samples 02 and 03. Leaching from UC-06 was significantly less than the other samples, apart from cobalt. For sample UC-02, both Cu, Y, and Pb are abundant in the leachate. Leaching of the radioactive element Th is low in PLS from samples UC-02 and 06, less than 3% of the total Th in the solid. Th leaching from sample UC-01 is nearly 8× that from the other samples. The concentration of Th in UC-01 starting solid material was lowest of the samples tested, 14 μg/g, compared to the concentration of Th in UC-02 and 06 which was 22 and 24 μg/g respectively (see Figure 5 and Table S3). The difference in Th leaching between the samples may be due to the mineral associations of Th in the different clay strata or formations. Notably, there is evidence that Th leaching is minimum, with less than 3% of the total leached from the clay when using citrate for leaching. Reducing Th in leachates, as well as other elements such as U, can help to reduce costs associated with waste disposal and remediation.

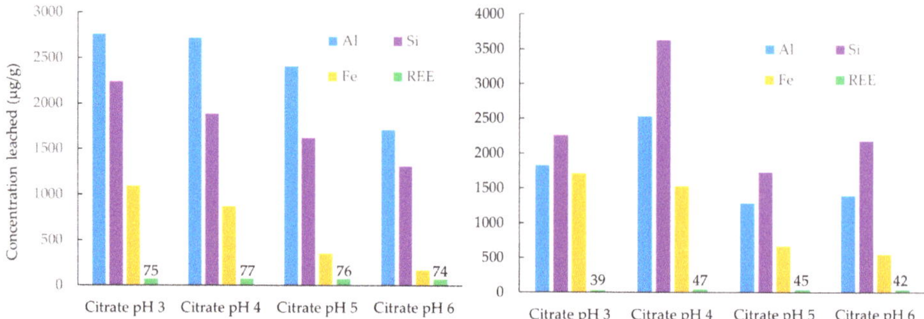

Figure 5. Concentration of Al, Si, Fe, and REE in PLS generated by citrate solution at different pH values for underclay sample UC-02 (**left**) and UC-03 (**right**). Values are converted to micrograms of element leached into PLS per gram of underclay (µg/g).

We evaluated how the pH of the leaching solution impacts the release of REE and gangue elements from samples UC-02 and UC-03 at a range of pH values from 3 to 6. Figure 5 shows the results from leaching of samples UC-02 and UC-03. For both samples tested, there was no significant change in the concentration of REE leached between pH 3–6. Decrease in the concentration of Al, Si, and Fe in the leachate was observed as pH increased (Figure 5, Table S7). Conversely, calcium in the leachate increased with increasing pH. Sequential acid digestions indicate that a majority of the REE are bound in mineral phases in the residual pool and are not extractable by $(NH_4)_2SO_4$, 1 M HCl, or 1.2 M H_2SO_4. The exceptions are for samples UC-02 and UC-03 which have a greater proportion of REE bound as exchangeable or extractable using hydrochloric and sulfuric acid.

At different pH values, range 3–6, there is a reduction in the concentration of gangue in the leachates (Figure 6). The solution can be pH buffered across a range of pH vales from 3–6 which allows for selective leaching of elements from the material. Leaching of non-REE base cations decreases with increasing pH and provides a system to selectively leach REE and minimize leaching of Al, Si, and Fe. Increasing pH should yield greater chelation recovery due to the increase in the number of deprotonated anion sites associated with pKa values for citric acid [12,13]. This is likely due to the monodentate bonding of the REE-citrate complex which is typical of the citrate-metal ion complex [12–14]. This bonding regime is the strongest of the bonds associated with chelation/complexation to carboxylic acid, the functional group on the citrate molecule [14]. There is little change in the concentration of REE in the citrate leachate across pH range 3–6 (see Figure 5, Table S7), which crosses two pKa boundaries for citric acid (e.g., 3.1, 4.7). At pH 4.7, a second carboxylic acid group becomes deprotonated. If this functional group could attract an additional RE ion, there should be a commensurate increase in REE concentration observed between pH 4 and 5 leachates. However, since there is little to no change in REE concentration across the range of pH it is likely the complexation of REE-citrate is more influenced by other factors such as saturation of the clay surface with the solution, dispersion of the clay grains, or amount of carboxylic acid group present (e.g., initial molarity of citrate leaching solution) for chelation and recovery of the metal ions.

The preliminary results from the bench-top experiments indicate that citrate is a chemically effective lixiviant that can be used to generate a PLS from clay-rich sedimentary rock. Presumably, REE and other ionically bound base cations that are present on clay surfaces are solubilized via complexation/chelation with the carboxylic functional group on the citrate anion. Simple chelation/complexation using the citrate anion in solution at pH 5 reduces the potential for dissolution of crystalline mineral phases and concurrent release of elements gangue elements. Such is the case with refractory minerals such as monazite and xenotime that are not only sources of REE but also radioactive elements Th and U. The process presented here is not aimed at leaching of crystalline bound, phosphatic REE. Rather, REE is sorbed to the surface of mineral grains that can be recovered with basic hydrometallurgical leaching.

The citrate PLS generated in the experiments contains a relatively low concentration of gangue elements because of the limited mineral dissolution. The overall economics of the process cannot be assessed at this stage of the research or at the scale of this work. However, future experiments and analysis will be aimed at evaluating the economics of the process including subsequent steps to purify REE from the PLS. Currently there are few processes that have been developed at a higher Technology Readiness Level (TRL) on the extraction and recovery of REE from secondary products and wastes.

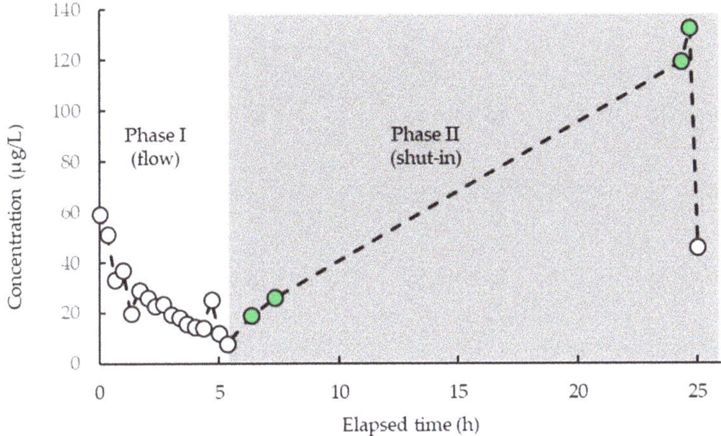

Figure 6. REE concentration (in µg/L) in effluent PLS collected during flow-through leaching of Middle Kittanning underclay (UC-02) powder. Black dots denote sampling points taken during continuous flow. Green dots indicate samples taken after various shut-in times.

3.4. Leaching of REE from Powdered Underclay Using Core Flow-Through

An additional set of leaching tests were conducted on the remaining unreacted powders from the benchtop test tube leaching tests. Flow-through leaching reactors (See Supplementary Materials) packed with rock powder were flooded and subjected to pressurized flow at ~1600–1800 psi to compensate for low porosity and correspondingly low permeability of clay. The highest concentration of REE in effluent leachates occurred during the initial 20–40 min of fluid flow through the powder. Powder reactor runs exhibited lower concentrations of REE released to the fluid and relative to amount of material in reaction vessel. Shut-in periods up to 5 days in some cases recovered equal amounts as continuous flow for 7 h (based on cumulative concentrations). The results indicate that shut-in time may be necessary to fully saturate the material for increased fluid interaction with the surfaces of the clay grains. Peak REE concentrations were also noted to occur within the first 20–40 min of fluid flow, denoted as phase I in Figure 6. Though absolute concentrations increased with increasing fluid shut-in times, the peak REE concentrations were consistently noted to occur within 20–40 min of initializing fluid flow (Figure 6).

Increased recoveries of the REE in the effluent solutions of the underclay powders were evident after increased shut-in times. Peak concentrations increased from the initial start of fluid flow to the samples flowed at 24 h (after ~17-h shut-in period) to the samples flowed at 5 days. An example of the effect shut-in time has on the liberation of REE into solution is shown in Figures 6 and 7.

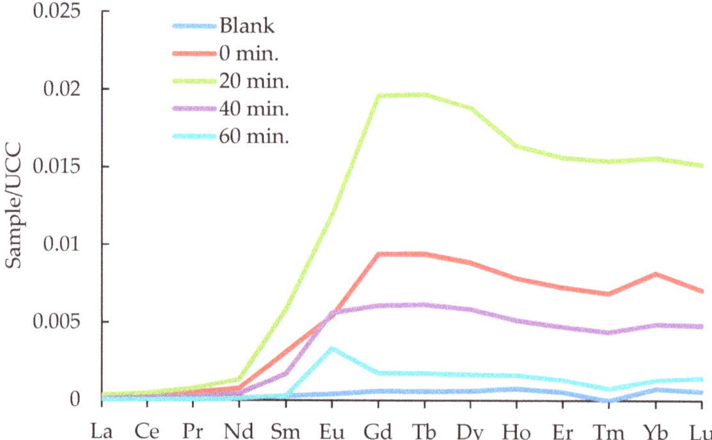

Figure 7. Concentration of REE, normalized to upper continental crust values, in flow-through PLS analyzed at 0, 20, 40, and 60 min.

The highest concentrations of REE were observed in effluents after a 5-day shut-in period. In Figure 6, the initial opening of the reactor, after 5 days, occurs at 0 min. Subsequent measurements were taken at 20, 40, and 60 min after flow was established. The highest concentration of REE in the leachate occurred at 20 min after reestablishing flow (Figure 7). Additionally, the pregnant leach solutions exhibit middle—to heavy—REE enrichments.

Results from these initial flow-through experiments highlight the considerations and parameters that need to be optimized to maximize the extraction of REE from underclay feedstocks using an organic lixiviant such as a sodium citrate solution. The flow-through experiments demonstrate that the initial 20–40 min of flow is the most critical for recovering the highest concentrations of the REE. Additional flow beyond that initial 20–40 min recovers significantly lower concentrations of the REE and may not be economical. The relatively quick release of REE in the first 20–40 min of fluid flow is consistent with our hypotheses that the sodium citrate solution targets the sorbed/colloidal components of the underclays. Increased shut-in times of the fluid with the underclay sample at pressure also led to increased concentrations in the PLS. The highest concentrations of REE in PLS produced came after a 5-day shut-in period. Notably, after these extended shut-in periods, the highest concentrations of the REE were still observed as occurring in the first 20–40 min of fluid flow. We suggest that the mechanism of REE extraction remains the same as in the powder benchtop tests, i.e., desorption from clay surfaces and/or complexation of ions from colloidal phases, but that the additional shut-in time is needed for wetting of micropores and packed grains. These observations are consistent between the powdered samples and the fractured core experiments. However, extraction efficiencies for the flow-through experiment were low, <1% of the total REE content, and likely due to the lower liquid–solid ratio and decreased contact between solution and material due to low transmissivity of the fluid through the material. Increasing the extraction efficiencies for flow-through applications remains an area of active research and further development and testing of a flow-through method that could eventually be employed at the field scale for in situ solution mining is necessary.

3.5. A Method for Recovery of REE from Citrate PLS

In this study a citrate-based al, the subsequent processing, and quantitative recovery of REE from the pregnant leach solution (PLS) is an essential consideration for mining operations and technology economic evaluation of the application. Sorbent capture is a promising technology for the recovery of a high purity REE fraction from PLS, including citrate-based solutions. In conventional mining operations of the REE, the PLS is subjected to a series of purification steps to remove impurities such as Fe, Al, P, and Th (i.e., gangue elements) before the REE can be recovered [33,34]. While these downstream processes are still an area of active research for the citrate solutions described herein, there are a number of potential pathways that include the use of a novel amine-based sorbent to selectively recover the REE. In order to process the pregnant leachate solution for the recovery of a high purity REE fraction, the leachate is passed over a bed of sorbent material [34]. After leaching, the PLS is passed over a bed of sorbent material [35]. During capture, alkali and alkaline earth elements, as well as the lixiviant, pass through the bed while gangue elements (e.g., Al, Si, Fe, and Th) and REE bind to the bed. Once bound to the sorbent bed, the REE is selectively eluted, away from the gangue elements to produce high purity REE fraction [33,35,36].

In a separate series of laboratory experiments we produced and tested the ability to capture REE from a citrate-based PLS and simultaneously remove gangue elements from the leachate. A citrate PLS—from the leaching of coal prep plant fines with the citrate leaching solution RS-2—was used. The initial, unoptimized sorbent capture test using the citrate PLS showed approximately 60–70% uptake of REE from the feed solution, up through holmium, with a bias toward the light and middle rare earth elements in the unoptimized recovery of bound REE from the sorbent. The overall recovery for the light and middle REE was found to be 80–100% for elements lanthanum to dysprosium, while the recovery ranged from 50–70% for MREE and HREE, with the greatest recovery in this subset occurring with holmium. These results show great promise toward the concentration of REE from a citrate-based PLS generated by leaching of clay-rich geologic material. Future work will be aimed at using the solid sorbent to remove all base metals and other elements from the citrate lixiviant in order to recycle the solution and reuse multiple cycles of leaching. Reuse of the lixiviant during sequential cycles will add cost savings to the process. When coupled with the use of solid polymer sorbents to recover and concentrate REE, citrate leaching may be a promising method for the leaching and concentration of REE from clay-rich coal mine waste rock and coal preparation plant refuse.

4. Conclusions

Underclay associated with coal seams in the Lower Freeport, Middle Kittanning, and Pittsburgh formations contain REE concentrations ranging from 250–353 ppm. Clay minerals such as illite, halloysite, and kaolinite are the predominant clay minerals that make up >55% of the total bulk mineralogy of the rock. The introduction of leaching solutions into underclay rock powder initiates chemical reactions such as ion exchange, hydrolysis, and mineral dissolution that result in the release of ion constituents into the citrate PLS. Bench-top leaching and flow-through experimental results indicate that citrate is a chemically effective molecule for leaching weakly bound REE or other elements of interest from clay-rich sedimentary rock. Rare earths and other metal elements may exist as water soluble, ion exchangeable, or ion/colloidal where the REE is adsorbed to clay or other minerals such as metal oxyhydroxides that are present in the rock. The properties of the citrate molecule provide an added benefit of pH-controlled selectivity against leaching gangue elements from the rock matrix. The process of using organic acid anions for chelation and complexation of target elements is a promising method to leach REE and other critical metals from a variety of different sedimentary lithologies (e.g., underclay, sandstone, and shale) and produced materials such as coal mining waste and coal preparation plant refuse. A chief advantage of the citrate leaching technology is the demonstrated ability to recover REE from feedstocks with minimal release of gangue and radioactive elements, using an environmentally benign and relatively cost-effective leaching solution.

Changes to the physical and/or chemical properties of the clay rock may have both a positive and negative outcome pertaining to the leaching of specific elements from the rock matrix. Bench top powder leach and the flow-through experiments conducted here provide first-order results on the efficacy of organic acid anions for leaching of REE and other elements. Additionally, the results provide observational data on the minimal impact the leaching has on the physical structure of the rock because of the likelihood that there is minimal dissolution of the rock matrix when leaching with citrate at a pH range of 3–6. Continuing work will be aimed toward maximizing the extraction efficiency of the organic acid-based citrate leaching solution. As described in the methods, a total accounting by mass of the extracted REE from flow-through experiments was not possible but estimates of the extraction efficiency of the system remain low, <1% of the total REE content for the underclay samples. Comparatively, benchtop experiments where powdered underclay samples were reacted with sodium citrate solution at 1% and 10% solids (e.g., Table S5) demonstrated leaching up to ~30% of the total REE. Future work using flow-through experiments will explore parameters such as increasing the leach solution ratio of fluids reacting with the underclay samples, the concentration of the sodium citrate solution, and sampling schemes such as a step-wise shut-in reaction. Information gained from these results and the guided work of future studies will aid in developing a technology economic evaluation of the process to determine costs associated with upscaling this technology for use in larger scale operations.

Supplementary Materials: The following are available online at http://www.mdpi.com/2075-163X/10/6/577/s1, Figure S1: Photograph of underclay reaction vessel used in flow-through leaching experiments, Figure S2: Schematic of flow-through apparatus used in powder leaching experiments, Table S1: Concentration of elements in bulk underclay powders used in pH tests, Table S2: Semiquantitative XRD results from random and oriented mounts, Table S3: Results of sequential digest of powdered underclay samples, Table S4: Concentration of trace elements in sequential digest PLS, Table S5: Concentration of elements in PLS from leaching of UC-02 (1 and 10% solids) with 0.1 mol/L citrate and 0.5 mol/L NaCl, Table S6: Concentration of elements (μg/g) leached into PLS from underclay samples using citrate solution RS-2, Table S7: Concentration of elements leached from underclay samples UC-02 and UC-03 using citrate solutions buffered to pH 3, 4, 5, and 6, Table S8: Tabulated parameters for particle size distributions of unreacted (initial) underclay samples UC-01, UC-02, UC-03, and UC-06.

Author Contributions: Conceptualization, S.N.M., M.M., and C.V.; methodology, S.N.M.; formal analysis, S.N.M., J.Y. and J.B.; investigation, S.N.M. and J.Y.; resources, J.B., M.M. and C.V.; data curation, S.N.M., and C.V.; writing—original draft preparation, S.N.M., J.Y., J.B., M.M. and C.V.; writing—review and editing, S.N.M., J.Y., J.B., and M.M.; supervision, M.M. and C.V.; project administration, C.V.; funding acquisition, M.M. and C.V. All authors have read and agreed to the published version of the manuscript.

Funding: This work was performed in support of the US Department of Energy's Fossil Energy Crosscutting Technology Research Program. The Research was executed through the NETL Research and Innovation Center's Rare Earth Elements Program. Research performed by Leidos Research Support Team staff was conducted under the RSS contract 89243318CFE000003. **Disclaimer:** This work was funded by the Department of Energy, National Energy Technology Laboratory, an agency of the United States Government, through a support contract with Leidos Research Support Team (LRST). Neither the United States Government nor any agency thereof, nor any of their employees, nor LRST, nor any of their employees, makes any warranty, expressed or implied, or assumes any legal liability or responsibility for the accuracy, completeness, or usefulness of any information, apparatus, product, or process disclosed, or represents that its use would not infringe privately owned rights. Reference herein to any specific commercial product, process, or service by trade name, trademark, manufacturer, or otherwise, does not necessarily constitute or imply its endorsement, recommendation, or favoring by the United States Government or any agency thereof. The views and opinions of authors expressed herein do not necessarily state or reflect those of the United States Government or any agency thereof.

Acknowledgments: We thank Mary Anne Alvin (DOE Rare Earths Technology Manager) and Thomas Tarka (REE FWP Technical Portfolio Lead) for their support. We thank MacMahan Gray (USDOE-NETL) and Brian Kail (LRST) for their time and efforts with sorbent capture tests and analysis, Ward Burgess and Randal Thomas (LRST) for reviewing the manuscript prior to submission.

Conflicts of Interest: The authors declare no conflicts of interest.

References

1. DOE Report. *Report on Rare Earth Elements from Coal and Coal Utilization Byproducts, Report to Congress*; United States Department of Energy: Washington, DC, USA, 2017. Available online: https://www.energy.gov/sites/prod/files/2018/01/f47/EXEC-2014-000442%20-%20for%20Conrad%20Regis%202.2.17.pdf (accessed on 29 May 2020).
2. U.S. Geological Survey. *Mineral Commodity Summaries 2020*; U.S. Geological Survey: Reston, VA, USA, 2020; 200p. [CrossRef]
3. Rozelle, P.; Khadikar, A.; Pulati, N.; Soundarrajan, N.; Klima, M.; Mosser, M.; Miller, C.; Pisupati, S. A study on removal of rare earth elements from U.S. coal byproducts by ion exchange. *Metall. Mater. Trans.* **2016**. [CrossRef]

4. Montross, S.N.; Verba, C.A.; Chan, H.L.; Lopano, C. Advanced characterization of rare earth element minerals in coal utilization byproducts using multimodal image analysis. *Int. J. Coal Geol.* **2018**, *195*, 362–372. [CrossRef]
5. Zhang, W.; Yang, X.; Honaker, R.Q. Association characteristic study and preliminary recovery investigation of rare earth elements from fire clay seam coal middlings. *Fuel* **2018**, *215*, 551–560. [CrossRef]
6. Honaker, R.Q.; Zhang, W.; Werner, J. Acid leaching of rare earth elements from coal and coal ash: Implications for using fluidized bed combustion to assist in the recovery of critical materials. *Energy Fuels* **2019**, *33*, 5971–5980. [CrossRef]
7. Appalachian Region Independent Power Producers Association (ARIPPA). 2018 Coal Refuse Whitepaper. Available online: https://arippa.org/wp-content/uploads/2018/12/ARIPPA-Coal-Refuse-Whitepaper-with-Photos-10_05_15.pdf (accessed on 2 April 2019).
8. Lin, R.; Stuckman, M.; Howard, B.; Bank, T.; Roth, E.; Macala, M.; Lopano, C.; Soong, Y.; Granite, E. Application of sequential extraction and hydrothermal treatment for characterization and enrichment of rare earth elements from coal fly ash. *Fuel* **2018**. [CrossRef]
9. Taggart, R.; Hower, J.; Hsu-Kim, H. Effects of roasting additives and leaching parameters on the extraction of rare earth elements from coal fly ash. *Int. J. Coal Geo.* **2018**. [CrossRef]
10. Huang, Q.; Noble, A.; Herbst, J.; Honaker, R. Liberation and release of rare earth minerals from Middle Kittanning, fire clay, and west kentucky No. 13 coal sources. *Powder Technol.* **2018**. [CrossRef]
11. Burckhard, S.R.; Schwab, A.P.; Banks, M.K. The effects of organic acids on the leaching of heavy metals from mine tailings. *J. Hazard. Mater.* **1994**, *41*, 135–145. [CrossRef]
12. Kondoh, A.; Oi, T. Interaction of alkaline earth metal ions with carboxylic acids in aqueous solutions studied by 13CNMR spectroscopy. *Z. Für Nat. A* **1998**, *53*, 77–91.
13. Wyrzykowski, D.; Chmurzyński, L. Thermodynamics of citrate complexation with Mn^{2+}, Co^{2+}, Ni^{2+}, and Zn^{2+} ions. *J. Term. Anal. Calorim.* **2010**, *102*, 61–64. [CrossRef]
14. Zabiszak, M.; Nowak, M.; Taras-Goslinksa, K.; Kaczmarek, M.T.; Hnatejko, Z.; Jastrzab, R. Carboxyl groups of citric acid in the process of complex formation with bivalent and trivalent metal ions in biological systems. *J. Inorg. Biochem.* **2018**, *182*, 37–47. [CrossRef] [PubMed]
15. Wang, L.; Lioa, C.; Yang, Y.; Xu, H.; Xiao, Y.; Yan, C. Effects of organic acids on the leaching process of ion-adsorption type rare earth ore. *J. Rare Earths* **2017**, *35*, 1233–1238. [CrossRef]
16. Moldovean, G.A.; Papangelakis, V.G. Recovery of rare earth elements adsorbed on clay minerals: II. Leaching with ammonium sulfate. *Hydrometallurgy* **2013**, *131–132*, 158–166. [CrossRef]
17. Zhi Li, L.; Yang, X. China's rare earth ore deposits and beneficiation techniques ERES2014. In Proceedings of the 1st European Rare Earth Resources Conference, Milos, Greece, 4–7 September 2014; pp. 26–36.
18. Brisson, V.L.; Zhuang, W.; Alvarez, L. Bioleaching of rare earth elements from monazite sand. *Biotechnol. Bioeng.* **2015**, *113*, 339–348. [CrossRef]
19. Shan, X.Q.; Wen, J.J.B. Effect of organic acids on adsorption and desorption of rare earth elements. *Chemosphere* **2002**. [CrossRef]
20. Schumacher, B.; Shines, K.; Burton, J.; Papp, M. *Comparison of Soil Sample Homogenization Techniques*; Lewis Publishers: Chelsea, MI, USA, 1990.
21. Eberl, D.D. *User's Guide to RockJock—A Program for Determining Quantitative Mineralogy from Powder X-ray Diffraction Data, Open-File Report 03-78*; U.S. Geological Survey: Boulder, CO, USA, 2003.
22. Poppe, L.J.; Paskevich, V.F.; Hathaway, J.C.; Blackwood, D.S. *A Laboratory Manual for X-Ray Powder Diffraction, Open-File Report 01-041*; U.S. Geological Survey: Woods Hole, MA, USA, 2010.
23. Sperazza, M.; Moore, J.N.; Hendrix, M.S. High-resolution particle size analysis of naturally occurring very fine-grained sediment through laser diffractometry. *J. Sed. Res.* **2004**. [CrossRef]
24. Peelman, S.; Sun, Z.H.; Siestma, J.; Yang, Y. Leaching of rare earth elements: Past and present. In Proceedings of the 1st European Rare Earth Resources Conference, ERES2014, Milos, Greece, 4–7 September 2014.
25. Bank, T.; Roth, E.; Tinker, P.; Granite, E. *Analysis of Rare Earth Elements in Geologic Samples using Inductively Coupled Plasma Mass Spectrometry*; US DOE Topical Report-DOE/NETL-2016/1794[R]; National Energy Technology Lab. (NETL): Pittsburgh, PA, USA, 2016.
26. Taylor, A.; Blum, J. Relation between soil age and silicate weathering rates determined from chemical evolution of a glacial chronosequence. *Geology* **1995**, *23*, 979–982. [CrossRef]

27. Yang, J.; Montross, S.N.; Britton, J.; Stuckman, M.; Lopano, C.; Verba, C. Microanalytical approaches to characterizing REE Appalachian basin underclays. *Minerals* **2020**, *10*. [CrossRef]
28. Komnitsas, K.; Modis, K. Geostatistical risk estimation at waste disposal sites in the presence of hot spots. *J. Haz. Mat.* **2009**, *164*, 1185–1190. [CrossRef]
29. Montross, S.; Verba, C.; Falcon, A.; Poston, J.; McKoy, M. Characterization of rare earth element minerals in coal utilization byproducts and associated clay deposits from Appalachian basin coal resources. In Proceedings of the 34th International Pittsburgh Coal Conference, Pittsburgh, PA, USA, 3–9 September 2019.
30. Hower, J.C.; Groppo, J.G.; Joshi, P.; Dai, S.; Moecher, D.P.; Johnston, M. Location of cerium in coal-combustion fly ashes: Implications for recovery of lanthanides. *Coal Combust. Gasificat. Prod.* **2013**, *5*, 73–78. [CrossRef]
31. Allen, J.R.L. *Principles of Physical Sedimentology*; Springer: New York, NY, USA, 2012; p. 212.
32. Norrström, A.-C.; Bergstedt, E. The impact of road de-icing salts (NaCl) on colloid dispersion and base cation pools in roadside soils. *Water Air Soil Pollut.* **2001**, *127*, 281–299. [CrossRef]
33. Gray, M.L.; Kail, B.W.; Wang, Q.; Wilfong, W.C. Stable immobilized amine sorbents for REE and heavy metal recovery from liquid sources. *Environ. Sci. Water Res. Technol.* **2018**. [CrossRef]
34. Wilfong, C.W.; Kail, B.W.; Wang, Q.; Shi, F.; Shipley, G.; Tarka, T.J.; Gray, M.J. Stable immobilized amine sorbents for heavy metal and REE removal from industrial wastewater. *Environ. Sci. Water Res. Technol.* **2020**, *6*, 1286–1299. [CrossRef]
35. Wang, Q.; Kail, B.W.; Wilfong, W.C.; Shi, F.; Tarka, T.J.; Gray, M.L. Amine sorbents for selective recovery of heavy rare-earth elements (Dysprosium, Ytterbium) from aqueous solution. *ChemPlusChem* **2020**, *85*, 130–136. [CrossRef]
36. Wang, Q.; Wilfong, W.C.; Kail, B.W.; Yu, Y.; Gray, M.L. Novel polyethylenimine–acrylamide/SiO_2 hybrid hydrogel sorbent for rare-earth-element recycling from aqueous sources. *ACS Sustain. Chem. Eng.* **2017**, *5*, 10947–10958. [CrossRef]

Article

Effect of Sulfuric Acid Baking and Caustic Digestion on Enhancing the Recovery of Rare Earth Elements from a Refractory Ore

Rina Kim [1,*], Heechan Cho [2], Jinan Jeong [2], Jihye Kim [3], Sugyeong Lee [4], Kyeong Woo Chung [1], Ho-Sung Yoon [1] and Chul-Joo Kim [1]

1 Resources Recovery Research Center, Mineral Resources Division, Korea Institute of Geoscience and Mineral Resources (KIGAM), Daejeon 34132, Korea; case7@kigam.re.kr (K.W.C.); hsyoon@kigam.re.kr (H.-S.Y.); cjkim@kigam.re.kr (C.-J.K.)

2 Department of Energy Resources Engineering, Seoul National University, Seoul 08826, Korea; hccho@snu.ac.kr (H.C.); krenos@snu.ac.kr (J.J.)

3 Department of Chemical Engineering and Applied Chemistry, University of Toronto, Toronto, ON M5S 3E5, Canada; jhye.kim@mail.utoronto.ca

4 Robert M. Buchan Department of Mining, Queen's University, Kingston, ON K7L 3N6, Canada; sugyeong.lee@queensu.ca

* Correspondence: rkim@kigam.re.kr

Received: 28 April 2020; Accepted: 10 June 2020; Published: 12 June 2020

Abstract: To improve the recovery of rare earth elements (REEs) from a refractory ore, this study investigated two different chemical decomposition methods, namely sulfuric acid baking and caustic digestion, with their respective leaching processes. The studied lateritic ore contained goethite (FeOOH) as a major constituent with REEs scattered around and forming submicron grains of phosphate minerals, such as apatite and monazite. Therefore, despite the substantially high content of REEs (3.4% total rare earth oxide), the normal acidic leaching efficiency of REEs reached only 60–70%. By introducing sulfuric acid baking and caustic digestion, the REE-leaching efficiency was significantly improved. After sulfuric acid baking at 2.0 acid/solid ratio and 200 °C for 2 h, the leaching efficiency reached 97–100% in the subsequent water-leaching. When the ore was digested with a solid/liquid ratio of 100 g/L in a 30 wt% NaOH solution at 115 °C and 300 rpm for 3 h, the REE-leaching efficiency of 99–100% was attained at 80 °C using a 3.0 M HCl solution. The correlation between the REE and the Fe-leaching was determined. The improvements in REE-leaching in both methods were mostly attributed to the mineral phase and crystallinity changes of Fe-bearing minerals due to the ore pretreatments. Such findings were also supported by X-ray diffraction and scanning electron microscopy analyses.

Keywords: REE; sulfuric acid baking; caustic digestion; acid leaching; water leaching

1. Introduction

The rare earth elements (REEs) comprise 17 elements in the periodic table, namely the 15 lanthanide elements (La–Lu) plus Sc and Y. These 17 elements are generally divided into two groups: La–Eu (Nos. 57–63) are light REEs (LREEs), whereas Gd–Lu (Nos. 64–71) and Y (No. 39) are heavy REEs (HREEs). Sc is not classified in any of the two groups. The REEs are used in various high-tech industries, e.g., the manufacturing of computers, cell phones, optical glasses, phosphors, hybrid cars and batteries, because of their unique electrical and magnetic properties [1–4].

REEs can be recovered from different minerals. More than 250 rare earth minerals have been discovered to date [1], and the representative minerals are monazite ($REEPO_4$), bastnasite ($REECO_3F$) and xenotime (YPO_4). Monazite is an REE phosphate mineral that is very stable and difficult to leach under mild conditions. In contrast, bastnasite is an REE fluorocarbonate. When bastnasite is used as an REE resource, the fluorocarbonate should be converted first into an oxide or a sulfate to facilitate the succeeding leaching process. Xenotime is an yttrium phosphate mineral, which also has a very stable form; thus, similar to monazite, it is also difficult to leach under mild conditions. Other REE minerals include apatite, cheralite, eudialyte, loparite, phosphorite, secondary monazite and spent uranium solutions [4]. REEs are also found in ion-adsorbed clays; they are physically adsorbed onto the clays and are mostly HREEs [5–7].

Although there are several REE minerals in nature, only monazite, bastnasite and xenotime are commercially utilized [8]. General processing methods have been developed for these well-known REE minerals: acid baking (mostly by sulfuric acid) followed by water-leaching can be applied to treat them [8–13]. The REE phosphate minerals, i.e., monazite and xenotime, can also be effectively pretreated using a NaOH solution, and then, the converted REE hydroxide compounds are mostly leached in a weak acid solution [8,12,14,15]. Bastnasite can be leached by hydrochloric or sulfuric acid after a roasting process to decompose fluorine and carbonate [12]. The REEs from ion-adsorbed clays can be leached using an ammonium sulfate solution, the mechanism of which is similar to that of the ion exchange process [5–7].

However, the described methods are effective only when the minerals are recovered as high-grade concentrates and when the REE minerals exist as major minerals in the ores. Furthermore, the minerals contained within an ore vary depending on the mine; therefore, different processing methods need to be applied based on the ore type. Some researchers studied the leaching behavior of apatite containing considerable quantities of REEs. Habashi [16] reviewed the REE recovery from Kola apatite, and Jorjani et al. [17] conducted a leaching study using an apatite concentrate from the Chadormalu Plant in Iran. For the two studies, nitric acid was used as the leaching agent to prevent gypsum production. In addition, REEs in apatite can generally be recovered as a byproduct from calcium sulfate sludge from the wet processing of phosphoric acid [18,19]. REEs contained in apatite can also be treated by a two-step leaching method using hydrochloric acid as a leaching agent [20]. In the first step, approximately 80% of Ca and 60% of P are removed with almost no REEs loss, and the REEs are recovered at the second step with less impurities. Several metallurgical studies of refractory REE ores have also been conducted. In Brazil, Neumann and Medeiros [21] tried to separate the REE minerals from Araxá REE ore by physical methods; however, they decided to omit physical separation due to the complex mineralogy, fine REE crystals and particles having strong intergrowth with the gangue mineral. In a pilot plant, Araxá REE ore was introduced into direct leaching after sulfuric acid baking [15]. The REE ores in Chuktukon deposits from Russia showed similar mineralogical properties as the Araxá ores and direct chemical processing was also considered, i.e., nitric acid-leaching in an autoclave [22]. The processes mentioned above are summarized in Table 1.

As indicated in Table 1, appropriate processing methods based on the ore properties have to be developed to handle gangue minerals as well as REE minerals. This study investigated the enhancement in REE recovery from a refractory ore containing 3.4% total rare earth oxide (TREO) by applying two different chemical decomposition methods followed by leaching processes, i.e., sulfuric acid baking—water-leaching and caustic digestion–acid-leaching.

Table 1. Rare earth elements (REEs) recovery metallurgical processes.

REE Mineral Type	Origin	Method	REE-Leaching Efficiency	Reference
Monazite	Unknown	H_2SO_4 baking (180–250 °C)/ Water-leaching (70 °C)	85–100%	Berry et al. [9]
Monazite	Australia	H_2SO_4 baking (250 °C)/ Acid-leaching (0.9 M H_2SO_4, 20–25 °C)	96%	Demol et al. [10]
Red mud	Greece	H_2SO_4 baking (700 °C)/ Water-leaching (room temperature)	>80% (Sc 60%)	Borra et al. [11]
Apatite residue	Iran	H_2SO_4 baking (270 °C)/ Water-leaching	95%	Soltani et al. [13]
Bastnasite, monazite	China	Caustic digestion ($NaOH + Ca(OH)_2$ roasting, 700 °C)/ Acid washing (0.5 M HCl + 0.05 M citric acid, 40 °C)/ Acid-leaching (3 M HCl + 0.5 M $AlCl_3$, 70 °C)	86%	Huang et al. [14]
Monazite, xenotime, bastnasite	Kyrgyz Republic	Caustic digestion (Na_2CO_3 fusion)/Acid-leaching (HNO_3)	-	Buchanan et al. [15]
Apatite, allanite, chevkinite, titanite	Turkey	Caustic digestion (600 °C)/ Water–HCl-leaching	-	Buchanan et al. [15]
Ion-adsorbed clays	Unknown	Acidic-leaching (0.5 M $(NH_4)_2SO_4$, pH 3 by H_2SO_4, 25 °C)	80–90%	Moldoveanu and Papangelakis [7]
Apatite	Iran	Acid-leaching (60% HNO_3, 60 °C)	59–74%	Jorjani et al. [17]
Apatite	South Africa	Acid-leaching (1.0 M HNO_3, 0.5 M $Ca(NO_3)_2$)	85%	Preston et al. [18]
Apatite	Mongolia	Two-step acid-leaching (1st: 1.0 M HCl/2nd: 2.0 M HCl)	92–94%	Kim et al. [20]

2. Materials and Methods

2.1. Ore Characteristics

The ore used in this study was a lateritic ore containing goethite as a major mineral. Figure 1 shows an X-ray diffraction (XRD) analysis of the REE ore, where the goethite, ettringite, muscovite, maghemite, and apatite peaks can be observed. However, no REE mineral peaks appeared because their concentrations were much lower than those of Fe-bearing minerals, and the REEs were scattered throughout the ore. REE-containing particles were observed through a scanning electron microscopy (SEM) analysis, as displayed in Figure 2. In the subsequent energy-dispersive X-ray (EDX) analysis (Figure 3), Ce oxide and mixed REE grains were observed, and some REEs were likely to substitute Ca in the apatite lattice. In addition, considering the high intensity of REEs and P in the mixed REE particle presented in Figure 3c, the presence of an REE phosphate mineral, i.e., monazite, was also expected.

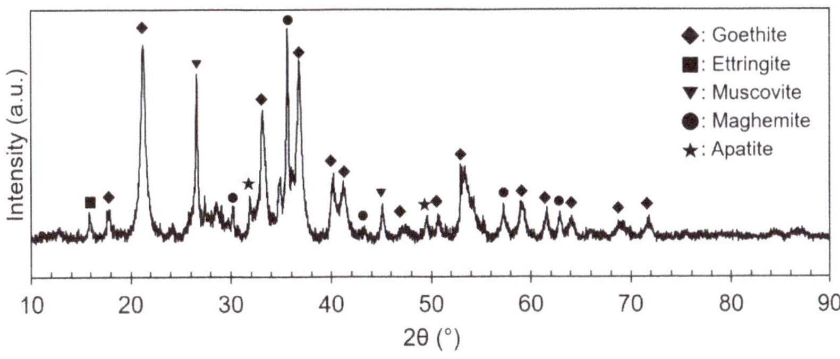

Figure 1. XRD pattern of the REE ore.

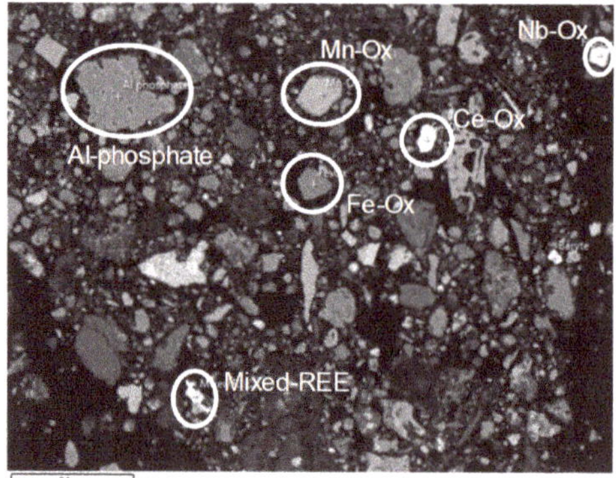

Figure 2. SEM image of the REE ore.

Because the ore was lateritic, the iron content was significantly high (more than 50% as Fe_2O_3) and 10% of manganese was present as an oxide as presented in Table 2. The TREO grade was 3.4%, which is relatively high as an REE ore. Ce was the most abundant element, followed by La, Nd and Y (Table 2). The ore was obtained as a ground material, and the 80% passing size was 30 μm (Figure 4).

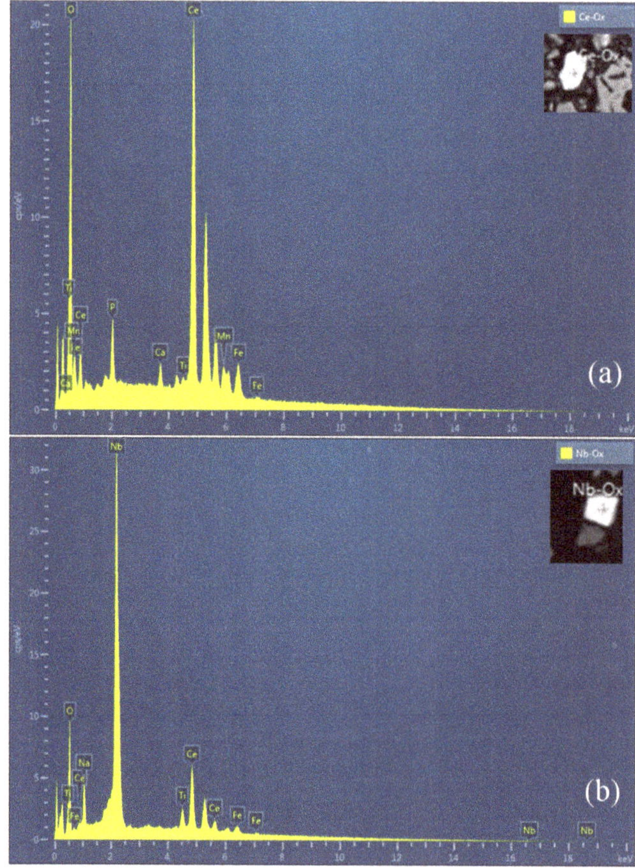

Figure 3. *Cont.*

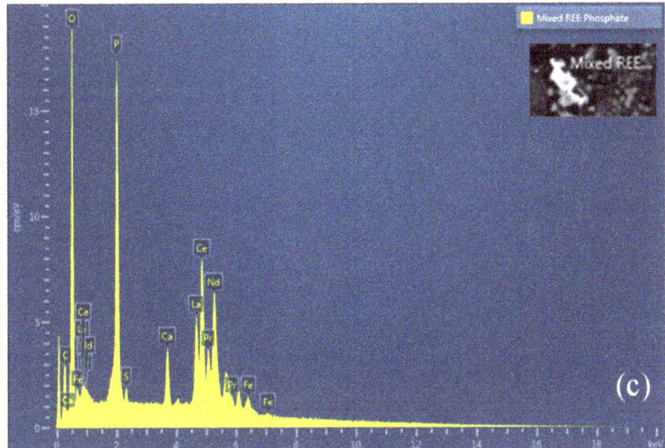

Figure 3. EDX spectra of REE-containing grains: (**a**) Ce oxide grain; (**b**) Nb oxide grain, which is host to minor Ce; (**c**) mixed REE oxide grain with Ca and P.

Table 2. Chemical composition of the REE ore.

Formula	Fe_2O_3	MnO	SiO_2	P_2O_5	Al_2O_3	TiO_2	CaO	BaO	CeO_2	MgO	La_2O_3
wt%	50.1	9.9	7.8	6.9	6.7	4.4	3.6	2.2	1.7	1.2	0.9
Formula	K_2O	SrO	Na_2O	Nd_2O_3	SO_3	Nb_2O_5	ZnO	Cl	ZrO_2	Y_2O_3	MoO_3
wt%	0.8	0.7	0.6	0.6	0.5	0.4	0.3	0.2	0.2	0.2	0.07

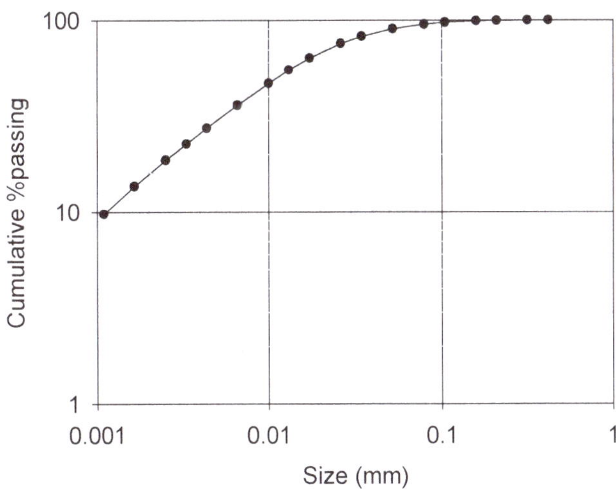

Figure 4. Cumulative particle size distribution of the ground REE ore (P_{80} of 30 μm).

2.2. Preliminary Leaching

Generally, REE oxide can dissolve at a slightly acidic pH (pH of less than 4.0–5.0), and REE phosphate dissolves at a highly acidic pH of approximately −1.0 [20,23], respectively. The purpose of preliminary leaching was to determine if the REE minerals in the ore could really dissolve at those pH ranges. The suggested pH ranges can be achieved by hydrochloric or nitric acid concentrations of 2.0 M or less. Using sulfuric acid for the preliminary tests was avoided because it can cause the REEs to co-precipitate with Ca minerals and interfere with the analysis of the REE-leaching behavior.

In the preliminary leaching, 1.0–3.0 M of nitric acid was used as a leaching reagent as it has less effect on the REE-leaching behavior since nitrate anions rarely react with any metal cations. The leaching temperature was varied as 25, 50 and 80 °C, with leaching times of 10 h, 3 h and 3 h, respectively. All leaching tests were conducted with a pulp density of 10% (*w*/*w*). The leaching efficiency and final metal recovery were calculated as follows, respectively:

$$\text{Leaching efficiency}(\%) = \frac{C_L \times V}{M_R \times m_R + C_L \times V} \times 100 \quad (1)$$

$$\text{Final recovery}(\%) = \frac{C_L \times V}{M_F \times m_F} \times 100 \quad (2)$$

Here, C_L is the metal concentration dissolved in the leaching solution (mg/L), V is the volume of the leaching solution (L), M_R and M_F are the metal contents in the leaching residue and ore (mg/g), respectively, and m_R and m_F are the amounts of the leaching residue and the ore used in the leaching test (g), respectively.

The leaching solution was sampled at regular intervals, and the samples were analyzed using inductively coupled plasma optical emission spectrometry (ICP-OES) to measure the metal concentration in the solution. The leaching residue and the ore were first fused by NaOH, with an ore to NaOH ratio of 1:10 (*w*/*w*) at 400 °C for 1 h. The fused products were then dissolved in aqua regia to measure the metal contents in the residue and the ore. They were also analyzed by ICP-OES. This solution analysis procedure was applied to all leaching tests in this study.

2.3. *Sulfuric Acid Baking–Water-Leaching*

To decompose the ore to a more soluble form, sulfuric acid baking was applied to the ore treatment. Various acid/solid ratios (*w*/*w*) were tested, i.e., 0.5, 1.0 and 2.0, and the ore and acid mixture was baked in a muffle furnace at 200 °C for 2 h. The baked ore was subsequently introduced to the water-leaching step. Water-leaching was conducted at 25, 50 and 80 °C for 3 h with a pulp density of 10% (*w*/*w*).

2.4. *Caustic Digestion–Acid-Leaching*

For caustic digestion, sodium hydroxide was used as a caustic chemical, and NaOH solutions with concentrations of 20, 30, 40 and 50% (*w*/*w*) were prepared. For the caustic digestion step, the highest temperature without boiling was chosen for each NaOH solution concentration, i.e., 105 °C at 20%, 115 °C at 30%, 120 °C at 40% and 145 °C at 50%. The digestion was conducted for 3 h, and the digested solid was then prepared for the subsequent acid-leaching reaction after solid/liquid separation and washing. The metal contents in the solid before and after digestion were measured by X-ray fluorescence (XRF). The digested solid was leached by hydrochloric acid (1.0–3.0 M) at 80 °C and 10% (*w*/*w*) pulp density for 3 h.

3. Results and Discussion

3.1. *Preliminary Leaching Results*

As a baseline, ore-leaching tests in nitric acid solution at different HNO_3 concentrations and temperatures were conducted. In the preliminary leaching, the leaching efficiency was equal to the final recovery as it was a single-step process.

Prior to the leaching tests, Eh–pH diagrams of REEs were considered. According to the Eh–pH diagrams (Figure 5), REE(III) hydroxide can simply dissolve in an acidic pH region. Cerium shows a slightly different characteristic because it can exist as two valence states, i.e., +3 and +4. Ce(III) hydroxide behaves in a similar manner to other REE(III) hydroxides; however, if it oxidizes to a higher valence state, Ce(IV), its leaching can be inhibited under mild acidic conditions. The leaching behavior of Ce(IV) will be discussed thoroughly in later sections.

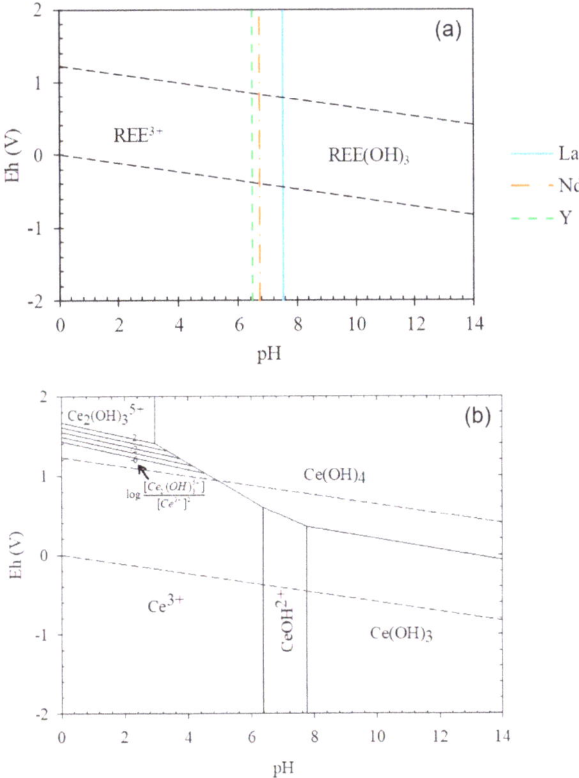

Figure 5. Eh–pH diagrams of REE systems: (**a**) (La, Nd and Y)–OH; (**b**) Ce–OH ([Ce], [La], [Nd], [Y] = 0.01 M).

As depicted in Figure 6 and Table 3, the REE-leaching efficiency generally increased with increasing acid concentration from 1.0 to 2.0 M at 25 °C (Ce = 8.4–17.3%, La = 10.2–18.6%, Nd = 13.8–26.3% and Y = 23.2–31.3%) and temperature from 25 to 80 °C in 2.0 M HNO_3 solution (Ce = 17.3–43.6%, La = 18.6–49.9%, Nd = 26.3–57.3% and Y = 31.3–48.2%). At 3.0 M HNO_3 and 80 °C, the REE recovery further improved to 63.0% Ce, 61.4% La, 72.3% Nd and 55.9% Y. However, even though the REE-leaching efficiency was improved by the high acid concentration and temperature, the maximum efficiency only ranged from 61% to 73% and approximately 30–40% of the REEs remained in the solid phase.

The leaching levels of the major impurities, namely Fe, Ca and P, were also examined. The leaching levels of Fe and P were not significant because their maximum values were only 20.0% and 41.2%, respectively, at 3.0 M HNO_3 and 80 °C. However, the Ca-leaching level was as high as 94–100% even at the most moderate condition of 1.0 M HNO_3 and 25 °C. Considering the mineralogy of the ore, Ca was mostly derived from apatite, and the pH of the solution was sufficiently low to dissolve apatite (less than 0.25, Table 3), which is available at pH values between 1.0 and 3.3 [20]. The impurity-leaching behavior will be discussed in more detail in later sections related to the REE-leaching behavior.

In the succeeding residue analysis by SEM-EDX, some REE oxide particles or REE-containing grains were trapped in other mineral matrices (Figure 7). The size of REE grains was in the range between 1.5 and 12 μm, which suggests that further liberation was essential. However, the ore particle size was already very fine at P_{80} of 30 μm; therefore, the authors decided to conduct chemical decomposition methods to free the REE-containing grains from the trapping matrices and improve the leaching recovery.

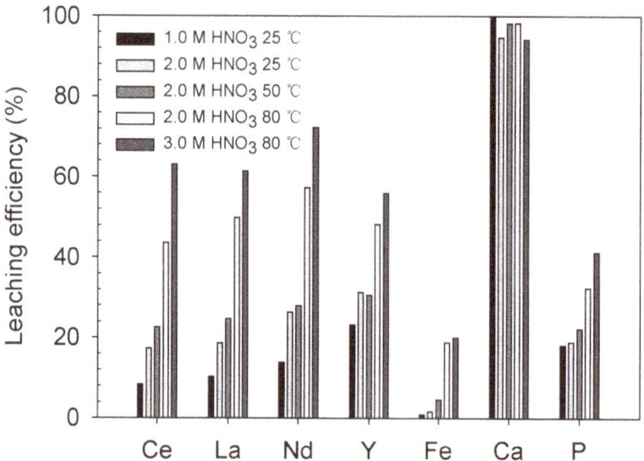

Figure 6. Preliminary leaching results: leaching efficiency (%) of REEs and impurities at different HNO_3 concentrations and temperatures.

Table 3. Preliminary leaching results: leaching efficiency (%) of REEs and impurities at different HNO_3 concentrations and temperatures.

HNO_3 Concentration (M)	Temperature (°C)	Initial pH	Leaching Efficiency (%)						
			Ce	La	Nd	Y	Fe	Ca	P
1.0	25	0.25	8.4	10.2	13.8	23.2	0.9	100.0	18.0
2.0	25	−0.37	17.3	18.6	26.3	31.3	1.6	94.6	18.9
2.0	50	−0.24	22.5	24.7	27.8	30.5	4.5	98.1	22.2
2.0	80	−0.13	43.6	49.9	57.3	48.2	18.8	98.2	32.3
3.0	80	−0.36	63.0	61.4	72.3	55.9	19.9	94.1	41.2

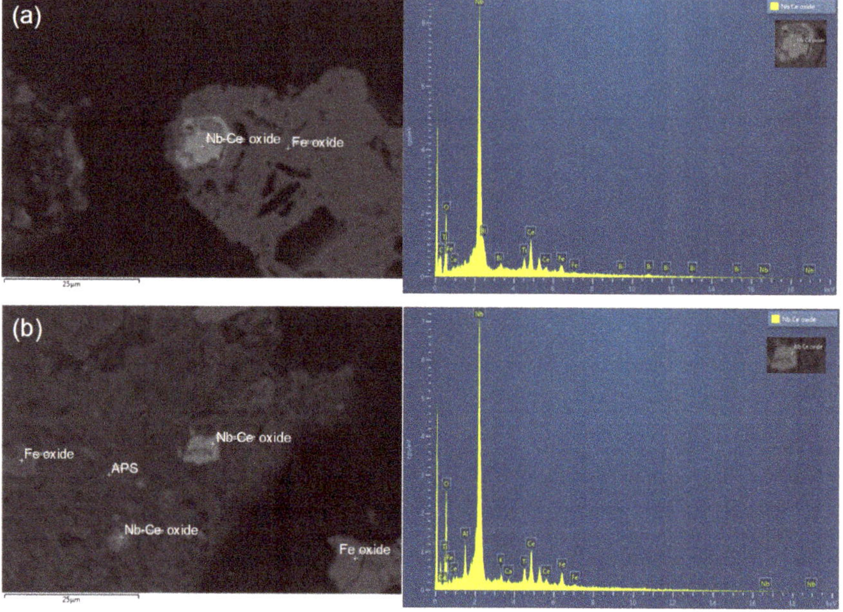

Figure 7. SEM-EDX analysis of preliminary leaching residue (leached at 3.0 M HNO_3 and 80 °C): (**a**) Fe oxide grain trapping Nb–Ce oxide; (**b**) micro-breccia of alumino-phosphate-sulfate (APS) bearing Fe oxide and Nb–Ce oxide.

3.2. Sulfuric Acid Baking–Water-Leaching

3.2.1. Effect of Acid/Solid Ratio

Sulfuric acid baking was expected to decompose the minerals trapping the REEs and the REE phosphate mineral itself (monazite and REE-substituted apatite). Because the entire baking product was sent to the next water-leaching step, it was considered that the leaching efficiency was equal to the final recovery. The detailed recovery and mass balance data are provided in Table S1 of a Supplemental data.

Through the sulfuric acid baking, the expected chemical reactions are likely to be those indicated in Equations (3)–(5). The ore and acid mixture before and after baking are displayed in Figure 8. According to Equation (3) and other previous studies [11,24], FeOOH, an Fe(III)-containing mineral, can only convert to Fe(III) sulfate species during sulfuric acid baking. Similarly, in this study, it was confirmed that most of the FeOOH was converted to $Fe_2(SO_4)_3$ after the acid baking, regardless of the acid amount, at 1.0 and 2.0 acid/solid ratios (Figure 9). Although it was not detected from the XRD analysis, Fe(II) sulfate species seemed to be also formed, because the baked product color was pale green. This implied that a reductive reaction occurred during the baking process, but the mechanism was not clear. The authors assumed that it was the effect of unknown reductive materials in the ore and lack of oxygen sources in the closed muffle furnace.

$$2FeOOH + 3H_2SO_4 \leftrightarrow Fe_2(SO_4)_3 + 4H_2O \quad (3)$$

$$REEPO_4 + H_2SO_4 \leftrightarrow REE_2(SO_4)_3 + H_3PO_4 \quad (4)$$

$$Ca_{10}(PO_4)_6X_2 + 10H_2SO_4 = 10CaSO_4 + 6H_3PO_4 + 2HX \ (X = OH,\ Cl,\ F) \quad (5)$$

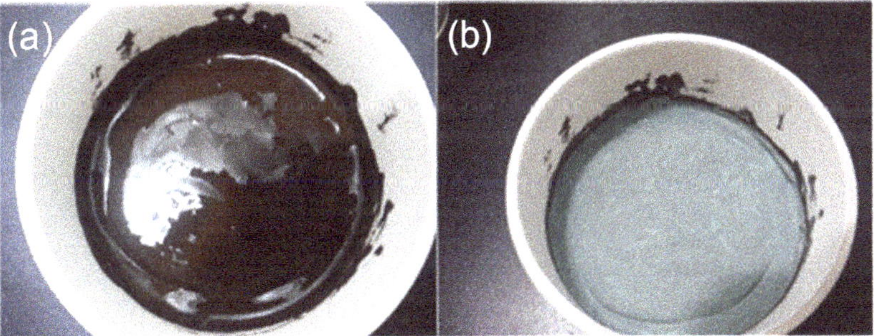

Figure 8. The ore and sulfuric acid mixture (**a**) before and (**b**) after baking at 200 °C for 2 h.

In the succeeding water-leaching step, the REE recovery was significantly improved as shown in Figure 10, as the maximum levels were Ce 98.4%, La 99.1%, Nd100% and Y 96.5% after baking at an acid/solid ratio of 2.0. The sulfuric acid concentration in the leaching step was calculated simply from the amount added for the baking step. The corresponding sulfuric acid concentrations in leaching solutions with acid/solid ratios of 0.5, 1.0 and 2.0 were 0.57, 1.13 and 2.27 M, respectively, although some part of the acid was consumed during the baking step. In the preliminary leaching results, when 0.5, 1.0 and 2.0 M of nitric acid were used, the REE-leaching efficiency was significantly less than that of the baked sample. Thus, it can be concluded that the sulfuric acid baking effectively decomposed the minerals in the ore.

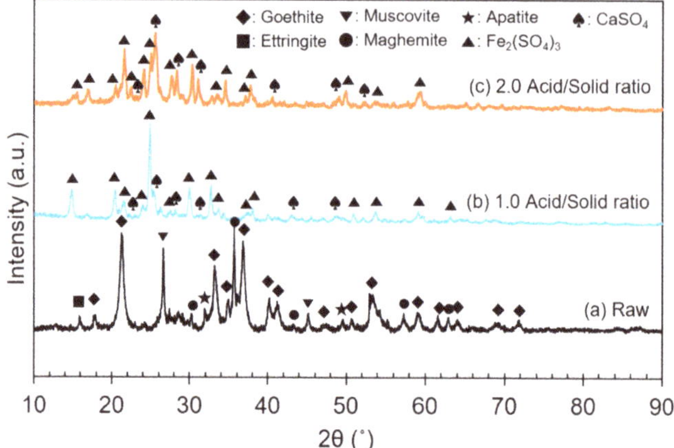

Figure 9. XRD patterns before and after sulfuric acid baking at different acid/solid ratios: (**a**) raw ore; (**b**) acid/solid ratio = 1.0; (**c**) acid/solid ratio = 2.0.

In terms of impurity-leaching, the Fe and P recoveries were considerably increased to 38.0% and 30.3%, 65.9% and 60.6% and 81.1% and 87.5% after baking at 0.5, 1.0 and 2.0 acid/solid ratios, respectively, compared with the maximum Fe and P recoveries of 20.0% and 41.2%, respectively, at 3.0 M nitric acid and 80 °C. This also implied that the minerals were effectively decomposed via the acid baking process. In addition, the Fe recovery was much higher in this study than in the previous one [24], which was approximately 20% at 1.5 acid/ore ratio. This was because the Fe(II) species formed in the current study were more soluble, whereas in the previous study, only less soluble Fe(III) species were present even after sulfuric acid baking.

In contrast, the Ca acted in a different manner from other metals. Its recovery remained at the same level of 57–58% regardless of the acid/solid ratio. This level was 1.62–1.65 times less than the recovery of the preliminary leaching results of 94.1% at 3.0 M nitric acid and 80 °C. Interestingly, the Ca recovery was considerably improved to 61.6, 74.4 and 97.7% after washing the leaching residues of 0.5, 1.0 and 2.0 acid/solid ratios, respectively. This is because calcium can easily form Ca-sulfate precipitates in a leaching solution in the presence of sulfate ions [20,25]; thus, the Ca recovery was not so high initially. After washing the residue with 3–4 times more water than the leaching solution amount, i.e., 300–400 mL, some of the Ca-sulfate precipitates were solubilized. More Ca dissolved at the higher acid/solid ratio because more residual acid remained in the solid residue pore. The precipitation problem will be further discussed in Section 3.2.2.

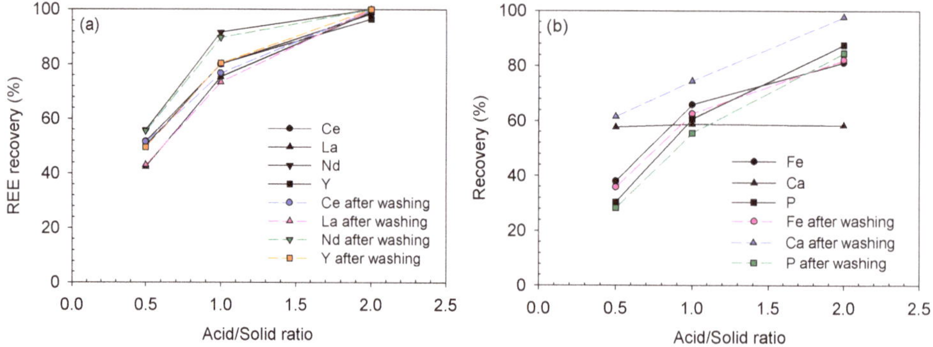

Figure 10. Leaching recoveries of (**a**) REEs and (**b**) impurities after water-leaching of the baked ore at different acid/solid ratios of sulfuric acid baking (baked at 200 °C for 2 h and leached at 25 °C for 3 h).

3.2.2. Effect of Leaching Temperature

The effect of temperature on the leaching process was examined for the ore treated with an acid/solid ratio of 2.0, which was the best condition for the highest recovery. As shown in Figure 11, the REE recovery decreased with increasing temperature from 25 to 80 °C. Between 25 and 50 °C, only a slight decrease in recovery occurred, but the REE recovery significantly decreased to 60–72% at 80 °C, compared with 96–100% at 25 °C. This trend was also observed in the recovery of impurities: the Fe, Ca and P recoveries decreased to 57.9, 51.2% and 73.2% at 80 °C, respectively, from 81.1, 58.3% and 87.5% at 25 °C, respectively.

The decrease in recovery was because of Ca-sulfate precipitation. Calcium can easily form precipitates with sulfate ions owing to its low solubility as presented in Table 4. The solubility of Ca sulfate, i.e., calcium sulfate dihydrate ($CaSO_4{\cdot}2H_2O$, CSD), is only 2.68 g/L at 25 °C (equivalent to 0.62 g/L Ca); then, it increases between 25 and 40 °C, and reduces to 2.59 g/L (0.60 g/L Ca) at 50 °C and 2.30 g/L (0.54 g/L Ca) at 80 °C [26,27]. The REE sulfates also showed a decrease in solubility with increasing temperature, but their solubility concentrations were significantly higher than their concentrations in the leaching solution (1.01 g/L Ce, 0.61 g/L La, 0.36 g/L Nd and 0.09 g/L Y at 2.0 acid/solid ratio and 25 °C). Despite these characteristics, the recoveries of REEs, Fe and P decreased at 80 °C. It seemed that these metals were adsorbed onto CSD and complexly reacted with each other to form co-precipitates. A similar behavior of REE co-precipitation and adsorption onto CSD was also observed in several previous studies [20,25,28]. They showed that the REE removal from the solution occurred at pH regions lower than the thermodynamic REE precipitation pH; thus, the adsorption onto the CSD surface would be a major process.

Moreover, it is known that the solubility of REE phosphates also decreases with increasing temperature [29–31]; therefore, the REE phosphate precipitation also seemed to affect the recoveries of REEs and impurities. After leaching at 80 °C, the recoveries of REEs and impurities were significantly improved by the washing process because of the solubilization of the precipitated CSD, which contained the adsorbed metals.

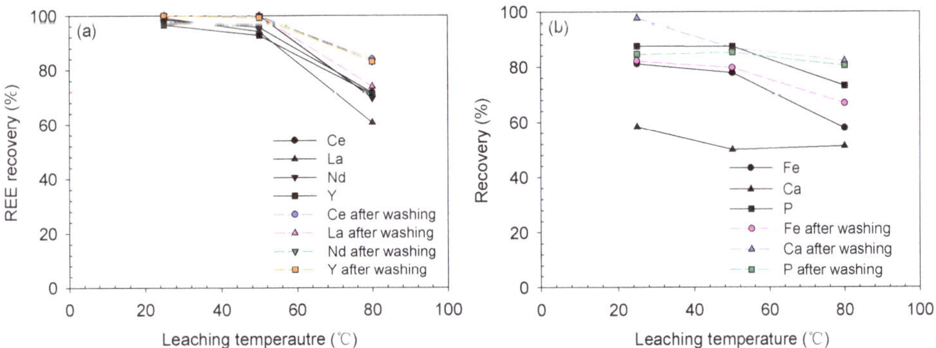

Figure 11. Leaching recoveries of (**a**) REEs and (**b**) impurities after water-leaching of the baked ore at 25–80 °C for 3 h (baked at 2.0 acid/solid ratio and 200 °C for 2 h).

Table 4. Solubilities of Ca sulfate and REE sulfate compounds at 25, 50 and 80 °C (the numbers inside the parentheses are the solubility of Ca and REEs).

Compound	Solubility (g/L)		
	25 °C	50 °C	80 °C
$CaSO_4{\cdot}2H_2O$ [1]	2.68 (0.62)	2.59 (0.60)	2.30 (0.54)
$Ce_2(SO_4)_3$ [2]	85.2 (42.0)	46.9 (23.1)	32.2 (15.9)
$La_2(SO_4)_3$ [2]	20.9 (10.3)	14.7 (7.22)	9.53 (4.68)
$Nd_2(SO_4)_3$ [2]	48.3 (24.2)	28.8 (14.4)	19.7 (9.87)
$Y_2(SO_4)_3$ [2]	61.4 (23.4)	43.1 (16.5)	29.0 (11.1)

[1] Dutrizac, 2002 [26]; [2] Das et al., 2019 [32].

3.3. Caustic Digestion–Acid-Leaching

3.3.1. Caustic Digestion Behavior

The second decomposition method, i.e., caustic digestion, was conducted at different concentrations of NaOH solution, i.e., 20–50 wt%. The significant chemical reactions of caustic digestion are shown in Equations (6)–(9).

$$REEPO_4 + 3NaOH \leftrightarrow REE(OH)_3 + Na_3PO_4 \quad (6)$$

$$Ca_{10}(PO_4)_6X_2 + 20NaOH \leftrightarrow 10Ca(OH)_2 + 6\,Na_3PO_4 + 2NaX\ (X = Cl,\ F) \quad (7)$$

$$Ca_{10}(PO_4)_6(OH)_2 + 18NaOH \leftrightarrow 10Ca(OH)_2 + 6\,Na_3PO_4 \quad (8)$$

$$AlPO_4 + 4NaOH \leftrightarrow Al(OH)_4^- + 4\,Na^+ + PO_4^{3-} \quad (9)$$

It was expected that due to caustic digestion, the Al and P would be dissolved into the solution because they are very soluble in strong basic solutions, whereas the REEs would remain in the solid phase as a hydroxide. Actually, as the NaOH concentration was increased, the REEs were concentrated in the digested product, except Y (Y is slightly amphoteric [33]), and the Al and P contents decreased in the solid phase and finally reached 1.0 wt% and 0.1 wt%, respectively, in the 50 wt% NaOH digested product (Figure 12). Iron is not soluble in basic solutions; thus, it was also concentrated in the digested product.

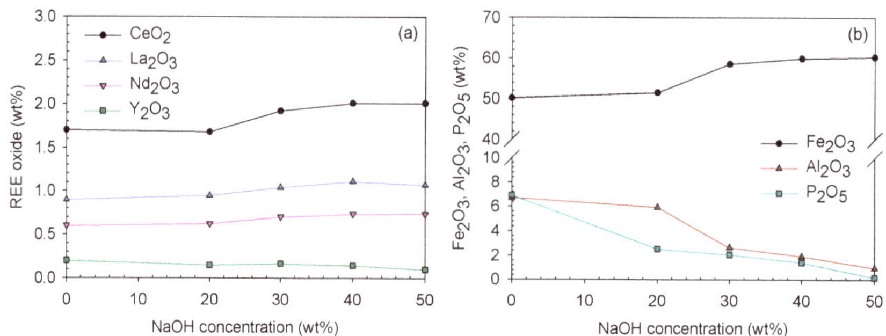

Figure 12. Oxide contents (wt%) of (**a**) REE and (**b**) Fe, Al and P of the digested product at different digesting NaOH concentrations.

Figure 13 shows that the intensity of goethite peaks in the digested product decreased. Representatively, at 2θ value of 21.3°, the peak intensity of goethite was 1019 and 363 before and after the digestion, respectively. Even some of the goethite peaks disappeared after digestion, although the Fe concentration increased. In contrast, the intensities of gangue mineral peaks, such as those of muscovite and ettringite, were relatively improved. These phenomena implied that the crystallinity of the goethite weakened, and it may be possible to free the trapped REE particles. Actually, as indicated in Figure 14, the surface of ore particles were chemically affected and eroded after caustic digestion.

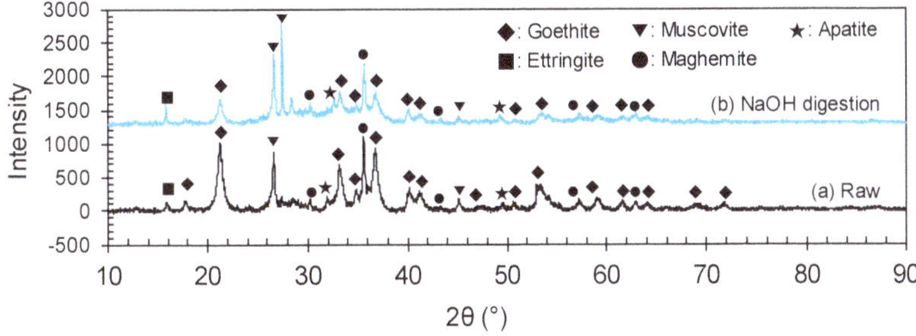

Figure 13. XRD patterns before and after caustic digestion using 50 wt% NaOH solution.

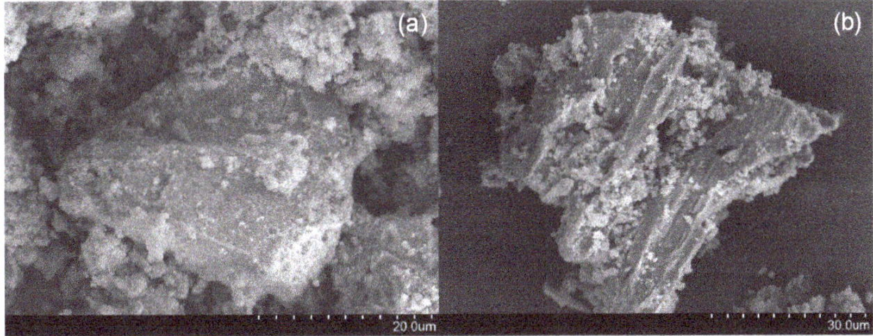

Figure 14. SEM images of the ore (**a**) before and (**b**) after caustic digestion (digestion condition: 50 wt% NaOH and 145 °C for 3 h).

3.3.2. Acid-Leaching Behavior

The caustic digested product was introduced to the succeeding acid-leaching step. During the digestion step, some elements of the ore were lost to the liquid phase, e.g., Al and P; thus, the leaching efficiency and final recovery were different in this test. The leaching efficiency was calculated based on the amount of elements in the digested product, whereas the recovery was based on the amount in the raw ore. The detailed recovery and mass balance data are provided in Table S2 of the Supplemental data.

The effect of acid concentration was initially investigated using 1.0–3.0 M hydrochloric acid. The digested ore was produced using a 50 wt% NaOH solution. As presented in Figure 15, the REE-leaching efficiency generally increased as the acid concentration increased, while Ce behaved in a slightly different manner.

In the case of Ce, its leaching efficiency only reached 12.1% at 1.0 M HCl. It was improved at elevated HCl concentrations of 2.0 and 3.0 M as 75.5% and 79.7%, respectively; however, these values were still lower than those of other REEs at 2.0 and 3.0 M HCl, which were in the range of 87–92% and 94–100%, respectively. It seemed that some part of Ce was converted from Ce(III) to Ce(IV) compound during the digestion step, probably $Ce(OH)_4$, which led to the low recovery of Ce. $Ce(OH)_4$ is less soluble than $Ce(OH)_3$ as the solubility product of $Ce(OH)_4$ is $10^{-51.56}$ and that of $Ce(OH)_3$ is $10^{-26.15}$ [34]. This can be also confirmed from the Eh–pH diagram of the Ce-OH system in Figure 5; $Ce(OH)_3$ can simply be leached at a slightly acidic pH, but $Ce(OH)_4$ needs a highly acidic pH. Although $Ce(OH)_4$ was not detected from the XRD analysis due to its low concentration, it was probable that $Ce(OH)_4$ was formed during the caustic digestion because of the significantly lower Ce-leaching efficiency than those of the other REEs.

The final recoveries of Ce, La and Nd showed almost similar trends with their leaching efficiencies. However, that of Y was significantly low, i.e., 26.2–38.4% at 1.0–3.0 M HCl. This was because Y was dissolved and lost during the digestion step, and only 40% of Y from the raw ore was introduced into the acid-leaching step.

In terms of impurity-leaching, the Fe-leaching efficiency and recovery almost linearly increased as the acid concentration increased. The Al-leaching efficiency ranged from 95% to 100%; hence, its final recovery was similar to that in the digestion step, which was in the range of 10–12%. Phosphorus was not detected in any leaching solutions because it was removed during the digestion step.

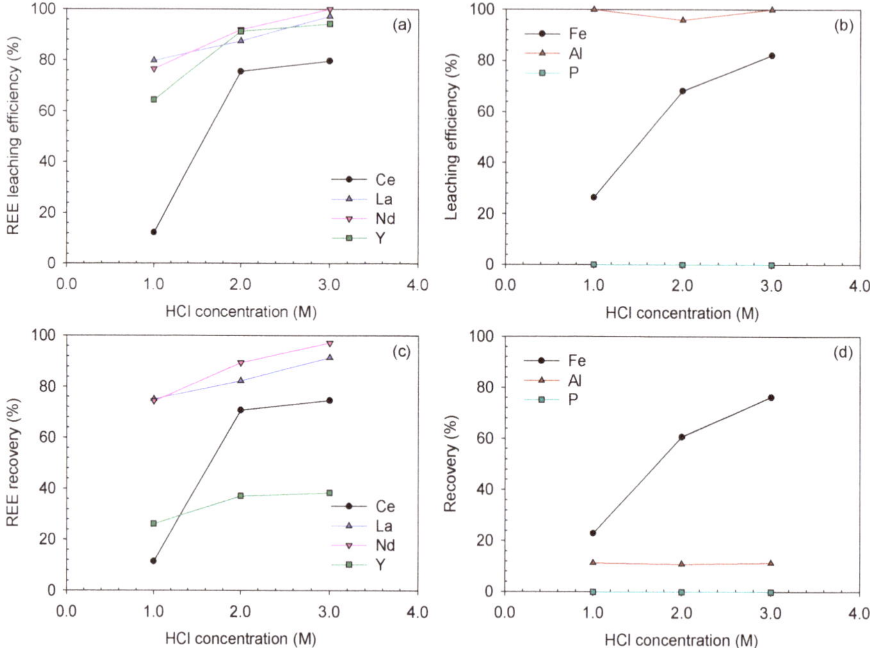

Figure 15. Leaching efficiencies and recoveries after acid-leaching of the caustic digested ore at HCl concentrations of 1.0–3.0 M and 80 °C for 3 h (digested using 50 wt% NaOH solution for 3 h): (**a**) REE and (**b**) Fe-, Al- and P-leaching efficiencies (%); (**c**) REE and (**d**) Fe, Al and P recoveries (%).

Figure 16 illustrates variation of leaching results with the digesting NaOH concentration. In terms of leaching efficiency and recovery, both were maximized at 30 wt% NaOH concentration, and those of La and Nd were not much changed at higher NaOH concentrations. However, Ce showed an evident decrease in the leaching efficiency and recovery at NaOH concentrations higher than 30 wt%, which was caused by the less soluble $Ce(OH)_4$ formation at higher NaOH concentrations. The final recovery of Y also significantly decreased as the NaOH concentration increased from 30 wt% to 50 wt%, although the leaching efficiency remained high at 93–94%. This was because its recovery from the digestion step decreased from 68% to 40% as the NaOH concentration increased from 30 wt% to 50 wt%.

Concerning the impurities, the Fe-leaching efficiency had a similar trend to that of the REEs (except Ce), which increased between 20 wt% and 30 wt% in the range of 74–82%. The Fe recovery was similar to its leaching efficiency, which ranged from 70% to 78%. However, the Al and P recoveries considerably decreased from 64.2% to 11.3% and 27.0% to 0%, respectively, as they were removed from the solution phase during the digestion step. In terms of leaching efficiency, that of P monotonously decreased, while that of Al fluctuated. It seemed that the residual NaOH in the digested product using 40 wt% NaOH solution suppressed the Al-leaching. A small amount of Al remained in the digested product using 50 wt% NaOH solution; thus, the Al-leaching efficiency reached 100%, although a higher NaOH concentration was applied.

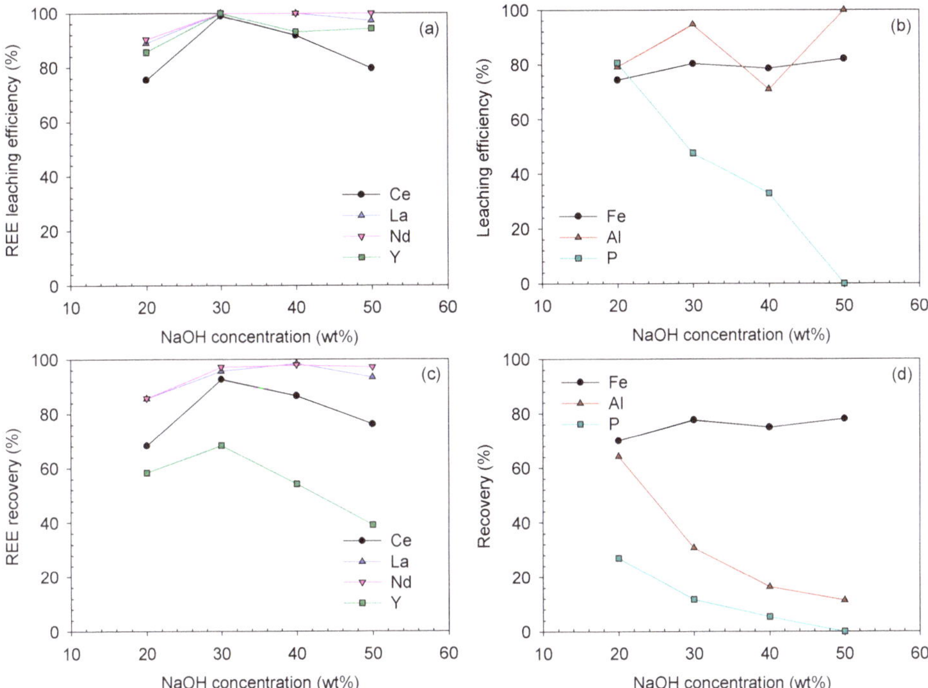

Figure 16. Leaching efficiencies and recoveries after acid-leaching of the caustic digested ore at digesting NaOH concentrations of 20–50 wt% (leached using 3.0 M HCl at 80 °C for 3 h): (**a**) REE and (**b**) Fe-, Al- and P-leaching efficiencies (%); (**c**) REE and (**d**) Fe, Al and P recoveries (%).

3.4. Determination of the Optimum Condition for Analyzing the REE–Fe-Leaching Efficiency Correlation

To statistically analyze the REE-leaching behavior, correlation curves of the REE–Fe-leaching efficiency are plotted in Figures 17 and 18 and the calculated R^2 values are listed in Table 5.

According to the analysis, the REE and Fe-leaching efficiencies were generally positively correlated. Specifically, the correlation was stronger during sulfuric acid baking–water-leaching (because the R^2 values were much closer or even equal to 1.0000) than during caustic digestion–acid-leaching. This means that (i) REE-leaching highly depends on the Fe-bearing mineral decomposition, which traps the REE constituents during the sulfuric acid baking–water-leaching process and (ii) REEs and Fe co-precipitation by CSD affected a good R^2 value; from REEs and Fe recovery trend before/after the residue washing process shown in Figures 10 and 11, their leaching behavior was somewhat governed by CSD formation as both the dissolved REEs and Fe can adsorb and desorb from CSD in a similar extent.

The curves showed a weaker REE–Fe correlation for the caustic digestion–acid-leaching test results, and the Ce–Fe curve based on the NaOH concentration showed the weakest correlation with an R^2 value of 0.1657. This is because of the Ce oxidation to the Ce(IV) species of $Ce(OH)_4$ as discussed earlier. In addition, the La and Nd leaching efficiencies showed weak correlations with that of Fe based on the digestion condition (Figure 18a) because they reached 100% recovery at 80% Fe-leaching efficiency, which is slightly lower than the maximum Fe recovery of 82%. If the points corresponding to the 50 wt% NaOH digestion are removed, the REE–Fe correlations are improved: 0.9998 Ce–Fe, 0.9158 La–Fe, 0.9158 Nd–Fe and 0.9575 Y–Fe. This implies that both the Fe-bearing mineral decomposition and digesting NaOH concentration can considerably affect the succeeding acid-leaching step in the caustic digestion–acid-leaching process. Based on the leaching condition, i.e., HCl concentration, the REE–Fe correlation was satisfactory regardless of the elements (Figure 18b).

Consequently, under the applied decomposition methods and leaching conditions in this study, the REE-leaching was more affected by Fe-leaching during sulfuric acid baking–water-leaching. At the Fe-leaching efficiencies of 68% and 66%, the REE-leaching efficiencies (except Ce) were 88–92% and 75–80% from the caustic digestion–acid-leaching and sulfuric acid baking–water-leaching, respectively. Therefore, caustic digestion was a better decomposition method in terms of REE-leaching, and the REE-leaching was inhibited by CSD co-precipitation in the sulfuric acid baking–water-leaching process as the leaching efficiency was lower than in caustic digestion–acid-leaching at the same corresponding Fe-leaching level.

Overall, as the Fe-leaching reaches the maximum, the REE-leaching efficiency also attains its highest value; among the factors, the Fe-bearing mineral decomposition has the most significant effect on REE-leaching. Thus, considering the REE-leaching efficiency only, the optimum decomposition-leaching conditions are as follows: 2.0 acid/solid ratio sulfuric acid baking +25 °C water-leaching for the sulfuric acid baking–water-leaching process and 30 wt% NaOH digestion + 3.0-M HCl-leaching at 80 °C for the caustic digestion–acid-leaching.

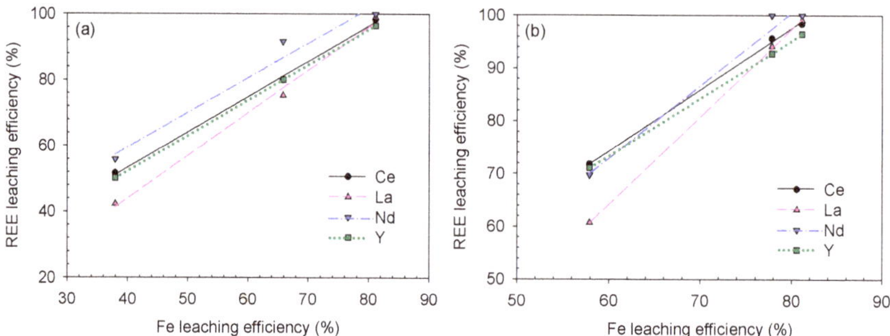

Figure 17. REE–Fe-leaching efficiency correlation for the sulfuric acid baking–water-leaching results based on (**a**) acid/solid ratio and (**b**) leaching temperature.

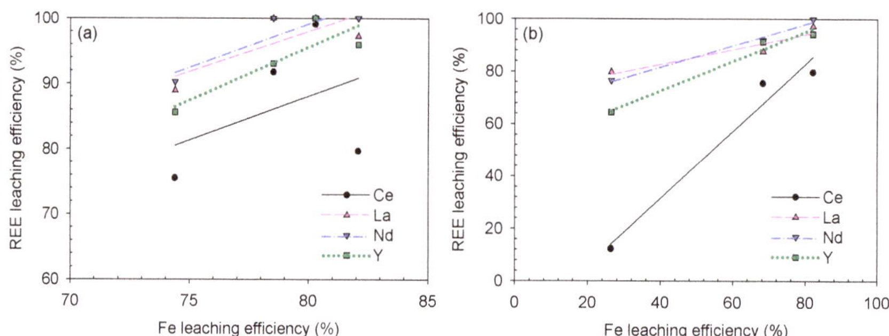

Figure 18. REE–Fe efficiency correlation for the caustic digestion–acid-leaching results based on (**a**) digesting NaOH solution concentration and (**b**) leaching HCl concentration.

Table 5. R^2 values of REE–Fe correlation curves based on the decomposition method and leaching condition.

Correlated Metals	Sulfuric Acid Baking-Water-Leaching		Caustic Digestion–Acid-Leaching	
	Acid/Solid Ratio	Leaching Temperature	NaOH Concentration	HCl Concentration
Ce–Fe	0.9980	0.9984	0.1657	0.9652
La–Fe	0.9946	0.9997	0.5990	0.8849
Nd–Fe	0.9692	0.9824	0.8062	0.9899
Y–Fe	1.0000	1.0000	0.7706	0.9772

4. Conclusions

In this study, the effects of two different chemical decomposition methods, namely sulfuric acid baking and caustic digestion, to enhance the REE recovery from a refractory ore were investigated. The REE-containing particles were scattered throughout the ore, and some of them were even trapped in other minerals such as Fe oxide, although the total REE concentration was relatively high at 3.4% TREO. From the preliminary leaching tests using nitric acid, the maximum REE recovery only reached 61–73%; thus, a significant amount of REEs was still left in the residue.

Using sulfuric acid baking–water-leaching, the REE recovery was improved to 97–100% after the ore was baked at 2.0 acid/solid ratio and 200 °C for 2 h. The REE recovery reduced with increasing leaching temperature because of the CSD solid formation at elevated temperatures and the accompanying REE co-precipitation.

During the caustic digestion reaction, the REEs were concentrated in the solid phase as Al and P were removed to the solution phase by the decomposition reaction. In the succeeding acid-leaching, all REE-leaching efficiencies reached the maximum after 30 wt% NaOH digestion for 3 h and 3.0 M HCl-leaching at 80 °C for 3 h. With increasing NaOH concentration, Ce became oxidized to Ce(IV) species of $Ce(OH)_4$, and its leaching efficiency from the acid-leaching step decreased.

From the REE–Fe correlation analysis, a stronger correlation between the REEs and Fe was observed for the sulfuric acid baking–water leaching results because of not only Fe-bearing mineral decomposition to iron sulfate, but also CSD precipitation with REEs and Fe. In addition, comparing the sulfuric acid baking–water-leaching and caustic digestion–acid-leaching, the caustic digestion showed a more powerful effect on enhancing the REE-leaching as it resulted to a higher REE-leaching efficiency at the same Fe-leaching level than that of sulfuric acid baking. It can be concluded that in terms of REE-leaching efficiency, as the Fe-leaching reached the maximum, the REE-leaching efficiency also attained its highest value. Furthermore, the Fe-bearing mineral decomposition significantly affected the REE-leaching of the studied ore.

Supplementary Materials: The following are available online at http://www.mdpi.com/2075-163X/10/6/532/s1, Table S1. Mass balance of sulfuric acid baking–water-leaching process; Table S2. Mass balance of caustic digestion–acid-leaching process.

Author Contributions: Conceptualization, R.K. and H.C.; methodology, R.K. and H.C.; validation, R.K., J.J., J.K. and S.L.; formal analysis, R.K., J.J., J.K. and S.L.; investigation, R.K.; resources, H.C., K.W.C., H.-S.Y. and C.-J.K.; data curation, R.K.; writing—original draft preparation, R.K.; writing—review and editing, H.C., K.W.C., H.-S.Y. and C.-J.K.; visualization, R.K.; supervision, H.C.; funding acquisition, H.C. and K.W.C. All authors have read and agreed to the published version of the manuscript.

Funding: This research was supported by the Energy Efficiency and Resources Core Technology Program of the Korea Institute of Energy Technology Evaluation and Planning (KETEP) granted financial resource from the Ministry of Trade, Industry and Energy (MOTIE) of the Korean government (No. 20122010300041). This research was also supported by the R&D Center for Valuable Recycling (Global-Top R&BD Program) of the Ministry of Environment of the Korean government (No. 2019002220001/KIGAM 20-9894).

Conflicts of Interest: The authors declare no conflicts of interest.

References

1. Jordens, A.; Cheng, Y.P.; Waters, K.E. A review of the beneficiation of rare earth element bearing minerals. *Miner. Eng.* **2013**, *41*, 97–114. [CrossRef]
2. Massari, S.; Ruberti, M. Rare earth elements as critical raw materials: Focus on international markets and future stategies. *Resour. Policy* **2013**, *38*, 36–43. [CrossRef]
3. Golev, A.; Scott, M.; Erskine, P.D.; Ali, S.H.; Ballantyne, G.R. Rare earths supply chains: Current status, constraints and opportunities. *Resour. Policy* **2014**, *41*, 52–59. [CrossRef]
4. Xie, F.; Zhang, T.A.; Dreisinger, D.; Doyle, F. A critical review on solvent extraction of rare earths from aqueous solutions. *Miner. Eng.* **2014**, *56*, 10–28. [CrossRef]
5. Tian, J.; Chi, R.; Yin, J. Leaching process of rare earths from weathered crust elution-deposited rare earth ore. *Trans. Nonferrous Met. Soc. China* **2010**, *20*, 892–896. [CrossRef]
6. Chen, Z. Global rare earth resources and scenarios of future rare earth industry. *J. Rare Earths* **2011**, *29*, 1–6. [CrossRef]
7. Moldoveanu, G.A.; Papangelakis, V.G. Recovery of rare earth elements adsorbed on clay minerals: II. Leaching with ammonium sulfate. *Hydrometallurgy* **2013**, *131–132*, 158–166. [CrossRef]
8. Sadri, F.; Nazari, A.M.; Ghahreman, A. A review on the cracking, baking and leaching processes of rare earth element concentrates. *J. Rare Earths* **2017**, *35*, 739–752. [CrossRef]
9. Berry, L.; Agarwal, V.; Galvin, J.; Safarzadeh, M.S. Decomposition of monazite concentrate in sulphuric acid. *Can. Metall. Quart.* **2018**, *57*, 422–433. [CrossRef]
10. Demol, J.; Ho, E.; Senanayake, G. Sulfuric acid baking and leaching of rare earth elements, thorium and phosphate from a monazite concentrate: Effect of bake temperature from 200 to 800 °C. *Hydrometallurgy* **2018**, *179*, 254–267. [CrossRef]
11. Borra, C.R.; Mermans, J.; Blanpain, B.; Pontikes, Y.; Binnemans, K.; Gerven, T.V. Selective recovery of rare earths from bauxite residue by combination of sulfation, roasting and leaching. *Miner. Eng.* **2016**, *92*, 151–159. [CrossRef]
12. Jha, M.K.; Kumari, A.; Panda, R.; Kumar, J.R.; Yoo, K.; Lee, J.Y. Review on hydrometallurgical recovery of rare earth metals. *Hydrometallurgy* **2016**, *165*, 2–26. [CrossRef]
13. Soltani, F.; Abdollahy, M.; Petersen, J.; Ram, R.; Javad Koleini, S.M.; Moradkhani, D. Leaching and recovery of phosphate and rare earth elements from an iron-rich fluorapatite concentrate: Part II: Selective leaching of calcium and phosphate and acid baking of the residue. *Hydrometallurgy* **2019**, *184*, 29–38. [CrossRef]
14. Huang, Y.; Dou, Z.; Zhang, T.; Liu, J. Leaching kinetics of rare earth elements and fluoride from mixed rare earth concentrate after roasting with calcium hydroxide and sodium hydroxide. *Hydrometallurgy* **2017**, *173*, 15–21. [CrossRef]
15. Buchanan, J.; Reveley, S.; Forrester, K.; Cummings, A. Review of current rare earth processing practice. In Proceedings of the ALTA 2014, Perth, Australia, 24–31 May 2014.
16. Habashi, F. The recovery of the lanthanides from phosphate rock. *J. Chem. Technol. Biotechnol. Chem. Technol.* **1985**, *35*, 5–14. [CrossRef]
17. Jorjani, E.; Bagherieh, A.H.; Chelgani, S.C. Rare Earth Elements Leaching from Chadormalu Apatite Concentrate: Laboratory Studies and Regression Predictions. *Korean J. Chem. Eng.* **2011**, *28*, 557–562. [CrossRef]
18. Preston, J.S.; Cole, P.M.; Craig, W.M.; Feather, A.M. the recovery of rare earth oxides from a phosphoric acid by-product. Part 1: Leaching of rare earth values and recovery of a mixed rare earth oxide by solvent extraction. *Hydrometallurgy* **1996**, *41*, 1–19. [CrossRef]
19. Wang, L.; Long, Z.; Huang, X.; Yu, Y.; Cui, D.; Zhang, G. Recovery of rare earths from wet-process phosphoric acid. *Hydrometallurgy* **2010**, *101*, 41–47. [CrossRef]
20. Kim, R.; Cho, H.; Han, K.N.; Kim, K.; Mun, M. Optimization of acid leaching of rare-earth elements from mongolian apatite-based ore. *Minerals* **2016**, *6*, 63. [CrossRef]
21. Neumann, R.; Medeiros, E.B. Comprehensive mineralogical and technological characterisation of the Araxá (SE Brazil) complex REE (Nb-P) ore, and the fate of its processing. *Int. J. Miner. Process.* **2015**, *144*, 1–10. [CrossRef]
22. Kuzmin, V.I.; Pashkov, G.L.; Lomaev, V.G.; Voskresenskaya, E.N.; Kuzmina, V.N. Combined approaches for comprehensive processing of rare earth metal ores. *Hydrometallurgy* **2012**, *129–130*, 1–6. [CrossRef]

23. Kim, R.; Cho, H.; Han, K.N. Behavior of anions in association with metal ions under hydrometallurgical environments: Part I–OH- effect on various cations. *Min. Metall. Explor.* **2014**, *31*, 34–39. [CrossRef]
24. Guo, X.; Li, D.; Park, K.; Tian, Q.; Wu, Z. Leaching behavior of metals from a limonitic nickel laterite using a sulfation–roasting–leaching process. *Hydrometallurgy* **2009**, *99*, 144–150. [CrossRef]
25. Sadri, F.; Kim, R.; Yang, Z.; Ghahreman, A. The effect of calcium sulfate crystallization and the crystal modification on aqueous REE stability in Ca saturated REE-Ca-SO_4-H_2O systems. *Hydrometallurgy* **2018**, *182*, 82–96. [CrossRef]
26. Dutrizac, J.E. Calcium sulphate solubilities in simulated zinc processing solutions. *Hydrometallurgy* **2002**, *65*, 109–135. [CrossRef]
27. Azimi, G.; Papangelakis, V.G.; Dutrizac, J.E. Modelling of calcium sulphate solubility in concentrated multi-component sulphate solutions. *Fluid Phase Equilibr.* **2007**, *260*, 300–315. [CrossRef]
28. Sadri, F.; Kim, R.; Ghahreman, A. Substitution of Calcium with Ce, Nd, Er, and Tb in the Structure of Microcrystals of Calcium Sulfates with Controlled Hydration Water: A Proposed Mechanism. *Cryst. Growth Des.* **2019**, *19*, 2621–2631. [CrossRef]
29. Firsching, F.H.; Brune, S.N. solubility products of the trivalent rare-earth phosphates. *J. Chem. Eng. Data* **1991**, *36*, 93–95. [CrossRef]
30. Poitrasson, F.; Oelkers, E.; Schott, J.; Montel, J. Experimental determination of synthetic $NdPO_4$ monazite end-member solubility in water from 21 °C to 300 °C: Implications for rare earth element mobility in crustal fluids. *Geochim. Cosmochim. Acta* **2004**, *68*, 2207–2221. [CrossRef]
31. Cetiner, Z.S.; Wood, S.A.; Gammons, C.H. The aqueous geochemistry of the rare earth elements. Part XIV. The solubility of rare earth element phosphates from 23 to 150 °C. *Chem. Geol.* **2005**, *217*, 147–169. [CrossRef]
32. Das, G.; Lencka, M.M.; Eslamimanesh, A.; Wang, P.; Anderko, A.; Riman, R.E.; Mavrotsky, A. Rare earth sulfates in aqueous systems: Thermodynamic modeling of binary and multicomponent systems over wide concentration and temperature ranges. *J. Chem. Thermodyn.* **2019**, *131*, 49–79. [CrossRef]
33. Ivano-Emie, B.N.; Nisel'son, L.A.; Ivolgina, A.T. Solubility of yttrium hydroxide in sodium hydroxide solution. *Zhur. Neorg. Khim.* **1961**, *6*, 1483–1484.
34. Joshi, S.; Kulp, E.A.; Fahrenholtz, W.G.; O'Keefe, M.J. Dissolution of cerium from cerium-based conversion coatings on Al 7075-T6 in 0.1 M NaCl solutions. *Corros. Sci.* **2012**, *60*, 290–295. [CrossRef]

Article

Concentration of Rare Earth Elements (Sc, Y, La, Ce, Nd, Sm) in Bauxite Residue (Red Mud) Obtained by Water and Alkali Leaching of Bauxite Sintering Dust

Leonid Chaikin [1], Andrei Shoppert [1,*], Dmitry Valeev [2,3], Irina Loginova [1] and Julia Napol'skikh [1]

1 Department of Non-Ferrous Metals Metallurgy, Ural Federal University Named after the First President of Russia B.N. Yeltsin, 620002 Yekaterinburg, Russia; l.i.chaikin@urfu.ru (L.C.); i.v.loginova@urfu.ru (I.L.); anapolskikh512@gmail.com (J.N.)

2 Laboratory of New Metallurgical Processes (#24), A.A. Baikov Institute of Metallurgy and Materials Science, Russian Academy of Sciences, 49, Leninsky Prospect, 119334 Moscow, Russia; dvaleev@imet.ac.ru

3 Scientific Research Centre "Thermochemistry of Materials", National University of Science & Technology "MISIS", 4, Leninsky Prospect, 119049 Moscow, Russia

* Correspondence: a.a.shoppert@urfu.ru; Tel.: +7-922-024-3963

Received: 31 March 2020; Accepted: 28 May 2020; Published: 30 May 2020

Abstract: One of the potential sources of rare-earth elements (REE) is the industrial waste known as red mud (bauxite residue), in which the majority of REE from the initial bauxite are concentrated via the Bayer process. Therefore, the studies of the subject, both in Russia and outside, focus almost exclusively on red mud processing. This article looks into the possibility of REE concentration into red mud by leaching an intermediate product of the bauxite sintering process at Russian alumina refineries, namely electrostatic precipitator (ESP) dust. The experimental works were performed by X-ray diffraction (XRD)and electron probe microanalysis (EPMA) of the sinter and sinter dust. The determination of major and rare-earth elements in the sinter from the rotary kilns and in the ESP dust before and after leaching was carried out by X-ray fluorescence (XRF) and plasma mass spectrometry (ICP-MS). The study showed that it is possible to obtain red mud that contains three times more REE than traditional waste red mud after two-stage leaching ESP dust in the water at 95 °C followed by leaching in an alkaline-aluminate liquor at 240 °C. The shrinking core model was used to study the kinetics of leaching of the original ESP dust and water-treated dust in alkaline-aluminate liquor. The study showed the change in the limiting stage of the alkaline leaching process after water treatment, with the activation energy growing from 24.98 to 33.19 kJ/mol.

Keywords: rare-earth elements; scandium; alumina production; sinter processes; red mud; electrostatic precipitation dust; alkali leaching; kinetics

---

1. Introduction

The volumes of production/consumption of rare and rare-earth metals, which are critical for modern industries, are steadily increasing [1–4]. In this group of metals—scandium, yttrium, lanthanum, and lanthanides—scandium is the most scarce and commercially attractive one. The increased interest in scandium is associated with its application in various industries [5–7]. Small amounts of scandium are found in ores of aluminum, cobalt, iron, molybdenum, nickel, phosphate, titanium, tungsten, uranium, and others [8–16].

At the moment, phosphogypsum—waste from the processing of apatite concentrates [17,18]—and red mud (bauxite residue) are considered to be the most promising sources of scandium [19–24]. Phosphogypsum is also considered to be the most promising source of other rare earth metals [6], as it may contain up to 1% of rare earth elements in total [25]. More than 250 million tons of phosphogypsum has been accumulated in Russia [26] with an annual increase by over 14 million tons. However, despite the large number of works devoted to the processing of phosphogypsum, there is still no commercially viable technology for its processing [27], which is explained by isomorphic co-crystallization of REE with gypsum, and therefore the need for its complete release [28].

Bauxite residue generated in the recovery and processing of bauxites is a source of environmental pollution [29,30] on the one hand and a promising object for obtaining valuable elements [31] on the other hand. Bauxite residue contains a high amount of scandium (70–260 ppm) that is close to its primary resources [32]. Large amounts of this waste form what may be called man-made deposits, which can be recycled into building materials, pigments, coagulants for wastewater treatment, and which can be used to re-extract alumina, extract iron concentrates, titanium, rare and rare earth metals [33–35]. In the industrial production of alumina, almost all Sc is concentrated in red mud. According to available estimates, the global reserves of scandium in the industrial waste range from 1 to 3 million tons, with 70–80% of it being contained in bauxite processing waste [36]. The full processing of the bauxite residue currently generated makes it possible to extract 6600–20,400 t of scandium per year. Therefore, a large number of studies have been dedicated to the recovery of rare-earth metals from various types of bauxite residue [37] obtained by processing various bauxites in the Bayer method and its variations. However, the low content of scandium in the red mud, the high content of alkali and alkaline earth metals, and the simultaneous recovery of iron together with REE [38] or low extraction efficiency [39,40] render the existing methods for extracting scandium from this type of industrial waste non-viable.

It should be noted that most studies of the separation of REE from alumina products have focused on the processing of red mud. In this article, we made an attempt to study the possibility of obtaining the REE concentrate from electrostatic precipitator (ESP) dust, which is an intermediate product of kiln sintering in alumina production. The dust is generated in a rotary kiln (Figure 1), where low-quality bauxite (the mass ratio of Al_2O_3 to SiO_2 (silicon modulus) is less than 7) is sintered with soda ash in order to convert the alumina-containing mineral into a water-soluble sodium aluminate. Iron and silicon-containing minerals react similarly with soda to form ferrite and sodium silicate, respectively. The technological process, especially at the drying and decarbonization stages, generates a large number of fine particles less than 100 μm (25–40% of the charge), which are carried away with the exhaust gases. To clean the exhaust gases of kilns from dust particles, a three-stage capture system is used. The first stage is the dust chamber where most coarse dust particles precipitate; the second stage is cyclones, where medium dust particles are removed from the exhaust gases through vortex separation; and the final stage is electrostatic precipitator. The last stage captures the finest particle fraction (less than 20 μm) using the force of an induced electrostatic charge.

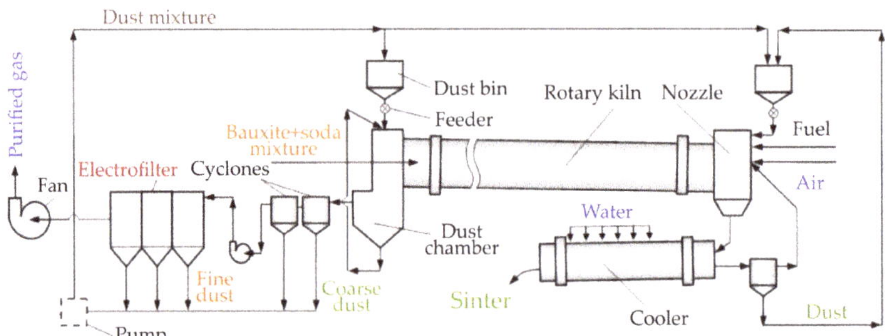

Figure 1. Flow chart of the sintering process on the RUSAL-Kamensk-Uralsky alumina refinery.

It is impossible to remove all dust from the technological process since that would upset the heat balance of the furnace and would greatly complicate the movement of the charge in the drying zone.

On the other hand, the ESP dust acts mostly as a dead weight, since a number of dust circulations through the kiln are larger than through all other units of the gas treatment system. Electrostatic precipitators capture only the smallest particles, which often pass very quickly through the hot zones of the furnace, and the sintering process fails to complete. Ultimately, they acquire the charge of the electrostatic precipitator and are released into the atmosphere. The physicochemical properties of sintering dust and the possibility of their leaching together with bauxite had already been investigated in our previous work [41]. It was shown that ESP dust can be highly reactive and can be removed from the process with further leaching to extract useful components. Also, the high recovery rate of ESP dust means significant losses of secondary heat. The proportion of ESP dust in the total mass of dust is up to 15%, which means that alumina refineries in Russia produce more than 30 thousand tons of ESP dust per year. At the same time, ESP dust contains about 50 ppm of scandium [42]. Therefore, more than 1.5 t of scandium and even more of other rare earth metals can be obtained annually from this intermediate product.

In this work we studied the possibility of leaching ESP dust with water and mother liquor of the Bayer process in order to extract rare-earth elements into valuable components such as alumina, caustic alkali, and soda, and, at the same time, concentrate them in the red mud resulting from leaching.

2. Materials and Methods

2.1. Solid Phase Characterization

The phase and quantitative composition of the ESP dust from sintering kilns of the Kamensk-Uralsky alumina refinery and leaching products were determined by X-ray diffraction (XRD) on a Rigaku D/MAX-2200 diffractometer (Rikagu Corp., Tokyo, Japan) using the PDF-2 database (International Center for Diffraction Data) and by X-ray fluorescence (XRF) using an Axios MAX X-ray fluorescence spectrometer (Malvern Panalytical Ltd., Almelo, The Netherlands). The content of REE micro impurities in the feedstock and leachate was determined using inductively coupled plasma mass spectrometry (ICP-MS) on a PerkinElmer NexION 300S instrument (PerkinElmer Inc., Waltham, MA, USA).

Electron probe micro-analysis (EPMA) was performed using a Cameca SX 100 microanalyzer (CAMECA Instruments, Inc., Madison, WI, USA) equipped with an energy-dispersive X-ray spectroscopy analysis (EDS) module Bruker XFlash 6 (Bruker Nano GmbH, Berlin, Germany).

2.2. Reagents

All reagents used in this study were of analytical purity. Distilled water was obtained using a GFL distiller (GFL mbH, Burgwedel, Germany). A model alkali liquor obtained was obtained by dissolving NaOH in distilled water. Then, the necessary amount of $Al(OH)_3$ was dissolved in the resulting hot caustic alkali liquor to achieve the desired ratio of NaOH and Al_2O_3 in the model liquor. After complete dissolution of aluminum hydroxide, the liquor was brought to the required concentration in caustic alkali. The aluminate liquor had the following chemical composition: Na_2O_k (caustic alkali) = 240 g/dm^3, Al_2O_3 = 120 g/dm^3. The sinter and ESP dust were obtained from the RUSAL-Kamensk-Uralsky alumina refinery (56.304530, 61.980334; Kamensk-Uralsky, Russia) where low-grade bauxite is processed via combined Bayer-Sintering process. Bauxite with the high silica content is sent in rotary kiln together with sodium carbonate to obtain water-soluble sinter at 1150 °C. The chemical and phase composition of the ESP dust and the sinter are different, since dust passes through the hot zone of the sintering kiln faster and many processes fail to complete; also, the difference may be caused by the difference in the distribution of minerals into different fractional classes. Table 1 shows the chemical composition of the ESP dust of the Kamensk-Uralsky alumina refinery, against the chemical composition of the bauxite sinter.

Table 1. The chemical composition of electrostatic precipitator (ESP) dust and bauxite sinter, wt. %.

Products	Al_2O_3	SiO_2	Na_2O	MgO	CaO	Fe_2O_3	LOI [1]	μ_{Si} [2]	ΣREE
ESP dust	25.5	2.5	28.3	0.4	2.5	12.2	24.8	10.2	755
Bauxite sinter	35.0	5.0	33.2	0.8	4.8	16.7	0.5	7.0	567

[1] LOI—loss on ignition at 1000 °C. [2] μ_{Si} (silicon modulus)—mass ratio of Al_2O_3 to SiO_2.

The results of X-ray diffraction analysis of the electrostatic precipitator dust of the RUSAL-Kamensk-Uralsky alumina refinery is shown in Figure 2.

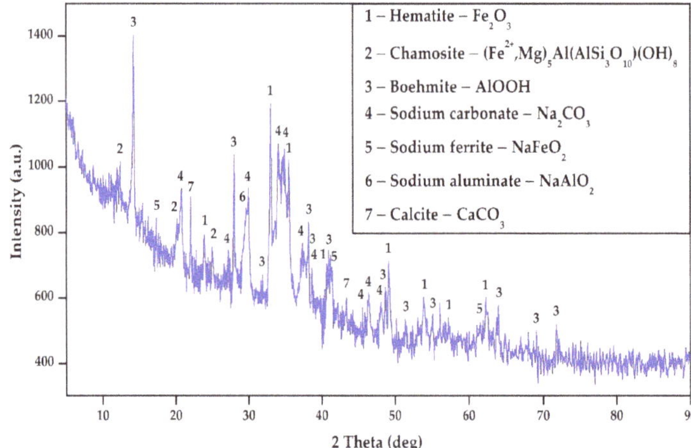

Figure 2. X-ray diffraction analysis of the electrostatic precipitator dust of the RUSAL-Kamensk-Uralsky alumina refinery in Russia.

2.3. *Experimentation*

The leaching of the required weight of ESP dust in distilled water was carried out at L:S = 10:1 at 95 °C for 60 min in a 0.5 dm^3 Lenz Minni thermostatic reactor (Lenz Laborglas GmbH & Co. KG, Wertheim, Germany) fitted with an overhead stirrer and ports for a refrigerator and samplers.

Autoclave leaching of the ESP dust sample at L:S = 3:1 was carried out at 160, 200, and 240 °C in an aluminate mother liquor for 15–90 min in a 1 dm^3 Parr autoclave (Parr Instrument, Moline, IL, USA). Leaching in an aluminate liquor at 90 °C also used a thermostatic Lenz Minni reactor.

After water and alkaline leaching, the resulting red mud was separated from the aluminate liquor by filtration on a Buchner funnel. After washing and drying of the bauxite residue for 8 h at 110 °C, the content of rare-earth elements and other components was measured. The solid phase obtained after water leaching of ESP dust is designated as ESPDW, and that obtained by autoclave leaching with an alkaline aluminate liquor at 240 °C for 90 min is designated as ESPDA.

3. Results and Discussion

It can be seen from Table 1 that the chemical composition of ESP dust and bauxite sinter are different. The high LOI in the dust is associated with the incomplete decomposition of soda and calcite; however, there is also a significant decrease in the content of silica, magnesium, and calcium in dust relative to alumina and an increase in sodium content. A decrease in silica content in dust leads to a significant increase in the silicon modulus (μ_{Si}). The results of X-ray diffraction analysis (Figure 2) confirm the presence of sodium aluminate in the ESP dust; however, most of the soda remains unreacted.

According to the previous research [19,21,43,44], the REE and Sc are possibly adsorbed on goethite, hematite and in the channels of aluminosilicates in the typical Bayer bauxite residue. The primary container of these elements could be cancrinite [45]. However, during the sintering process iron minerals are transformed into sodium ferrite and aluminosilicate—into sodium silicate and sodium aluminate (Figure 2). Therefore, REE and Sc could be liberated through the sintering process, which could lead to an increase in their concentration in the by-products and increased leachability than in typical Bayer bauxite residue.

Figure 3 shows the content of rare-earth elements in ESP dust and sinter. The obtained data show that the ESP dust from the sintering process contains more REE than the sinter itself. Moreover, if we detract LOI (consisting of water and CO_2 carbonates, which are removed from sinter at 1000 °C and which remain in ESP dust, according to Table 1), the REE content in the calcined ESP dust will be approximately 70–80% higher than in the sinter. The obtained data is in good agreement with the results presented previously in the patent [42], where it was shown that ESP dust contains about 50 ppm of scandium. This may be due to the repeated circulation of fine dust in the process and mineralogical transformation. Consequently, ESP dust can be a source of REE, and we can consider the option of removing it from the cycle in order to extract valuable components and increase the efficiency of the sintering kiln.

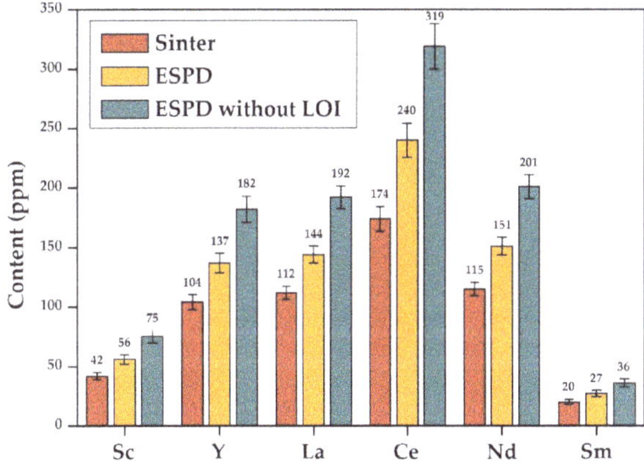

Figure 3. Content of rare earth elements in electrostatic precipitator dust and bauxite sinter, ppm (LOI—loss on ignition at 1000 °C).

3.1. Water Leaching of ESP Dust

The XRD analysis of ESP dust (Figure 2) shows a high content of soluble minerals (such as sodium silicate and sodium aluminate). Tests were carried out to leach ESP dust with water for 1 h at 95 °C, which is necessary for the complete conversion of sodium salts into liquor. The output of the solid phase (ESPDW) after leaching was 40% of the initial weight of ESP dust. Table 2 shows the chemical composition of the resulting red mud, and the degree of recovery of the main components into the liquor is shown in Figure 4. Part of the aluminum in electrostatic precipitator dust was converted to sodium aluminate; it is therefore easily leached with water. Notably, aqueous leachate has a higher silicon modulus than the original dust, as silicon has a degree of recovery higher than aluminum. This indicates that the degree of transformation of silica to sodium silicate is higher than alumina to sodium aluminate. Na_2O is almost completely leached, which indicates that with water leaching for less than 1 h the components in the resulting liquor do not have time to form a disilication product according to the following Equation (1):

$$6Na_2SiO_3 + 6NaAl(OH)_4 + Na_2X \rightarrow Na_6[Al_6Si_6O_{24}]\cdot Na_2X + 12NaOH + 6H_2O, \quad (1)$$

where X represents various inorganic anions, most often sulfate, carbonate, chloride, aluminate, etc. [46].

Table 2. The chemical composition of electrostatic precipitator dust after water leaching, wt. %.

Product	Al_2O_3	SiO_2	Na_2O	MgO	CaO	Fe_2O_3	LOI	μ_{Si}
ESPDW	39.8	3.1	1.7	1.0	6.2	30.5	18.3	12.8

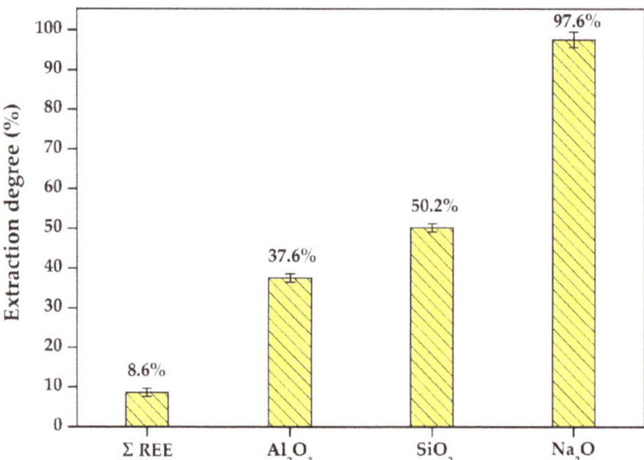

Figure 4. The extraction degree of the main components into the liquor by water leaching of electrostatic precipitator dust.

It can also be seen in Figure 4 that REE (Sc, Y, La, Ce, Nd) are slightly extracted during water leaching, because of their interaction with alkali and soda [47]. As shown in some studies [39,40], the degree of scandium recovery from red mud to soda liquor usually does not exceed 20%.

As it was stated in the introduction section, scandium is the most valuable REE in red mud. It was previously shown that scandium in red mud is mainly associated with iron minerals [48]. However, some researches show [45] that cancrinite could accumulate Sc as well. Therefore, to study the association of rare-earth metals with different phases, we performed ESPD and ESPDW surface mapping using EPMA (Figure 5).

It can be seen from Figure 5 that scandium in ESP dust is mainly associated with iron minerals, and to a lesser extent—with silicon compounds. It could be associated with the fact that iron minerals (hematite) are not fully transformed into sodium ferrite during sintering because small particles of dust pass very quickly through the hot zones of the furnace and a disilication product is not formed yet. Almost the same picture is observed in ESPDW; however, the distribution of scandium is more uniform than in ESP dust. This may be due to the formation of a disilication product (sodalite or cancrinite), which can adsorb the REE. This suggests that destruction of the hematite matrix is required at first for the complete recovery of REE from ESPD, as is the case with the typical Bayer process red mud. This also explains the low degree of REE recovery at the water leaching stage, since hematite cannot be dissolved by sodium carbonate liquor.

3.2. *Kinetics of ESP Dust Leaching by Alkaline Aluminate Liquor*

To study the mechanism and effect of temperature on the leaching of initial ESPD and ESPDW with the alkaline aluminate mother liquor, experiments were carried out to measure the aluminum extraction degree from the liquor with variable duration and temperature of the process. Aluminum was chosen as an indicator of leaching efficiency as it was the only element extracted from the liquor during alkali leaching of ESPDW. The temperature range was selected, taking into account that, after the extraction of readily soluble sodium salts, the alumina in the material is mainly represented by boehmite, the leaching of which requires a temperature of more than 160 °C [49]. The experimental results are shown in Figure 6.

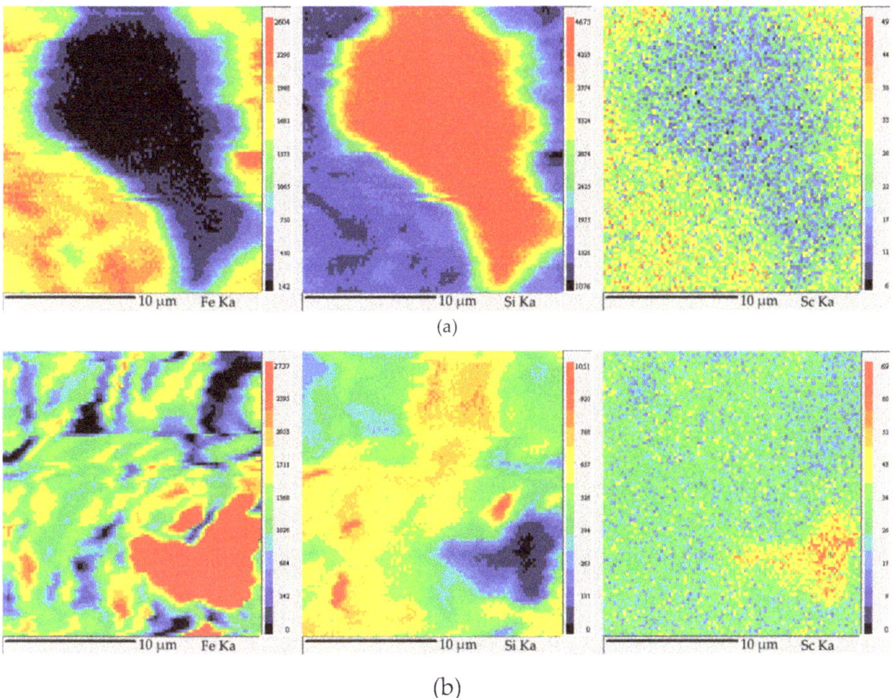

Figure 5. Mapping of electrostatic precipitator dust surface (**a**) and water-leached electrostatic precipitator dust surface (**b**) using the electron probe micro-analyzer.

It can be seen from Figure 6 that the kinetics of aluminum extraction into the liquor is quite high at all temperatures for the first 30 min for ESP dust. Even at 90 °C, more than 40% of aluminum is extracted from ESP dust after 30 min, apparently because of a certain degree of conversion of boehmite to sodium aluminate in the sintering kiln. However, after pre-leaching in water (ESPDW), sodium aluminate had already been extracted into the first-stage liquor; therefore, the rate of leaching of aluminum from ESPDW at 90 °C is significantly lower. Although at higher temperatures, due to a lower silica content, the efficiency of alumina recovery from ESPDW increases, reaching 90% after 90 min of leaching at 240 °C.

To study the leaching mechanism, the obtained kinetic curves were processed using the shrinking core model [50]. We studied six kinetic equations [51] describing the process in various modes, from kinetic to intra-diffusion; however, the models shown below (Equations (2) and (3)) proved to be most promising for the process description:

$$1 - 3(1 - X)^{2/3} + 2(1 - X) = k_1 t, \tag{2}$$

$$1/3\ln(1 - X) + ((1 - X)^{-1/3} - 1) = k_2 t, \tag{3}$$

where X is the degree of aluminum recovery into the liquor at a time t, k_i is the apparent rate constant. Equation (2) describes the process in the intra-diffusion area, while Equation (3) describes the process limited by interfacial transfer and diffusion through the product layer. The plot of $1/3\ln(1 - X)$ +

($(1-X)^{-1/3}-1$) versus t for ESP dust leaching gives a straight line (Figure 7a) with the determination coefficient R^2 the highest among all models used (more than 0.98 for all temperatures except 90 °C), which indicates that leaching, in this case, is most likely limited by interfacial transfer and diffusion through the product layer. It can be concluded that during leaching, a disilication product is formed around the core of the boehmite (Figure 8), which slows down the leaching process. Also, a film of sodium titanate can form on the surface of boehmite, which is known [52] to reduce the rate of dissolution of aluminum hydroxides significantly. For water-leached dust, the highest determination coefficient is observed, when using Equation (2) (Figure 7c), which implies that the process is limited by diffusion through the layer of the product or unreacted matter.

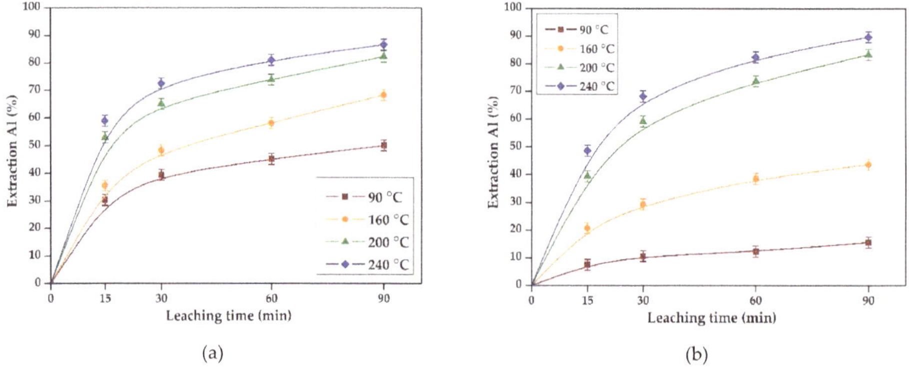

Figure 6. Effect of leaching time and temperature for Al extraction from raw ESP dust (**a**), from water-leached ESP dust (**b**).

Using the obtained values of the apparent rate constant (k_i) in Figure 7b,d and the Arrhenius equation (Equation (4)), we determined the values of the apparent activation energy for leaching ESPD and ESPDW (Figure 6) to be 24.98 kJ/mol and 33.19 kJ/mol, respectively.

$$k_i = A\exp(-E_a/RT), \tag{4}$$

where A is the Arrhenius constant, R is the universal rate constant (8.314 J/mol·K), T is the temperature (K), E_a is the apparent activation energy (J/mol).

The obtained values of the activation energy also confirm diffusion limitation. However, for an intradiffusion stage, the activation energy should be in the range of 8–22 kJ/mol. The higher values, in this case, maybe due to the fact that higher activation energy is required for the dissolution of the boehmite since a temperature of more than 160 °C is required for its extraction into alkali liquor. In the second case, the activation energy is higher, since there is no easily soluble phase of aluminum left after water leaching and a low degree of leaching efficiency of boehmite is observed at 90 °C.

Figure 8 shows XRD patterns of ESPDW and ESPDA, from which it can be seen that alkaline leaching results in disappearing of boehmite peaks and appearing instead of the peaks corresponding to cancrinite; iron, in contrast to the Bayer red mud, is represented by both hematite and hydroxide phases.

The yield of solid residue (red mud) from the leaching of ESPD in alkaline aluminate liquor at 240 °C for 90 min was 29.8% of the initial weight of dust. At the same time, the yield of red mud (ESPDA), after ESPDW leaching with alkali liquor at 240 °C for 90 min was 21.0%. As a result, the degree of REE enrichment of red mud in the second case was higher. Figure 9 shows the content of rare earth metals in these products.

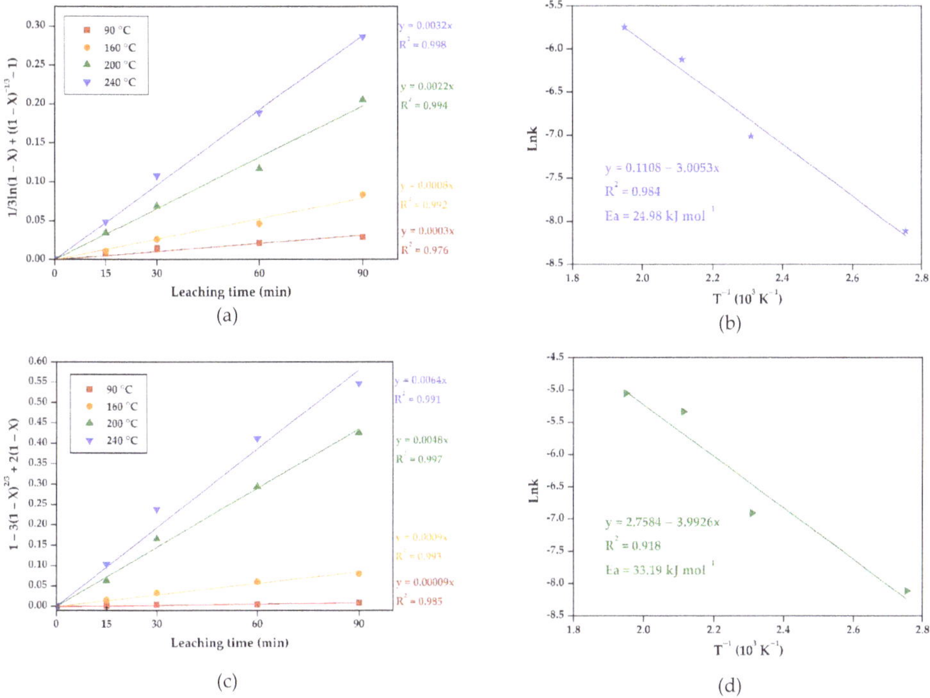

Figure 7. Results of substituting the data for ESPD leaching to Equation (3) (**a**); substituting the data for ESPDW leaching to Equation (2) (**c**); dependence ln$k$ – T^{-1} for leaching the original ESP dust (**b**) and water-treated ESP dust (**d**).

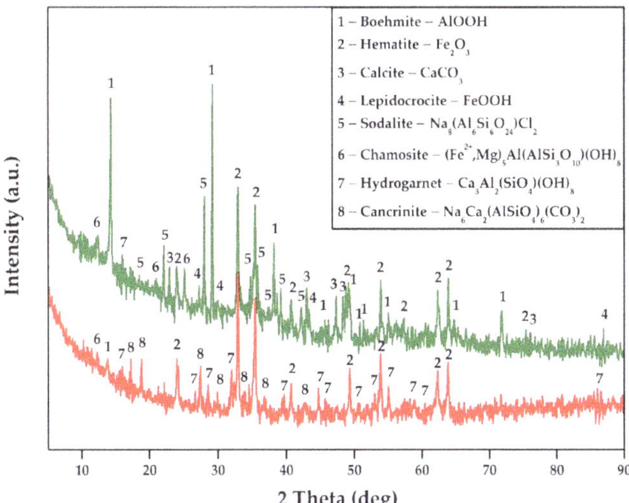

Figure 8. XRD pattern of water-leached electrostatic precipitator dust (**green**) and alkali-leached electrostatic precipitator dust (**red**).

The data in Figure 9 show that the total amount of REE in the red mud after ESP dust leaching was more than 1700 ppm in the first process and over 3200 ppm in the second process. Thus, we demonstrated that, in principle, it is possible to concentrate rare-earth elements in red mud by leaching ESP dust with water and alkaline aluminate liquor. The REE content in the red mud obtained thereby is three times higher than in conventional red mud of the alumina refinery, which can significantly reduce the cost of obtaining the REE concentrate in the future. Moreover, the recovery of additional components from natural raw materials can improve the efficiency of processing of bauxite raw materials in general. Based on these findings, we propose the following ESPD processing scheme (Figure 10), which enables efficient recovery of both alumina and rare earth elements. The stage of leaching REE from ESPDA will be discussed in the next paper.

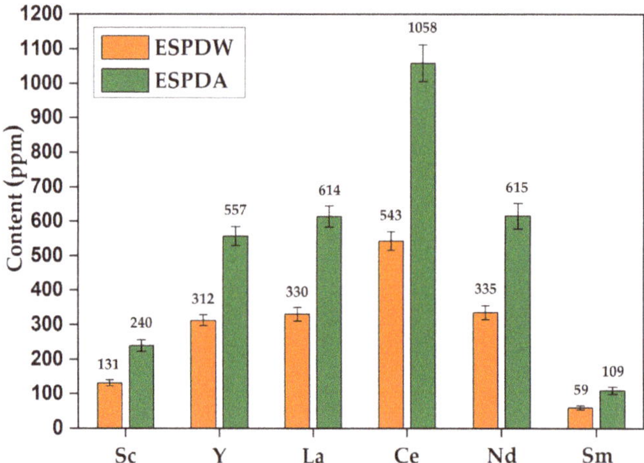

Figure 9. Content of rare earth elements in water-leached electrostatic precipitator dust and alkali-leached electrostatic precipitator dust.

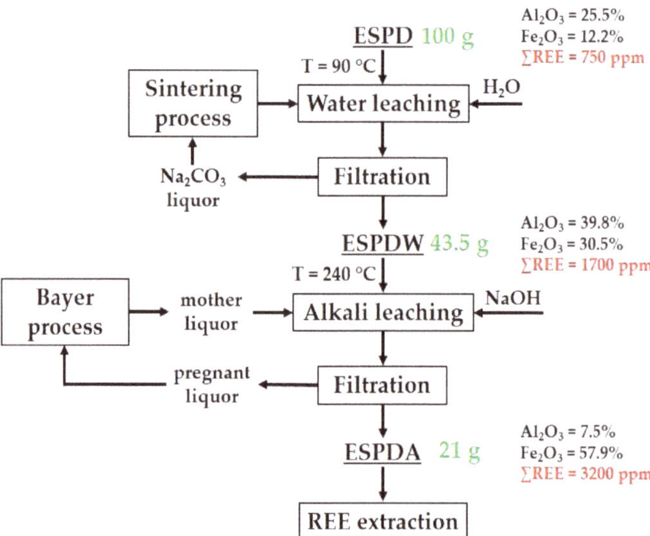

Figure 10. A flowsheet of rare-earth elements (REE) concentration from electrostatic precipitator dust by water/alkali leaching.

4. Conclusions

In this article, we investigated the possibility of concentrating rare-earth elements into red mud obtained by leaching dust from electrostatic precipitators used in bauxite charge sintering. We compared the chemical composition of the sinters and ESP dust formed during bauxite sintering in tubular rotary kilns. The sinter and the dust were analyzed for the content of rare earth elements. We also studied their association with various minerals of the materials used. The study of leaching of electric precipitation dust allows us to draw the following conclusions:

1. The chemical and phase composition of the electrostatic precipitation dust and sinter obtained in the same technological process are different. The content of rare-earth elements in the dust of electrostatic precipitators is 70–80% higher than in the sinter, which may be explained by a large number of cycles of ESP dust return to the kiln due to the very small size of particles.
2. Similar to the red mud of the Bayer process, scandium in ESP dust and in the solid residue after ESP dust leaching is associated to a greater extent with hematite, because of an insufficient transformation of latter into sodium ferrite in the sintering process and to a lower extent with disilication product (sodalite and cancrinite); the complete recovery of REE requires the destruction of the hematite at first.
3. The following conditions allow to obtain red mud from ESP dust that contains three times more REE than traditional waste red mud of the Ural Alumina Refinery: pre-leaching with water at 90 °C followed by autoclave leaching with an alkaline-aluminate liquor at 240 °C for 90 min.
4. The kinetic patterns of leaching alumina with the alkali liquor have been studied for the original electrostatic precipitation dust and water-leached dust. The change in the limiting stage of the process after water-treatment has been shown by an increase of the activation energy from 24.98 to 33.19 kJ/mol; changes in the form of the kinetic curves have also been demonstrated.

A technological scheme has been proposed for the removal of ESP dust from the cycle and its further comprehensive processing. Recovery of additional components from ESP dust will improve the efficiency of processing of bauxite raw materials in general and reduce the environmental footprint.

Author Contributions: Conceptualization, I.L. and L.C.; methodology, I.L.; validation, L.C., I.L.; formal analysis, A.S., D.V.; investigation, L.C., J.N.; resources, L.C., I.L.; data curation, I.L.; writing—original draft preparation, A.S. I.L.; writing—review and editing, A.S., D.V.; visualization, J.N., D.V.; supervision, L.C.; project administration, I.L.; funding acquisition, A.S. All authors have read and agreed to the published version of the manuscript.

Funding: The research was funded by the Russian Science Foundation, grant number 18-19-00186.

Acknowledgments: RUSAL-Kamensk-Uralsky are acknowledged for providing materials. Technicians at Ural Branch of Russian Academy of Sciences are acknowledged for their assistance with XRD, XRF, EPMA, and ICP-MS analysis.

Conflicts of Interest: The authors declare no conflict of interest.

References

1. Qi, D. *Hydrometallurgy of Rare Earths: Extraction and Separation*; Elsevier: Amsterdam, The Netherlands, 2018; ISBN 978-0-12-813921-9.
2. Savchenkov, S.A.; Bazhin, V.Y.; Brichkin, V.N.; Kosov, Y.I.; Ugolkov, V.L. Production Features of Magnesium-Neodymium Master Alloy Synthesis. *Metallurgist* **2019**, *63*, 394–402. [CrossRef]
3. Wübbeke, J. Rare earth elements in China: Policies and narratives of reinventing an industry. *Resour. Policy* **2013**, *38*, 384–394. [CrossRef]
4. Savchenkov, S.; Bazhin, V.; Brichkin, V.; Povarov, V.; Ugolkov, V.; Kasymova, D. Synthesis of magnesium-zinc-yttrium master alloy. *Lett. Mater.* **2019**, *9*, 339–343. [CrossRef]
5. Røyset, J.; Ryum, N. Scandium in aluminium alloys. *Int. Mater. Rev.* **2005**, *50*, 19–44. [CrossRef]
6. Binnemans, K.; Jones, P.T.; Blanpain, B.; Van Gerven, T.; Yang, Y.; Walton, A.; Buchert, M. Recycling of rare earths: A critical review. *J. Clean. Prod.* **2013**, *51*, 1–22. [CrossRef]

7. Wang, W.; Cheng, C.Y. Separation and purification of scandium by solvent extraction and related technologies: A review. *J. Chem. Technol. Biotechnol.* **2011**, *86*, 1237–1246. [CrossRef]
8. Li, Z.; Din, J.; Xu, J.; Liao, C.; Yin, F.; Lŭ, T.; Cheng, L.; Li, J. Discovery of the REE minerals in the Wulong–Nanchuan bauxite deposits, Chongqing, China: Insights on conditions of formation and processes. *J. Geochem. Explor.* **2013**, *133*, 88–102. [CrossRef]
9. Ou, Z.; Li, J.; Wang, Z. Application of mechanochemistry to metal recovery from second-hand resources: A technical overview. *Environ. Sci. Process. Impacts* **2015**, *17*, 1522–1530. [CrossRef]
10. Rychkov, V.; Koukkari, P.; Kirillov, S.; Kirillov, E. Best Practices of Russia and Finland in Extracting REE from Fertilizer Waste. *KnE Mater. Sci.* **2017**, *2*, 168–173. [CrossRef]
11. Diallo, M.S.; Baier, G.; Moyer, B.A.; Hamelers, B. Critical Materials Recovery from Solutions and Wastes: Retrospective and Outlook. *Environ. Sci. Technol.* **2015**, *49*, 9387–9389. [CrossRef]
12. Ding, Y.; Wang, J.; Wang, G.; Xue, Q. Innovative Methodology for Separating of Rare Earth and Iron from Bayan Obo Complex Iron Ore. *ISIJ Int.* **2012**, *52*, 1772–1777. [CrossRef]
13. Wang, W.; Pranolo, Y.; Cheng, C.Y. Metallurgical processes for scandium recovery from various resources: A review. *Hydrometallurgy* **2011**, *108*, 100–108. [CrossRef]
14. Ayora, C.; Macías, F.; Torres, E.; Lozano, A.; Carrero, S.; Nieto, J.-M.; Pérez-López, R.; Fernández-Martínez, A.; Castillo-Michel, H. Recovery of Rare Earth Elements and Yttrium from Passive-Remediation Systems of Acid Mine Drainage. *Environ. Sci. Technol.* **2016**, *50*, 8255–8262. [CrossRef] [PubMed]
15. Ramasamy, D.L.; Puhakka, V.; Repo, E.; Sillanpää, M. Selective separation of scandium from iron, aluminium and gold rich wastewater using various amino and non-amino functionalized silica gels—A comparative study. *J. Clean. Prod.* **2018**, *170*, 890–901. [CrossRef]
16. Smirnov, A.L.; Titova, S.M.; Rychkov, V.N.; Bunkov, G.M.; Semenishchev, V.S.; Kirillov, E.V.; Poponin, N.N.; Svirsky, I.A. Study of scandium and thorium sorption from uranium leach liquors. *J. Radioanal. Nucl. Chem.* **2017**, *312*, 277–283. [CrossRef]
17. Rychkov, V.N.; Kirillov, E.V.; Kirillov, S.V.; Semenishchev, V.S.; Bunkov, G.M.; Botalov, M.S.; Smyshlyaev, D.V.; Malyshev, A.S. Recovery of rare earth elements from phosphogypsum. *J. Clean. Prod.* **2018**, *196*, 674–681. [CrossRef]
18. Lokshin, E.P.; Tareeva, O.A.; Elizarova, I.P. A study of the sulfuric acid leaching of rare-earth elements, phosphorus, and alkali metals from phosphodihydrate. *Russ. J. Appl. Chem.* **2010**, *83*, 958–964. [CrossRef]
19. Rivera, R.M.; Ounoughene, G.; Malfliet, A.; Vind, J.; Panias, D.; Vassiliadou, V.; Binnemans, K.; Van Gerven, T. A Study of the Occurrence of Selected Rare-Earth Elements in Neutralized–Leached Bauxite Residue and Comparison with Untreated Bauxite Residue. *J. Sustain. Metall.* **2019**, *5*, 57–68. [CrossRef]
20. Loginova, I.V.; Shoppert, A.A.; Chaikin, L.I. Extraction of Rare-Earth Metals During the Systematic Processing of Diaspore-Boehmite Bauxites. *Metallurgist* **2016**, *60*, 198–203. [CrossRef]
21. Borra, C.R.; Pontikes, Y.; Binnemans, K.; Van Gerven, T. Leaching of rare earths from bauxite residue (red mud). *Miner. Eng.* **2015**, *76*, 20–27. [CrossRef]
22. Mordberg, L.E. Patterns of distribution and behaviour of trace elements in bauxites. *Chem. Geol.* **1993**, *107*, 241–244. [CrossRef]
23. Anawati, J.; Azimi, G. Recovery of Strategic Materials from Canadian Bauxite Residue by Smelting Followed by Acid Baking–Water Leaching. In *Rare Metal Technology 2020*; Azimi, G., Forsberg, K., Ouchi, T., Kim, H., Alam, S., Baba, A.A., Eds.; The Minerals, Metals & Materials Series; Springer International Publishing: Cham, Switzerland, 2020; pp. 139–150. ISBN 978-3-030-36757-2.
24. Sinha, S.; Sinha, M.K.; Pandey, B.D. Extraction of lanthanum and cerium from Indian red mud. *Int. J. Miner. Process.* **2014**, *127*, 70–73. [CrossRef]
25. Anhaeusser, C.R.; Geological Society of South Africa (Eds.) *Mineral Deposits of Southern Africa: In 2 Vol. 1*; Geological Society of South Africa: Johannesburg, Southern Africa, 1986; ISBN 978-0-620-09438-2.
26. Lokshin, E.P.; Tareeva, O.A. Production of high-quality gypsum raw materials from phosphogypsum. *Russ. J. Appl. Chem.* **2015**, *88*, 567–573. [CrossRef]
27. Binnemans, K.; Jones, P.T.; Blanpain, B.; Van Gerven, T.; Pontikes, Y. Towards zero-waste valorisation of rare-earth-containing industrial process residues: A critical review. *J. Clean. Prod.* **2015**, *99*, 17–38. [CrossRef]
28. Lokshin, E.P.; Tareeva, O.A. Activation of leaching of rare earth elements from phosphohemihydrate. *Russ. J. Appl. Chem.* **2013**, *86*, 1638–1642. [CrossRef]
29. Kovács, T.; Horváth, M.; Csordás, A.; Bátor, G.; Tóth-Bodrogi, E. Tobacco plant as possible biomonitoring tool of red mud dust fallout and increased natural radioactivity. *Heliyon* **2020**, *6*, e03455. [CrossRef]

30. Alam, S.; Das, B.K.; Das, S.K. Dispersion and Sedimentation Characteristics of Red Mud. *J. Hazard. Toxic Radioact. Waste* **2018**, *22*, 04018025. [CrossRef]
31. Liu, Z.; Li, H. Metallurgical process for valuable elements recovery from red mud—A review. *Hydrometallurgy* **2015**, *155*, 29–43. [CrossRef]
32. Rivera, R.M.; Xakalashe, B.; Ounoughene, G.; Binnemans, K.; Friedrich, B.; Van Gerven, T. Selective rare earth element extraction using high-pressure acid leaching of slags arising from the smelting of bauxite residue. *Hydrometallurgy* **2019**, *184*, 162–174. [CrossRef]
33. Klauber, C.; Gräfe, M.; Power, G. Bauxite residue issues: II. options for residue utilization. *Hydrometallurgy* **2011**, *108*, 11–32. [CrossRef]
34. Qu, Y.; Li, H.; Tian, W.; Wang, X.; Wang, X.; Jia, X.; Shi, B.; Song, G.; Tang, Y. Leaching of valuable metals from red mud via batch and continuous processes by using fungi. *Miner. Eng.* **2015**, *81*, 1–4. [CrossRef]
35. Shoppert, A.; Loginova, I.; Rogozhnikov, D.; Karimov, K.; Chaikin, L. Increased as Adsorption on Maghemite-Containing Red Mud Prepared by the Alkali Fusion-Leaching Method. *Minerals* **2019**, *9*, 60. [CrossRef]
36. Sabirzyanov, N.A.; Yatsenko, S.P. *Gidrokhimicheskie Sposoby Kompleksnoĭ Pererabotki Boksita*; IKhTT UrO RAN: Ekaterinburg, Russia, 2006; ISBN 978-5-7691-1629-2. (In Russia)
37. Akcil, A.; Akhmadiyeva, N.; Abdulvaliyev, R.; Abhilash; Meshram, P. Overview on Extraction and Separation of Rare Earth Elements from Red Mud: Focus on Scandium. *Miner. Process. Extr. Metall. Rev.* **2018**, *39*, 145–151. [CrossRef]
38. Borra, C.R.; Blanpain, B.; Pontikes, Y.; Binnemans, K.; Van Gerven, T. Smelting of Bauxite Residue (Red Mud) in View of Iron and Selective Rare Earths Recovery. *J. Sustain. Metall.* **2016**, *2*, 28–37. [CrossRef]
39. Petrakova, O.V.; Panov, A.V.; Gorbachev, S.N.; Klimentenok, G.N.; Perestoronin, A.V.; Vishnyakov, S.E.; Anashkin, V.S. Improved Efficiency of Red Mud Processing through Scandium Oxide Recovery. In *Light Metals 2015*; Hyland, M., Ed.; John Wiley & Sons, Inc.: Hoboken, NJ, USA, 2015; pp. 91–96. ISBN 978-1-119-09343-5.
40. Yatsenko, S.P.; Pyagai, I.N. Red mud pulp carbonization with scandium extraction during alumina production. *Theor. Found. Chem. Eng.* **2010**, *44*, 563–568. [CrossRef]
41. Loginova, I.V.; Shoppert, A.A.; Chaikin, L.I. Effect of Adding Sintering Furnace Electrostatic Precipitator Dust on Combined Leaching of Bauxites and Cakes. *Metallurgist* **2015**, *59*, 698–704. [CrossRef]
42. Diev, V.N.; Sabirzyanov, N.A.; Skryabneva, L.M.; Yatsenko, S.P.; Anashkin, V.S.; Aminov, S.N.; Zavadskij, K.F.; Sysoev, A.V.; Ustich, E.P. Method for Scandium Extraction from Bauxite Treatment for Alumina Production. Russia Patent 2,201,988, 10 April 2003.
43. Vind, J.; Malfliet, A.; Blanpain, B.; Tsakiridis, P.; Tkaczyk, A.; Vassiliadou, V.; Panias, D. Rare Earth Element Phases in Bauxite Residue. *Minerals* **2018**, *8*, 77. [CrossRef]
44. Borra, C.R.; Blanpain, B.; Pontikes, Y.; Binnemans, K.; Van Gerven, T. Recovery of Rare Earths and Other Valuable Metals from Bauxite Residue (Red Mud): A Review. *J. Sustain. Metall.* **2016**, *2*, 365–386. [CrossRef]
45. M Tóth, T.; Schubert, F.; Raucsik, B.; Fintor, K. Mineralogical and Geochemical Constraints of the REE Accumulation in the Almásfüzitő Red Mud Depository in Northwest Hungary. *Appl. Sci.* **2019**, *9*, 3654. [CrossRef]
46. Smith, P. The processing of high silica bauxites—Review of existing and potential processes. *Hydrometallurgy* **2009**, *98*, 162–176. [CrossRef]
47. Pasechnik, L.A.; Shirokova, A.G.; Koryakova, O.V.; Sabirzyanov, N.A.; Yatsenko, S.P. Complexing Properties of Scandium(III) in Alkaline Medium. *Russ. J. Appl. Chem.* **2004**, *77*, 1070–1073. [CrossRef]
48. Anawati, J.; Azimi, G. Recovery of scandium from Canadian bauxite residue utilizing acid baking followed by water leaching. *Waste Manag.* **2019**, *95*, 549–559. [CrossRef] [PubMed]
49. Alex, T.C.; Kumar, R.; Roy, S.K.; Mehrotra, S.P. Towards ambient pressure leaching of boehmite through mechanical activation. *Hydrometallurgy* **2014**, *144–145*, 99–106. [CrossRef]
50. Levenspiel, O. *Chemical Reaction Engineering*, 3rd ed.; Wiley: New York, NY, USA, 1999; ISBN 978-0-471-25424-9.
51. Rogozhnikov, D.A.; Shoppert, A.A.; Dizer, O.A.; Karimov, K.A.; Rusalev, R.E. Leaching Kinetics of Sulfides from Refractory Gold Concentrates by Nitric Acid. *Metals* **2019**, *9*, 465. [CrossRef]
52. Wang, Y.; Li, X.; Zhou, Q.; Qi, T.; Liu, G.; Peng, Z.; Zhou, K. Effects of Si-bearing minerals on the conversion of hematite into magnetite during reductive Bayer digestion. *Hydrometallurgy* **2019**, *189*, 105126. [CrossRef]

Article

Leaching Kinetics of Rare Earth Elements from Fire Clay Seam Coal

Xinbo Yang * and Rick Q. Honaker

Department of Mining Engineering, University of Kentucky, Lexington, KY 40506, USA; rick.honaker@uky.edu
* Correspondence: xinbo.yang@uky.edu

Received: 27 April 2020; Accepted: 25 May 2020; Published: 27 May 2020

Abstract: Recovery of rare earth elements (REEs) from coal samples collected from the Fire Clay coal seam using diluted mineral acid solutions was investigated. The initial processing step was coal recovery using conventional froth flotation which concentrated the REEs in tailing material resulting in an upgrade to values around 700 ppm on a dry whole mass basis. Leaching experiments were performed on the flotation tailings material using a 1.2 M sulfuric acid solution adjusted to a temperature of 75 °C to study the extractability of REEs from coal material. The effect of particle size, leaching time, leaching temperature, and solid concentration on REE leaching recovery were evaluated. The kinetic data obtained from leaching over a range of temperatures suggested that the leaching process follows the shrinking core model with possibly a mixed control mechanism that may be a result of several heterogenous materials leaching simultaneously. Leaching recovery increased rapidly at the beginning of the reaction then slowed as the system reached equilibrium. The apparent activation energy determined from test data obtained over a range of temperatures using 1 M sulfuric acid was 36 kJ/mol for the first 20 min of reaction time and 27 kJ/mol for the leaching period between 20 and 120 min. The leaching of light REEs during the initial stage was determined to be driven by a chemical reaction, followed by the formation of a product layer, which required lower activation energy in the later stage of leaching. In regards to the heavy REEs, the major mechanism for leaching is desorption and the product layer formation does not affect the heavy REEs significantly.

Keywords: rare earth elements; coal; leaching; kinetics; apparent activation energy

---

1. Introduction

Rare earth elements (REEs) exist in over 200 different mineral types. A portion of the REEs are associated with carbonates and oxides that are easy to extract while others are bonded with silicates and phosphates that are difficult to penetrate [1]. Commercially, REEs are produced from monazite, xenotime, bastnaesite, and clay adsorbed REEs, which is a unique source commonly associated with a commercial production operation in southern China [2,3]. Many studies have focused on developing REE extraction processes from secondary resources including mine waste streams, industrial wastes or byproducts, electronic waste recycling and magnet recycling industries, coal ash, slags and waste water, etc. [3–8]. The limitation of recovering REE from secondary resources is low grade, which limits the effort and cost that can be applied toward the extraction process. However, these resources take advantage of eliminating the mining cost as well as other associated costs such as the chemical cost already expended for treating the waste. Coal-based materials represent a potential secondary source for REEs which may be extracted and concentrated by the use of physical and/or chemical processes [9–11].

Several studies have investigated the REE mode of occurrences and distributions in coals of varying ranks, associated mineral matter, acid mine drainage and sludge, and coal combustion byproducts. Arbuzov et al. (2018) studied peat material collected from western Siberia and concluded that: (1) a maximum of 25% of the REEs had a mineral association, such as monazite, xenotime, and zircon, etc.; (2) 10%–30% of REEs was in humic substances, and (3) 40%–80% was in a water-soluble form [12]. Studies conducted on the ash product of pre-combustion coal material using a muffle furnace showed that 70% of the light REEs and 50% of the heavy REEs in bituminous coals are predominantly associated with phosphate minerals, and 50%–60% of the REEs in low rank coals are associated with clays [13]. The REEs in low rank coals (i.e., lignite and subbituminous) are primarily associated with the organic matter whereas the association with higher rank coals are with the associated mineral matter including grains of rare earth phosphate minerals that are micro-dispersed within the organic matrix [12,14,15].

The Fire Clay coal seam material appears to be a very suitable source for REEs extraction due to its high concentration of REEs on a dry ash basis as compared to other coal sources. The origin of the REEs in the Fire Clay coal has been associated with volcanic deposition during the coalification period [16]. Mineralogy studies performed on the bituminous coal source using scanning electron microscopy with energy dispersive spectroscopy (SEM-EDS) detected micro-dispersed rare earth phosphate mineral particles, which are generally difficult to dissolve in strong acid solutions [9]. The finely dispersed inorganic material contained in Fire Clay coal has relatively high concentrations of REEs Liberation of the rare earth minerals by size reduction provides an opportunity for recovering clean coal from middling streams which is generally discarded or partially blended into coal products to meet contract specifications. This study focused on the leaching characteristics of the REEs associated with various fractions of the Fire Clay coal and associated mineral matter.

2. Materials and Method

2.1. Materials

A representative sample (~200 kg) of a targeted fraction of the Fire Clay seam coal was collected from a coal preparation plant located in eastern Kentucky. The Fire Clay coal source is a high-volatile (30%–40% dry basis), bituminous coal well documented as being enriched with REEs and other critical elements. The sampled process stream was the product of a secondary dense medium vessel, which targeted the recovery of the 1.4 × 1.6 specific gravity (SG) fraction, commonly referred to as middlings, in the 150 × 9 mm particle size fraction of the plant feed.

Analysis of a representative sample of the bulk revealed that the middlings material contained 242 ppm REEs on a dry, whole sample basis and 1331 ppm on a dry, ash basis. The ash content of the representative sample was 18.2% on a dry basis. The distribution of the REEs shown in Figure 1 indicates that nearly 24% of the REEs are permanent magnet elements (Nd, Pr, Sm, and Dy) while scandium represents 3.1% of the total, which is valued as a component in high performance aluminum alloy. The high ash-based REE content value provided an indication that the use of grinding to liberate the associated mineral matter could potentially provide a means for significant pre-concentration of the REEs prior to the leaching process while producing a marketable clean coal product.

The 150 × 9 mm middling bulk sample was crushed using a laboratory jaw crusher and subsequently ground in a hammer mill to achieve a top particle size of 1 mm. The bulk sample was split into multiple representative samples by riffling the hammer mill product. For each test, a representative sample was pulverized to a top size of 80 mesh (177 μm) in a smaller hammer mill and then ground in an attrition mill at a pre-selected rotation speed and retention time to achieve the desired particle size reduction. The solids concentration used in the attrition mill was 40 g/L. The ground material was treated in an 8-L Denver conventional flotation system to recover clean coal material using diesel fuel No. 2 as the collector and 4-methyl-2-pentanol as the frother. The chemical concentrations varied from test-to-test based on the feed particle size distribution. As shown in Figure 2, a rougher stage of treatment was followed by three cleaner stages to ensure maximum recovery of the liberated mineral matter and to achieve an ultraclean coal product. The tailings material collected from the coal recovery process was further treated by a second flotation step using octanohydroxamic acid ($C_8H_{17}NO_2$) at a dosage of 1.5 kg/t and a pH value of 9.5 to remove acid consuming materials (e.g., calcite) [17]. No additional frother was used in the mineral flotation stage due to the frothing property of octanohydromaxic acid. A rougher-cleaner treatment process was used for the alkali mineral rejection to maximize recovery of non-floatable material to the tailings stream. The final tailings material was the material used as feed for the leaching studies.

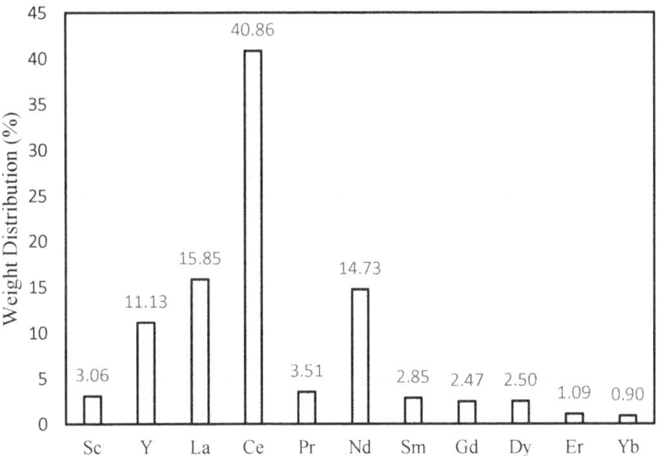

Figure 1. Rare earth element distribution in the nominal 1.4 × 1.6 Specific Gravity (SG) middling fraction of Fire Clay coal as collected from a secondary dense medium vessel product stream.

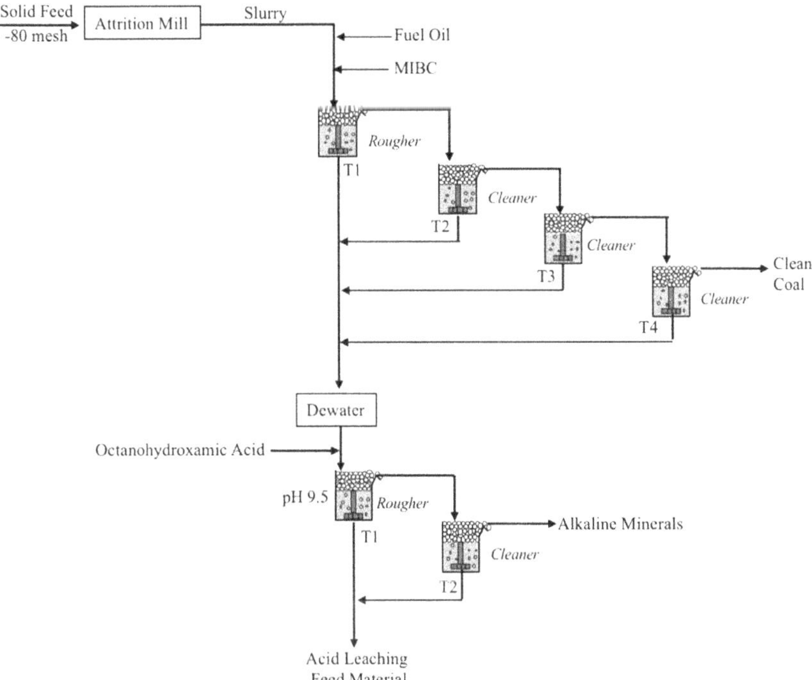

Figure 2. Sample preparation of the Fire Clay middlings for the leaching studies.

2.2. Methodology

Leaching experiments were conducted in a triple neck round bottom flask with the middle neck connected to a total reflux condenser, which ensured that a constant volume was maintained. The reactor was placed in a water bath where the solution was heated using an immersion heater and a precise temperature control system to maintain the temperature throughout the duration of the test. Agitation was provided using a magnetic stir at a speed that could be varied up to around 1200 rpm. The leaching experiments were conducted using deionized water and trace metal grade acid (purity > 99.99%).

The investigation involved the evaluation of the following parameters on leaching recovery and kinetics: i) lixiviant type (i.e., H_2SO_4, HCl, and HNO_3), ii) acid concentration (i.e., 0.1M, 0.5M, 1M, and 2M etc.), iii) solid-to-liquid ratio (i.e., S/L = 1/100, 1/50, 1/10, 1/5, etc.), and iv) solution temperature (i.e., 25 °C, 40 °C, 50 °C, 60 °C, and 75 °C). To assess leaching kinetics, samples were collected at time intervals established from the initial start of the test, i.e., 1 min, 3 min, 5 min, 10 min, 20 min, 30 min, 60 min, 90 min, and 120 min. Micro-filter (0.45 µm) plunger syringes were used to separate the leachate from the solids to immediately stop the solid–liquid reaction. The final solid residue was filtered and washed with deionized water.

The REE contents in addition to other elements of interest in the leachate and solid residue samples were determined using inductively coupled plasma optical emission spectrometry (ICP-OES). The results were used to calculate elemental and overall recovery of the REEs. Solid loss, REE content in the leach solid residue and leachate solution are presented in the supplementary materials (Table S1) that is accessible online. Leach recovery represents the amount of material in the test feed source that was solubilized into solution during the leaching process, which was quantified using the following expression:

$$\text{Leach recovery (\%)} = \frac{c_L * V_L}{c_L * V_L + c_{SR} * m_{SR}} \times 100\% \quad (1)$$

in which c_L is the elemental concentration in the leachate solution (µg/mL); V_L the volume of the analyzed leachate solution (mL); c_{SR} the elemental concentration in solid residue (µg/g); and m_{SR} the weight of solid residue (g).

2.3. Analytical Methods

REE content was determined by digestion and analysis of the resultant solution in an ICP-OES. The solid sample preparation procedure followed the ASTM D6357 method for ashing and digestion of coal and refuse samples with modifications made to the digestion to allow for use of a digestion block apparatus. The ICP-OES unit was calibrated using a standard solution identified as VHG-SM68 multi standard, which contained 48 elements. The REE recovery of these check standards was maintained at +/− 10% relative standard deviation (RSD). A duplicate sample was chosen at random and run through the entire process to verify repeatability at the frequency of not less than one every 40 samples. A certified coal ash sample (1633b) was utilized to ensure the digestion procedure and as a reference standard for peak selection. Three standard sample were repeated with each batch of digestion and the standard deviations for the rare earth elements are: <2% for Ce, Dy, Er, Eu, Gd, Ho, La, Lu, Nd, Pr, Sc, Th, Y, and Yb; <5% for Sm and Tm; <15% for Tb.

X-ray diffraction (XRD) analyses were conducted on feed samples using an Advance D8 instrument produced by the Bruker Company. The scanning was performed from 10° to 70° with a stepwise increase of 0.02° and a scanning speed of 0.5°/min. The XRD spectra were analyzed to estimate concentrations of major mineral components using the EVA software developed by the same company.

3. Results and Discussions

3.1. Particle Size Effect

A reduction in particle size may provide two significant benefits, i.e., (1) liberation of the clay particles and other mineral matter which exposes more surface area and exchangeable REEs for lixiviants to interact and extract the RE ions and/or (2) liberation of nano-sized RE minerals and RE oxides that may be dissolved in acid solutions. Acid leaching on finer size material can also provide faster kinetic rates and higher efficiency for REE extraction. On the negative side, by reducing the particle size to a micron level, the newly generated surface area is increased exponentially which escalates the consumption of hydrogen ions by dissolving more contaminate metal ions. As such, selectivity may be reduced, thereby elevating the cost of leaching and downstream concentration processes. Additional issues are the higher cost of energy for grinding and difficulties associated with thickening and dewatering ultrafine solid residuals.

To assess the effect of particle size on leaching performance, representative samples of the Fire Clay middlings material were ground for different lengths of time before the de-carbonization step to generate samples having a range of 80% passing sizes (P80). Acid leaching tests were conducted using 1.2 mol/L sulfuric acid solution with a solid concentration of 10 g/L at 75 °C. The REE recovery was quantified using test data generated after leaching for five hours and Equation (1).

The results shown in Table 1 indicated that reducing the particle size liberated mineral matter containing higher concentrations of REE. For example, by reducing the particle size to a P80 size of 150 μm (80 mesh top size) in the feed, the flotation tailings material contained 444 ppm of total REEs with a P80 size of 32 μm. The REE concentration of 444 ppm reflects the content of the coarser mineral matter dispersed in the middling particles. By grinding for greater lengths of time resulting in more applied energy, the P80 size was reduced to sub-micron level and the tailing material generated by flotation nearly doubled to 719 ppm of total REEs. This finding indicates that the finest mineral matter dispersed within the Fire Clay coal has the highest concentration of REEs. A previous study found that REEs in the Fire Clay coal were strongly associated with micro-dispersed kaolinite which may be liberated and released through size reduction [9].

Table 1. Liberation of REEs from the de-carbonized Fire Clay middlings material.

Attrition Mill		P80		Flotation		TREE Concentration		Acid Leaching	
Grinding Time (min)	Rotor Speed (rpm)	Flotation Feed (micron)	Flotation Tailing (micron)	Tailing Yield (%)	Ash Content (%)	Ash Basis (ppm)	Whole Mass Basis (ppm)	Solid Loss (%)	TREE Recovery (%)
Feed					18.2	1331	242		
0	0	150	32.0	3.7	89.1	499	444	11.9	71.2
2.5	150	45	14.5	7.4	85.9	581	499	9.9	74.8
5	200	37	10.2	9.5	87.1	609	531	9.2	79.0
20	200	13	8.7	12.8	90.8	685	622	16.2	83.7
40	200	11	7.3	14.2	91.0	768	699	17.2	84.3
60	200	10	6.5	13.9	91.4	771	705	20.4	83.6
90	200	9	5.0	13.6	90.9	788	716	24.2	82.3
120	200	8	0.9	15.4	91.0	791	719	27.9	82.0

TREE: total rare earth elements.

Reducing particle size resulted in a significant increase in leach recovery from 71.2% to 84.3% over the range of P80 sizes. As such, the size reduction increased the amount of REEs reporting in the leach feed and increased the percentage of the REEs being recovered through leaching. These two positive outcomes suggest that the REEs associated with micro-dispersed mineral matter in the Fire Clay middlings are more concentrated and more easily extractable by leaching relative to the coarser grain fractions. In addition, the finer mineral matter is, in general, more soluble as indicated by an increase in the amount of solid loss during the leaching process. As much as 20% of the solids in the finest sample tested was dissolved under the standard leaching conditions, which may reflect both the solubility of the mineral matter and surface area exposure. A negative impact is an increase in the amount of contaminates in the leachate due to the elevated level of dissolved solids.

Based on the liberation test results, 20 min of grinding time was selected to generate the acid leach feed material used in this study. A flow sheet of the sample preparation process is shown in Figure 3 along with weight yield and content data for each feed and product stream. The decarbonization step resulted in a high-quality clean coal product containing around 7% ash-forming material while the reject material was nearly pure mineral matter as indicated by an ash content of 90.81%. The mineral flotation rougher-cleaner treatment resulted in 0.74% of the total feed reporting to the concentrate having an ash content of 75.12%. The lower ash content in the flotation concentration was an indicator of calcite flotation ($CaCO_3$). The float product also contained 741 ppm of TREEs, which may be due to RE mineral flotation resulting from the use of octanohydroxamic acid. The flotation tailing material produced from the two stages of flotation represented 12.1% of the feed and contained 615 ppm of TREEs and 90.67% ash-forming material. The REE upgrade is 2.54:1 starting from a feed content of 242 ppm. This material was used as the acid leach feed in the subsequent leaching studies presented in this paper.

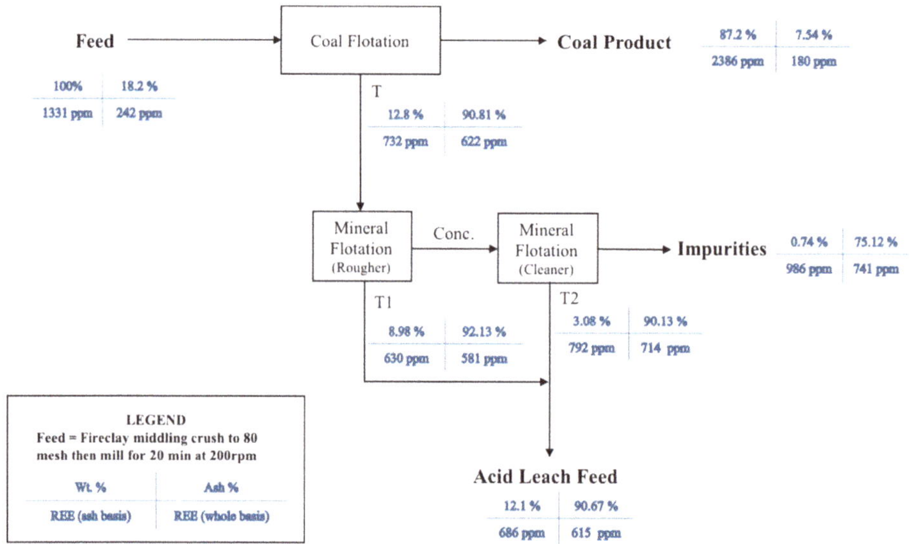

Figure 3. Schematic of sample preparation for the acid leach feed material using coal and mineral flotation.

Five representative samples of the acid leach feed were analyzed to assess the repeatability of the ICP-OES. The average TREE value was 607 ± 18 ppm (2.97% variation) with thorium content of 41 ± 0.6 ppm (1.46% variation). The Ce content accounted for 42% of the total REEs as shown in Figure 4a. The Fire Clay material was rich in light REEs (i.e., Ce, La, Pr, Nd, Sm, Eu, Gd, Sc) as indicated by a content of 534 ppm or 88.0% of the total. Yttrium was the most abundant heavy REE (i.e., Y, Tb, Dy, Ho, Er, Tm, Yb, and Lu) with a concentration of 47 ppm. The major minerals present in the sample were quartz, kaolinite, illite, and muscovite as shown in the XRD plot in Figure 4b.

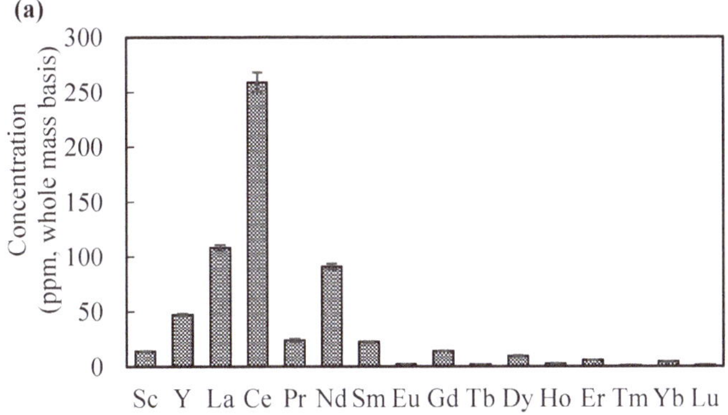

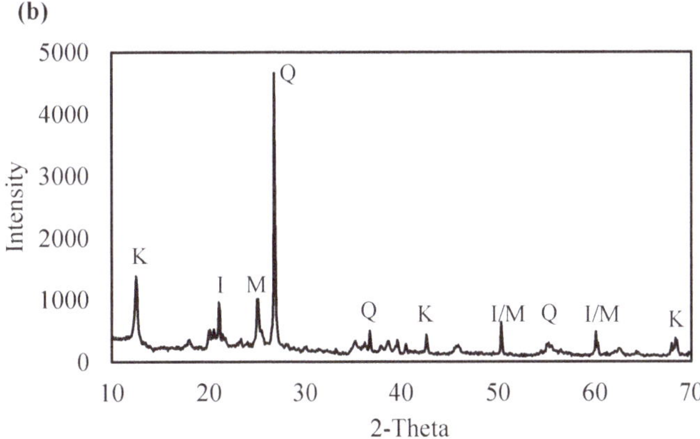

Figure 4. The composition of the acid leaching feed material used in this study on the basis of (**a**) rare earth content and (**b**) mineralogy as determined by X-ray diffraction analysis (Q—quartz, K—kaolinite, I—illite, M—muscovite).

3.2. *Effect of Major Variables on REE Leaching*

3.2.1. Effect of Acid Type

The lixiviant type affects the REE leaching characteristics by changing the solution speciation stabilities due to the existence of various anions in varying concentrations. Sulfate ions were reported to have a higher coordination ability with rare earths than chloride ions even in high monovalent concentration solutions [18]. Leaching experiments were conducted using different inorganic acids at an acid concentration of 1M, solid/liquid ratio of 10 g/L, and a temperature of 75 °C. Sulfuric acid (H_2SO_4), hydrochloric acid (HCl), and nitric acid (HNO_3) were used to study the effect on REE leaching recovery and reaction rate as shown in Figure 5. Total REE (TREEs) recovery values of 80%, 76%, and 74% were achieved after 3 h of leaching using HCl, HNO_3, and H_2SO_4 solution, respectively. The pH of the leachate solutions at the end of the tests were 0.105, 0.113, and 0.112, respectively.

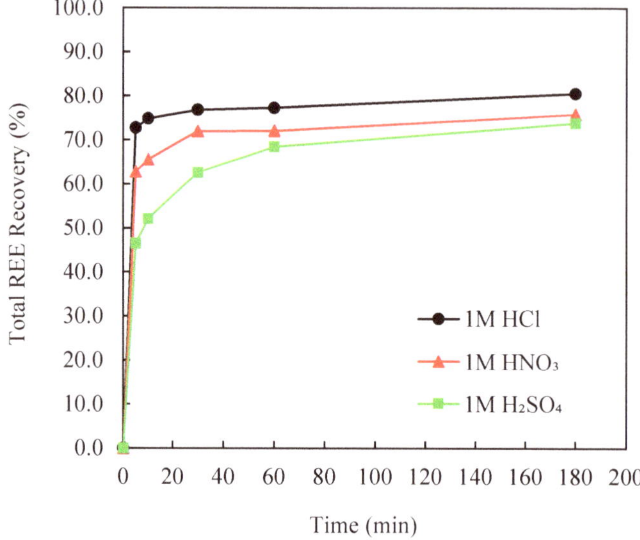

Figure 5. Effect of acid type on the leaching recovery of total rare earth elements contained in the Fire Clay coal middlings (75 °C, 530 rpm, solid-to-liquid ratio (S/L) = 10 g/L, d_{80} = 8.7 μm).

Hydrochloric acid provided the fastest leaching rate, which achieved 73% recovery after the first 5 min of leaching, and slowly reached equilibrium after 3 h. Nitric acid also provided fast leaching rate within the first 30 min. Sulfuric acid was the least effective under the leaching conditions and provided the slowest leaching rate. This finding was likely due to the fact that sulfate ions have a higher coordination ability with rare earths than chloride ions even in high monovalent concentration solutions [19].

The coal-based leachate contained high concentrations of trivalent ions that may coordinate with sulfate ions resulting in depression of the rare earth-sulfate coordination. In addition, sulfuric acid requires two steps of dissociation reaction to release H^+ into solution whereas hydrochloric acid and nitric acid dissociates more rapidly into solution. Viscosity of the sulfuric acid solution is another factor that could have resulted in the slower reaction rate as the wetting rate of the solid particle surfaces is reduced when the solution viscosity is high. Despite the negative aspects of sulfuric acid, the lixiviant is still considered a viable lixiviate due to its relatively low cost and the negative aspects of the other lixiviants including the volatility of hydrochloric acid and the decomposability of nitric acid under 75 °C [20].

3.2.2. Stirring Speed Effect

Stirring speed affects the thickness of the film layer surrounding a solid particle suspended in the lixiviate solution. A high stirring speed creates an enhanced shear rate in solution which reduces the film layer thickness thereby increasing the mass transfer rate through the film diffusion layer [21]. The effect of stirring speed was evaluated at 300 rpm, 530 rpm, 760 rpm, and 900 rpm as shown in Figure 6. The leaching condition included 1M H_2SO_4 solution and a solid/liquid ratio of 10 g/L at 75 °C. The test results indicated that a stirring speed of 300 rpm did not provide sufficient agitation due to inadequate suspension of the slurry based on visual observations, while stirring speeds of 530 rpm to 900 rpm provided nearly equal kinetics. The recovery achieved using a 900-rpm stirring speed was slightly lower than that obtained at 760 rpm. A stirring speed of 530 rpm was established as an adequate value for the standard test conditions.

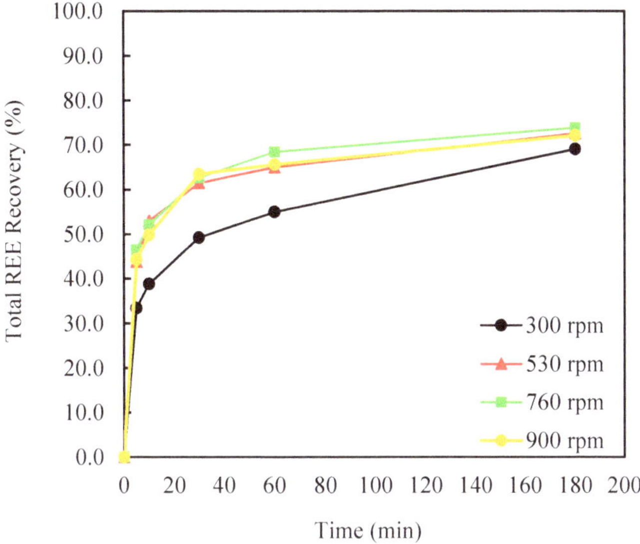

Figure 6. Effect of stirring speed on the leaching recovery of total rare earth elements contained in the Fire Clay coal middlings (75 °C, 1 M H_2SO_4, S/L = 10g/L, d_{80} = 8.7 µm).

3.2.3. Solid-to-Liquid Ratio Effect

The solid-to-liquid (S/L) ratio establishes to the stochiometric ratio of reactants, which directly affects the reaction equilibrium. The effect of the S/L ratio on rare earth leaching recovery was investigated in the range of 10 g/L to 200 g/L while maintaining the other parameters constant at 75 °C, 1 M H_2SO_4, and 530 rpm. The association between reactants decreased with an increase in the solid/liquid ratio, which resulted in a decrease in the extraction rate as shown in Figure 7. Leach recovery was reduced from 74% to 40% after increasing the S/L ratio from 10 g/L to 200 g/L. The magnitude of the recovery reduction is not commonly observed in other metal leaching operations. In the metallic copper leaching process, the leaching reaction was more effective when the Cu^{2+} concentration in solution was higher due to Cu^{2+} reacting with metallic Cu to Cu^{+} [22]. This type of reaction mechanism does not occur in a REE solution since the REEs exist mostly as a compound. Niobium leaching from titanium oxide residues did not show any effect from the S/L ratio on leaching recovery [21]. However, Li et al. (2013) reported on a rare earth concentrate leaching study that found the S/L ratio to have a negative effect on the leaching of a rare earth concentrate when the ratio was higher than 100 g/L [23]. Therefore, the solid/liquid ratio effect varies from source-to-source in different leaching environments.

3.2.4. Effect of Acid Concentration

The effect of sulfuric acid concentration on leaching recovery was studied using 0.1 M, 0.5 M, 1 M, and 2 M acid concentrations using the standard values for temperature, stirring speed, and solid-to-liquid ratio. The initial acid concentrations of 0.1 M, 0.5 M, 1 M, and 2 M resulted in ending pH values of 1.04, 0.38, 0.11, and −0.25, respectively, after 3 h of leaching. As shown in Figure 8, the total REE recovery increased substantially from 40% to 74% by elevating acid concentration from 0.1 to 1 M. However, from 1 M to 2 M, the increase was marginal at around 2.5 absolute percentage points. The optimal acid concentration was selected to be 1 M since higher concentrations of acid did not provide a significant increase in recovery of REEs.

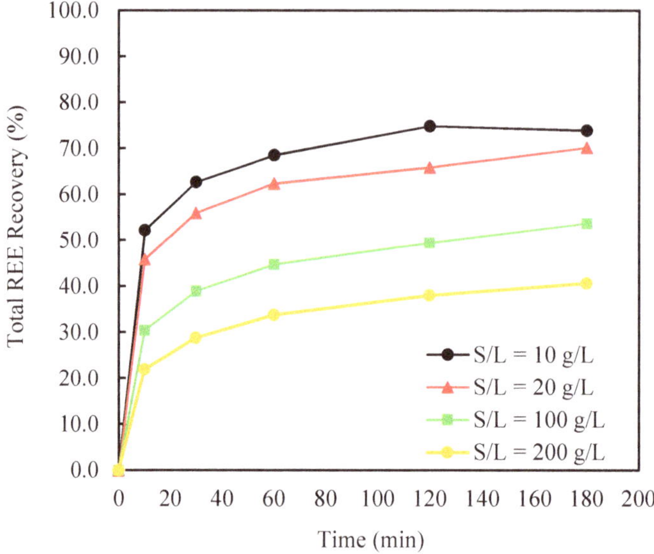

Figure 7. Effect of solid-to-liquid ratio on the leaching recovery of total rare earth elements contained in the Fire Clay coal middlings (75 °C, 1 M H_2SO_4, 530 rpm, D_{80} = 8.7 μm).

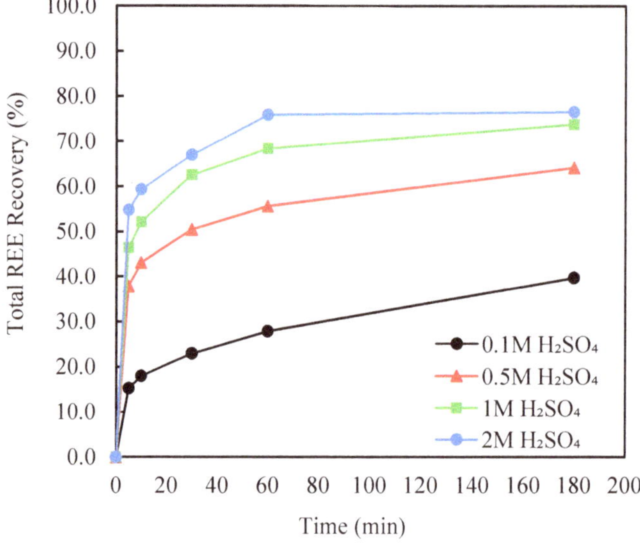

Figure 8. Effect of sulfuric acid solution concentration on the leaching recovery of total rare earth elements contained in the Fire Clay coal middlings (75 °C, 530 rpm, S/L = 10g/L, D_{80} = 8.7 μm).

The effect of acid concentration on individual RE leaching recovery is shown in Figure 9. Recovery of the light REEs significantly increased when acid concentration was elevated from 0.1 M to 1 M and very little improvement was realized when using a 2 M acid solution. Scandium was the least sensitive to acid concentration. The remaining un-leached portion of Sc is likely associated with a mineral structure that requires a higher level energy to break down. Yang et al. (2018) improved the scandium leaching recovery from 31% to 74% after treated by roasting with no chemical additives at 750 °C for 2 h from a coal-based material [24].

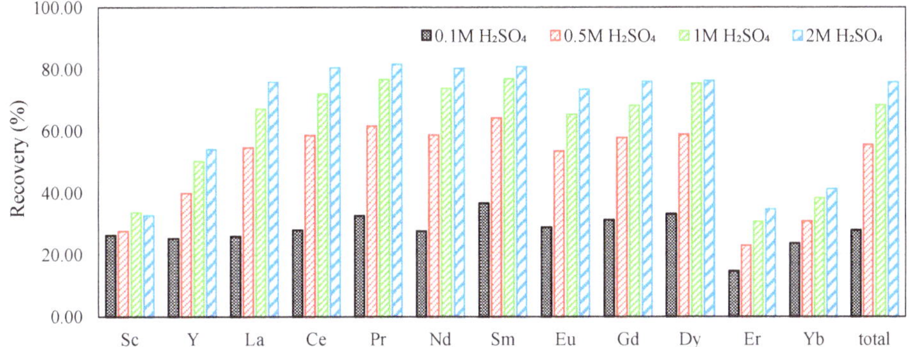

Figure 9. Effect of sulfuric acid solution concentration on individual rare earth element leaching recovery (2 h, 75 °C, 530 rpm, S/L = 10 g/L, D_{80} = 8.7 μm).

3.2.5. Effect of Temperature

A leaching process that is mainly controlled by a diffusion process is more dependent on mixing conditions whereas temperature has a more significant effect on chemical reaction controlled processes [25]. The effect of temperature on REE leaching using 1 M H_2SO_4 was investigated using a stirring speed of 530 rpm and S/L ratio of 10 g/L for 2 h. Samples were taken over shorter time increments due to the relatively fast kinetics during the first 20–30 min. Figure 10 shows that REE leach recovery increased significantly with an elevation in leaching temperature. When the temperature was increased from 298 K (25 °C) to 348 K (75 °C), leaching recovery increased from 35% to 75% after 2 h of leaching. The data suggests the existence of a relatively fast leaching process during the first 20 min followed by a slow process. As such, two or more reaction mechanisms may be occurring when leaching the coal source.

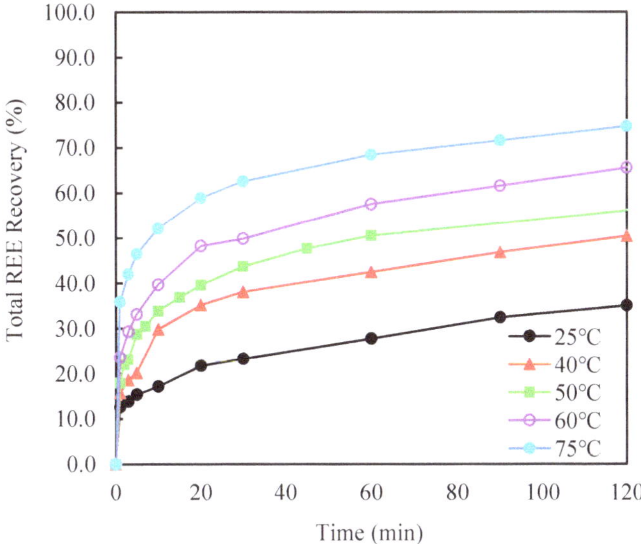

Figure 10. Effect of leaching reaction temperature on the leaching recovery of total rare earth elements contained in the Fire Clay coal middling (1 M H_2SO_4, 530 rpm, S/L = 10g/L, D_{80} = 8.7 μm).

The effect of temperature on individual REEs is shown in Figure 11. Most of the light REEs (i.e., La, Ce, Pr, Nd, Sm) appeared to be very sensitive to temperature, which indicated that the leaching mechanism of light REEs was mostly chemical reaction controlled. The recovery of Ce, Pr, and Nd increased from 36%, 39%, and 36% to 79%, 84%, and 80%, respectively, by increasing the temperature from 25 °C to 75 °C. The heavy REEs and scandium recovery improved with higher temperature, but the increase was not as significant. Scandium recovery rose from 29% to 36%. For the recovery of elements that were relatively insensitive to temperature, the activation energy is generally low and more likely to be a result of a diffusion controlled process [26].

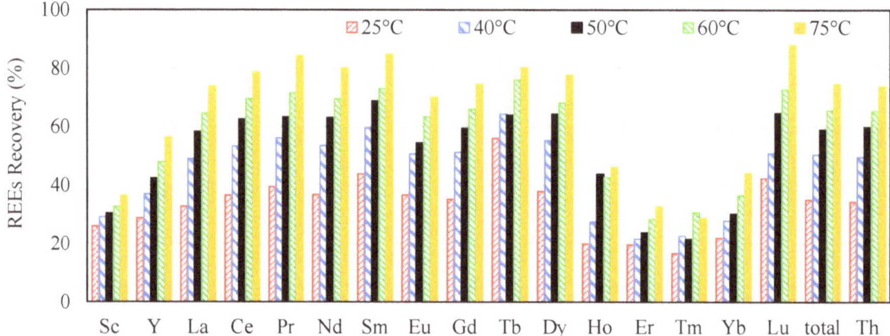

Figure 11. Effect of leaching reaction temperature on the leaching recovery of individual rare earth element (1 M H_2SO_4, 530 rpm, S/L = 10g/L, D_{80} = 8.7 μm, retention time of 120 min).

3.2.6. Morphology

The morphology of the Fire Clay coal middling particles before and after sulfuric acid leaching was studied using SEM. As shown in Figure 12a, the feed material consisted of heterogeneous particles comprised of mostly quartz and clay, which agrees well with the XRD analysis shown in Figure 12b. After 2 h of leaching at 50 °C, the particles were found to have a porous structure on the surface with a micro pore structure as shown in Figure 12b. After 2 h of leaching at 75 °C, the porous structure on some of the particle surfaces appeared larger in size as shown in Figure 12c. The images show no visible layer on the surface instead of porous structure due to dissolution. Therefore, the diffusion process in this reaction may be the results of interfacial transfer of the products and the reagent diffusion through the porous structure of solid particles.

Figure 12. SEM images of particles found in (**a**) leaching feed material; (**b**) solid residue after 2 h leaching at 50 °C; (**c**) solid residue after 2 h leaching at 75 °C (1 M H_2SO_4, 530 rpm, S/L = 10 g/L, D_{80} = 8.7 μm).

The leaching process involved several simultaneous reactions due to the mineral composition and the variety of REE associations. The REEs were found to exist in crystallized structures (mostly silicates and phosphate compounds), which usually require decomposition to be extracted under the current leaching conditions. A small portion of the REEs are present as RE ion substitution form in clays whereas most are associated with soluble RE containing minerals. Based on the results shown in Section 3.2.5, extraction of light REEs in this coal source was more sensitive to temperature thus the light REEs were more likely to be mineral associated, whereas the heavy REEs extraction was more independent to temperature thus more likely to be soluble metal oxides and adsorbed ions onto clay minerals.

3.3. *Kinetic Analysis*

The leaching process is classified as a fluid–particle heterogenous reaction in which a liquid reacts with a solid by contacting and transforms the solid into a product. A solid particle that reacts with a liquid and shrinks in size during the reaction can be described by a shrinking core model. The reaction is a five-step process, i.e., (1) diffusion through the film layer, (2) diffusion through the product layer, (3) chemical reaction on the surface, (4) product diffusion through the product layer and (5) product diffusion through the film layer to the solution. The slowest step is known as the rate determining process. The activation energy of a certain leaching step can be quantified by selecting the most accurate rate equation to represent the reactions [25].

A variety of rate equations have been developed and reported in literature that describe the leaching rate process [25,27,28]. Among the equations, the rate equation (Equation (2)) developed by Crank–Ginstling–Brounshtein, which describes the mass transfer across product layer, fits the experimental data well, i.e.,

$$k_d t = \left[1 - \frac{2}{3}\alpha - (1-\alpha)^{\frac{2}{3}}\right] \quad (2)$$

here α is the fraction that reacted; k_d is the kinetic constant.

The Crank–Ginstling–Brounshtein equation was used to linearize the extraction fraction (α) among all the temperatures using the experimental data for the first 20 min of leaching and the following 20–120 min of the reaction as shown in Figure 13. The correlation coefficient values (R^2) and the corresponding slopes (k) of the plots are listed in Table 2. Rate constants were calculated and the Arrhenius plots of ln(k) versus 1/K are as shown in Figure 14 for the two leaching stages. The activation energy determined for the first 20 min was 36 kJ/mol and 27 kJ/mol for the following 20–120 min of leaching. The activation energy values for both leaching periods were close to the energy barrier that is typically used to identify a diffusion controlled or chemical reaction controlled process, which is around 20 kJ/mol [26].

Table 2. Correlation coefficients of diffusion-controlled kinetics models at different temperatures for total REEs.

Temperature		Diffusion 1–20 min			Diffusion 20–120 min		
°C	K	k	a	R^2	k	a	R^2
25	298	0.0002	0.0017	0.9949	0.0001	0.0036	0.996
40	313	0.0007	0.0022	0.9648	0.0002	0.0129	0.9977
50	323	0.0009	0.0046	0.9616	0.0003	0.0196	0.9402
60	333	0.0014	0.0068	0.9919	0.0004	0.0262	0.9934
75	348	0.0019	0.0189	0.9627	0.0005	0.0487	0.9796

Note: k denotes the slope of the regression line, and a denotes the intercept of the regression line.

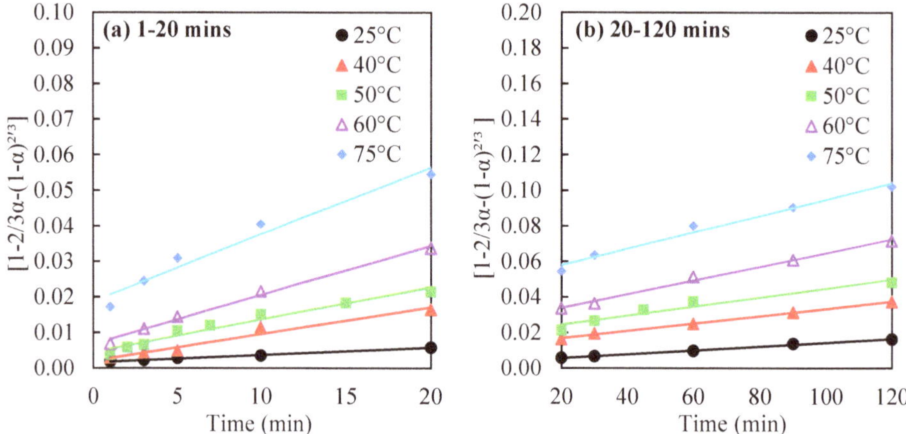

Figure 13. Kinetic modeling of total REEs recovery during the (**a**) first 20 min, and (**b**) 20–120 min of leaching at various temperatures for the Fire Clay middlings (1 M H_2SO_4, 530 rpm, S/L = 10 g/L, D_{80} = 8.7 μm, retention time of 120 min).

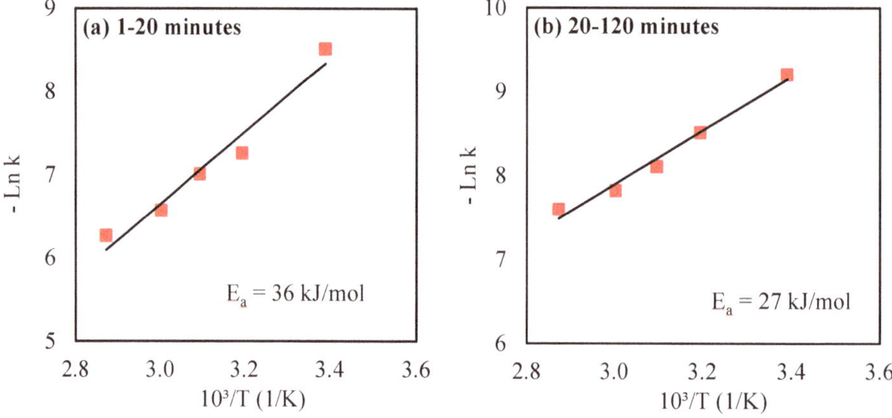

Figure 14. Arrhenius plot for the total REEs leached from the Fire Clay coal middlings during the (**a**) first 20 min, and (**b**) 20–120 min of leaching (1 M H_2SO_4, 530 rpm, S/L = 10g/L, d_{80} = 8.7 micron).

Since the coal tailing material is a heterogenous material that contains a number of potential modes of occurrence of REEs, the leaching process is not a single reaction. The resulting requirement for activation energy is a combination of the various forms of REEs. In addition, the material contains both calcite and pyrite among other soluble minerals that create a complex solution environment where the localized pH elevation on the solid particle surface could cause a product layer to be formed. The interfacial transfer of product through the porous structure of the solid particles requires high activation energies as reported by Li et al. (2010 and 2013), which can be as high as 40 kJ/mol [23,29].

To support the hypothesis, the activation energies for light and heavy REE groups were calculated using the data provided in Tables 3 and 4, respectively. The activation energy values for leaching the light REEs over the first 20 min and the period between 20 and 120 min were 41.8 kJ/mol and 28.1 kJ/mol, respectively. On the other hand, the activation energy values for the leaching of heavy REEs for the first 20 min and the 20–120 min of reaction were 24.2 kJ/mol and 26.1 kJ/mol, respectively. These values indicate that the leaching of the light REEs during the initial stage is more of a chemical reaction followed by the formation of a product layer and a reduced activation energy. The activation energy required for leaching the heavy REEs during the initial stage was significantly lower than that of the light REEs. This finding implies that the major rate controlling mechanism for heavy REEs leaching is diffusion.

Table 3. Correlation coefficients of diffusion-controlled kinetics models at different temperatures for light REEs.

Temperature		Diffusion 1–20 min			Diffusion 20–120 min		
°C	K	k	a	R^2	k	a	R^2
25	298	0.0002	0.0016	0.9975	0.0001	0.0034	0.9971
40	313	0.0005	0.0017	0.9963	0.0002	0.0115	0.9931
50	323	0.001	0.0042	0.9712	0.0003	0.0214	0.9542
60	333	0.0015	0.0068	0.9929	0.0004	0.0278	0.9928
75	348	0.0021	0.0198	0.9648	0.0005	0.0571	0.9888

Note: k denotes the slope of the regression line, and a denotes the intercept of the regression line.

Table 4. Correlation coefficients of diffusion-controlled kinetics models at different temperatures for heavy REEs.

Temperature		Diffusion 1–20 min			Diffusion 20–120 min		
°C	K	k	a	R^2	k	a	R^2
25	298	0.0002	0.0028	0.9593	0.00005	0.0049	0.9634
40	313	0.0004	0.0032	0.9943	0.00008	0.104	0.98
50	323	0.0005	0.0051	0.939	0.00009	0.0127	0.9704
60	333	0.0007	0.0068	0.9803	0.0002	0.016	0.9972
75	348	0.0008	0.013	0.941	0.0002	0.0251	0.9857

Note: k denotes the slope of the regression line, and a denotes the intercept of the regression line.

It was noted that the reaction kinetics was extremely fast within the first 1 min of the reaction. A possible explanation is that, due to the different modes of occurrence of REEs in coal-based material, the easy-to-leach REEs was instantaneously released at the beginning of the leaching process, and the hard-to-leach fraction controlled the reaction rate. The mode of occurrence of REEs can be categorized into different forms (i.e., ion exchange form, carbonate form, metal oxide form, acid soluble form, and insoluble form) with different levels of activation energy needed for extraction [24]. Zhang and Honaker (2020) studied the REE mode of occurrence in coal using the sequential chemical extraction method and quantified the REEs associated with each mode [30]. The REEs associated with ion exchangeable form and carbonates are likely to be released instantaneously at the beginning of the extraction process under the leaching conditions of the current study.

4. Conclusions

The Fire Clay coal is an excellent source for REE extraction due to its elevated REE contents and high leaching recovery values. Reducing the particle size of the middings material prior to de-carbonization resulted in the liberation of the associated micro-dispersed mineral matter that contained significantly higher concentrations of REEs. Decreasing the P80 particle size from 38 µm to 5 µm nearly doubled the concentration of REEs in the tailings of the flotation steps, which was the material used a feed for the leaching tests. The major minerals present in the sample were quartz, kaolinite, illite, and muscovite. Exponentially increasing the particle surface area through grinding resulted in higher solid dissolution which elevates the consumption of hydrogen ions and the concentration of contaminant metal ions.

The type of inorganic acid does not affect leaching recovery significantly but has an impact on the initial leaching rate. The mixing condition is sufficient at rotational speed values above 500 rpm.

Furthermore, solid concentration and acid concentration have a significant effect on leaching recovery of REEs. Scanning electron microscope (SEM) images showed no visible product layer or coating product on the particle surfaces. Therefore, the diffusion process in this reaction may be contributed by the interfacial transfer of the products and the reagent diffusion through the porous structure of solid particles.

The kinetic data obtained from leaching over a range of temperatures suggested that the leaching process follows the diffusion control shrinking core model. The activation energy determined from test data obtained over a range of temperatures using 1 M sulfuric acid was 36 kJ/mol for the first 20 min of reaction time and 27 kJ/mol for the leaching period between 20 and 120 min. The leaching of light REEs during the initial stage is much greater than the typical energy barrier of 20 kJ/mol, which indicates that a mixed control mechanism may occur as a result of several heterogenous materials leaching simultaneously. The energy required for later stage of leaching reduced significantly. The activation energy required for leaching the heavy REEs during the initial stage was significantly lower than that of the light REEs, which implies that the major rate controlling mechanism for heavy REE leaching is the diffusion.

Supplementary Materials: The following are available online at http://www.mdpi.com/2075-163X/10/6/491/s1, Table S1: Solid loss and REs content in leaching solid residue and leachate solution.

Author Contributions: Conceptualization, X.Y. and R.Q.H.; methodology, X.Y. and R.Q.H.; software, X.Y.; validation, X.Y.; formal analysis, X.X.; investigation, X.Y. and R.Q.H.; resources, R.Q.H.; data curation, X.Y.; writing—original draft preparation, X.Y.; writing—review and editing, R.Q.H.; visualization, X.Y.; supervision, R.Q.H.; project administration, R.Q.H.; funding acquisition, R.Q.H. All authors have read and agreed to the published version of the manuscript.

Funding: This research was funded by the U.S. Department of Energy, Grant Number DE-FE0029900.

Acknowledgments: This material is based upon work supported by the Department of Energy under Award Number DE-FE0029900. Disclaimer: This report was prepared as an account of work sponsored by an agency of the United States Government. Neither the United States Government nor any agency thereof, nor any of their employees, makes any warranty, express or implied, or assumes any legal liability or responsibility for the accuracy, completeness, or usefulness of any information, apparatus, product, or process disclosed, or represents that its use would not infringe privately owned rights. Reference herein to any specific commercial product, process, or service by trade name, trademark, manufacturer, or otherwise does not necessarily constitute or imply its endorsement, recommendation, or favoring by the United States Government or any agency thereof. The views and opinions of authors expressed herein do not necessarily state or reflect those of the United States Government or any agency thereof.

Conflicts of Interest: The authors declare no conflict of interest.

References

1. Kumari, A.; Panda, R.; Jha, M.K.; Lee, J.Y.; Kumar, J.R.; Kumar, V. Thermal treatment for the separation of phosphate and recovery of rare earth metals (REMs) from Korean monazite. *J. Ind. Eng. Chem.* **2015**, *21*, 696–703. [CrossRef]
2. Chi, R.; Zhang, X.; Zhu, G.; Zhou, Z.A.; Wu, Y.; Wang, C.; Yu, F. Recovery of rare earth from bastnasite by ammonium chloride roasting with fluorine deactivation. *Miner. Eng.* **2004**, *17*, 1037–1043. [CrossRef]
3. Jha, M.K.; Kumari, A.; Panda, R.; Rajesh Kumar, J.; Yoo, K.; Lee, J.Y. Review on hydrometallurgical recovery of rare earth metals. *Hydrometallurgy* **2016**, *165*, 2–26. [CrossRef]
4. Binnemans, K.; Jones, P.T.; Blanpain, B.; Van Gerven, T.; Yang, Y.; Walton, A.; Buchert, M. Recycling of rare earths: A critical review. *J. Clean. Prod.* **2013**, *51*, 1–22. [CrossRef]
5. Wu, G.; Li, J.; Xu, Z. Triboelectrostatic separation for granular plastic waste recycling: A review. *Waste Manag.* **2013**, *33*, 585–597. [CrossRef]
6. Peelman, S.; Sun, Z.H.I.; Sietsma, J.; Yang, Y. *Chapter 21—Leaching of Rare Earth Elements: Review of Past and Present Technologies*; Rare Earths Industry, Elsevier Inc.: Delft, The Netherlands, 2016; ISBN 9780128023280. [CrossRef]
7. Lin, R.; Bank, T.L.; Roth, E.A.; Granite, E.J.; Soong, Y. International Journal of Coal Geology Organic and inorganic associations of rare earth elements in central Appalachian coal. *Int. J. Coal Geol.* **2017**, *179*, 295–301. [CrossRef]

8. Lin, R.; Howard, B.H.; Roth, E.A.; Bank, T.L.; Granite, E.J.; Soong, Y. Enrichment of rare earth elements from coal and coal by-products by physical separations. *Fuel* **2017**, *200*, 506–520. [CrossRef]
9. Zhang, W.; Yang, X.; Honaker, R.Q. Association characteristic study and preliminary recovery investigation of rare earth elements from Fire Clay seam coal middlings. *Fuel* **2018**, *215*, 551–560. [CrossRef]
10. Honaker, R.Q.; Zhang, W.; Yang, X.; Rezaee, M. Conception of an integrated flowsheet for rare earth elements recovery from coal coarse refuse. *Miner. Eng.* **2018**, *122*, 233–240. [CrossRef]
11. Honaker, R.; Groppo, J.; Bhagavatula, A.; Rezaee, M.; Zhang, W. *Recovery of Rare Earth Minerals and Elements from Coal and Coal Byproducts*; International Coal Preparation Conference: Louisville, KY, USA, 2016.
12. Arbuzov, S.I.; Maslov, S.G.; Finkelman, R.B.; Mezhibor, A.M.; Ilenok, S.S.; Blokhin, M.G.; Peregudina, E.V. Modes of occurrence of rare earth elements in peat from Western Siberia. *J. Geochem. Explor.* **2018**, *184*, 40–48. [CrossRef]
13. Finkelman, R.B.; Palmer, C.A.; Wang, P. Quantification of the modes of occurrence of 42 elements in coal. *Int. J. Coal Geol.* **2018**, *185*, 138–160. [CrossRef]
14. Dai, S.; Graham, I.T.; Ward, C.R. A review of anomalous rare earth elements and yttrium in coal. *Int. J. Coal Geol.* **2016**, *159*, 82–95. [CrossRef]
15. Riley, K.W.; French, D.H.; Farrell, O.P.; Wood, R.A.; Huggins, F.E. Modes of occurrence of trace and minor elements in some Australian coals. *Int. J. Coal Geol.* **2012**, *94*, 214–224. [CrossRef]
16. Hower, J.C.; Ruppert, L.F.; Eble, C.F. Lanthanide, yttrium, and zirconium anomalies in the Fire Clay coal bed, Eastern Kentucky. *Int. J. Coal Geol.* **1999**, *39*, 141–153. [CrossRef]
17. Zhang, W.; Honaker, R.; Groppo, J. *Fundamental Study of the Monazite Calcite Flotation Separation*; Canadian Institute of Mining, Metallurgy and Petroleum: Montrel, QC, Canada, 2016.
18. Xiao, Y.F.; Chen, Y.Y.; Feng, Z.Y.; Huang, X.W.; Huang, L.; Long, Z.Q.; Cui, D.L. Leaching characteristics of ion-adsorption type rare earths ore with magnesium sulfate. *Trans. Nonferrous Met. Soc. China Engl. Ed.* **2015**, *25*, 3784–3790. [CrossRef]
19. Xiao, Y.; Feng, Z.; Huang, X.; Huang, L.; Chen, Y.; Liu, X.; Wang, L.; Zhiqi, L. Recovery of rare earth from the ion-adsorption type rare earths ore: II. Compound leaching. *Hydrometallurgy* **2016**, *163*, 83–90. [CrossRef]
20. Yu, J.; Guo, Z.; Tang, H. Dephosphorization Treatment of High Phosphorus Oolitic Iron Ore by Hydrometallurgical Process and Leaching Kinetics. *ISIJ Int.* **2013**, *53*, 2056–2064. [CrossRef]
21. Makanyire, T.; Jha, A.; Sutcliffe, S. Kinetics of hydrochloric acid leaching of niobium from TiO2 residues. *Int. J. Miner. Process.* **2016**, *157*, 1–6. [CrossRef]
22. Wang, Z.; Guo, S.; Ye, C. Leaching of copper from metal powders mechanically separated from waste printed circuit boards in chloride media using hydrogen peroxide as oxidant. *Procedia Environ. Sci.* **2016**, *31*, 917–921. [CrossRef]
23. Li, M.; Zhang, X.; Liu, Z.; Hu, Y.; Wang, M.; Liu, J.; Yang, J. Kinetics of leaching fluoride from mixed rare earth concentrate with hydrochloric acid and aluminum chloride. *Hydrometallurgy* **2013**, *140*, 71–76. [CrossRef]
24. Yang, X.; Werner, J.; Honaker, R.Q. Leaching of rare Earth elements from an Illinois basin coal source. *J. Rare Earths* **2019**, *37*, 312–321. [CrossRef]
25. Levenspiel, O. *Chemical Reaction Engineering-3rd edition*; John Wiley & Sons, Inc.: Hoboken, NJ, USA, 1999; Volume 38, ISBN 9780471530169.
26. Free, M.L. *Hydrometallurgy: Fundamentals and Applications*; John Wiley & Sons, Inc.: Hoboken, NJ, USA, 2013; ISBN 978-1-118-23077-0.
27. Salmi, T.; Grenman, H.; Warna, J.; Murzin, D.Y. New modelling approach to liquid-solid reaction kinetics: From ideal particles to real particles. *Chem. Eng. Res. Des.* **2013**, *91*, 1876–1889. [CrossRef]
28. Dickinson, C.F.; Heal, G.R. Solid–liquid diffusion controlled rate equations. *Thermochim. Acta* **1999**, *340–341*, 89–103. [CrossRef]
29. Li, M.; Wei, C.; Qiu, S.; Zhou, X.; Li, C.; Deng, Z. Kinetics of vanadium dissolution from black shale in pressure acid leaching. *Hydrometallurgy* **2010**, *104*, 193–200. [CrossRef]
30. Zhang, W.; Honaker, R. Characterization and recovery of rare earth elements and other critical metals (Co, Cr, Li, Mn, Sr, and V) from the calcination products of a coal refuse sample. *Fuel* **2020**, *267*, 117236. [CrossRef]

Article

Grain Size Distribution and Clay Mineral Distinction of Rare Earth Ore through Different Methods

Lingkang Chen [1,2,3,*], Xiongwei Jin [3], Haixia Chen [2], Zhengwei He [1,4,*], Lanrong Qiu [3] and Hurong Duan [5]

1 College of Earth Sciences, Chengdu University of Technology, Chengdu 610059, China
2 College of Sciences, Guangdong University of Petrochemical Technology, Maoming 525000, China; chenhaixia1975@126.com
3 School of Resource and Environmental Engineering, Jiangxi University of Science and Technology, Ganzhou 341000, China; xiongweijjx@126.com (X.J.); lrqiu183@163.com (L.Q.)
4 State Key Laboratory of Geohazard Prevention and Geoenvironment Protection, Chengdu University of Technology, Chengdu 610059, China
5 College of Geomatics, Xi'an University of Science and Technology, Xi'an 710054, China; duanhurong@126.com
* Correspondence: lkchen@jxust.edu.cn (L.C.); hzw@cdut.edu.cn (Z.H.); Tel.: +86-028-84078705 (L.C.)

Received: 23 March 2020; Accepted: 11 April 2020; Published: 15 April 2020

Abstract: Although clay mineral content in ion-absorbed rare earth ores is crucial for migrating and releasing rare earth elements, the formation, distribution, and migration of clay minerals in supergene rare earth ores have not been fully understood. Therefore, this study analyzes the characteristics of clay mineral type and content, soil particle size, pH value, leaching solution concentration, and leaching rate. This analysis was performed using different methods, such as regional rare earth mine soil surveys, in situ leaching profile monitoring, and indoor simulated leaching. The results showed that the grain size and volume curve of rare earth ore have unimodal and bimodal shapes, respectively. X-ray diffraction showed the differences in clay mineral types formed by different weathered bedrocks. The principal clay minerals were kaolinite, illite, chlorite, and vermiculite, with their relative abundance varying with parent rock lithology (granite and low-grade metamorphic rocks). In the Ganxian granite weathering profile, the kaolinite content increased from top to bottom. The decomposition of feldspar minerals to kaolinite was enhanced with an increase in the SiO_2 content during weathering. The in situ leaching profile analysis showed that the kaolinite content increased initially and then decreased, whereas the illite/mica content exhibited the opposite trend. Under stable leaching solution concentration and leaching rate, clay mineral formation is favored by lower pH. Low pH, low leaching rate, and highly-concentrated leaching solution (12 wt%) resulted in a slow increase in kaolinite content in the upper part of the profile (30 cm). A lower concentration of the leaching solution (4 wt%) resulted in rapid enrichment of kaolinite after 15 days. Low pH, leaching solution concentration, and leaching rate promoted the formation of distinct kaolinite horizons. We suggest that by disregarding other control factors, rare earth recovery of over 90% can be achieved through leach mining with solutions of 8 wt% and a pH of 5 at a leaching rate of 5 mL/min.

Keywords: clay minerals; grain size characteristics; in situ leaching; simulated leaching; ion-absorbed type rare earth ore

1. Introduction

Rare earth elements (REEs) are 16 chemical elements grouped by their atomic number, and classified as light (LREEs), middle (MREEs), and heavy (HREEs). The weathered crust elution-deposited REE ores in southern China have drawn much attention because of their abundance of granitic residuum, their simple extraction processes, and their well-distributed composition [1–3].

The migration and enrichment of REEs are controlled by several factors, such as parent rock lithology, pH value, intensity of weathering, and topography [4–6]. Previous studies have shown that, for chemical index of alteration (CIA) values of 65%–85% in granite, clay minerals increase rapidly with an increasing degree of weathering. There is a positive correlation between the loss on ignition (LOI) of 2%–6% in the weathering crust and REE content [7].

Clay minerals have a controlling effect on the migration and release of REE ore. The completely weathered layer of a weathering crust mainly comprises quartz, feldspar, and clay minerals. The clay mineral content decreases gradually from the weathering crust surface to the lower layer, where clay minerals are converted from hydromica and montmorillonite to halloysite, kaolinite, and gibbsite [8].

The distributions of REEs in the weathering crust are controlled by both the composition of the parent rock and the clay mineral content of the weathering crust. Halloysite, a clay mineral, plays a significant role in the differentiation of cerium [9]. Halloysite has a stronger effect than kaolinite in the adsorption of REEs; however, this adsorption mechanism is not yet fully understood. Previous studies found that the adsorption of REEs is controlled by the properties of the clay minerals rather than the electrolyte solution or dissolved carbon dioxide content [10]. The adsorption capacity of kaolinite increases linearly with increasing pH. A fractionation between HREEs and LREEs due to selective sorption is observed, with HREEs being more sorbed than LREEs at high ionic strengths [10]. For montmorillonite at pHs below 4.5, the REE adsorption capacity is constant, and is modeled by cation exchange [11]. Different clay minerals have different adsorption capacities for REEs. Chi et al. [12] showed that for three common clay minerals, the cation adsorption capacity follows the order: montmorillonite > halloysite > kaolinite. This result shows that different parent rock lithologies will result in different weathering crust structures and clay mineral compositions. Intimate grain-to-grain contacts promote a unique chemical environment at the microscale, bringing about the formation of transient clay mineral phases which quickly disappear in the overlying soil [13]. The bulk of illite in the weathering crust is due to the weathering of mica minerals. A study of unstable soil profiles found that illite is converted into vermiculites or interstratified illite-smectite [14].

Climatic and environmental change is one of the causes of compositional differentiation in clay minerals. Kaolinite and kaolinite interlayer minerals are dominant in strongly leached soil layers [15], while illite and montmorillonite represent a cold and humid climate with weak chemical weathering [16]. Clay minerals of different crystal characteristics differ in physical structure and properties [13]. Clay minerals that host ion-adsorbed REE ores have large specific surface areas and a strong capacity to adsorb REE ions. The clay mineral content thus controls the migration and enrichment of REEs—processes of great significance for REE mineralization.

Although the clay mineralogy of weathered crust elution-deposited REE ores varies, several studies have demonstrated that the clay minerals in these ores commonly comprise halloysite, kaolinite, some illite, and rare montmorillonite [17]. It is widely believed that the horizon enriched in REE generally contains abundant halloysite and kaolinite [18,19], and that clay mineral migration is controlled by soil particle size and specific leaching conditions [20,21]. The metallogenetic mechanism of weathered crust elution REE deposits could involve the weathering of granodiorite and volcanic rocks in warm and humid climates, with the transformation of their parent mineralogy into kaolinite, halloysite, and montmorillonite [22]. In weathering crust elution-deposited REE ores, REEs adsorbed on the clay minerals by ion-exchangeable phases account for more than 80% of the total REE content [3]. However, leaching is controlled by the properties of the REE ore, by the nature and concentration of the leaching reagent, and by the hydrodynamics, kinetics, and mass transfer of the leaching process [22]. We postulate that the weathered crust elution-deposit REE ore is associated with REE ion enrichment, which is dissociated with hydrated or hydroxyl hydrated minerals and adsorbed by clay minerals, which are subsequently deposited, and mineralized in the weathered crust over a long period. In contrast, this is not to say that all REE mineralization can be explained by a single model.

It is important to understand that the clay mineralogy in different environments of REE ore formation varies with different conditions of parent rock, pH values, degrees of weathering [4,5], and mining conditions [9]. This study investigates the types and changing characteristics of clay minerals in several ion-absorbed REE ores in southern Jiangxi Province, China, during weathering and in situ leaching, with an aim of improving the recovery rate. To achieve this, soils on the surface of the weathered crust in a typical rare earth mining area in southern Jiangxi Province were sampled. Then, methods such as in situ leaching profile monitoring and indoor leaching simulation experiments were used to study the characteristics of the clay mineral properties and soil particle size.

2. Background

2.1. Study Site

The study sites are located in the REE mining regions of Longnan County, Anyuan County, Ganxian District, Ganzhou, Jiangxi Province, China (Figure 1). The region is situated in the subtropical monsoon climatic zone, with an average annual precipitation of 1461.2 mm and an annual average temperature of 19.4 °C [23]. The topography is high in the south and east and low in the north and west [23]. The central region consists mostly of basins between hills. REE mines are mainly distributed in these basins.

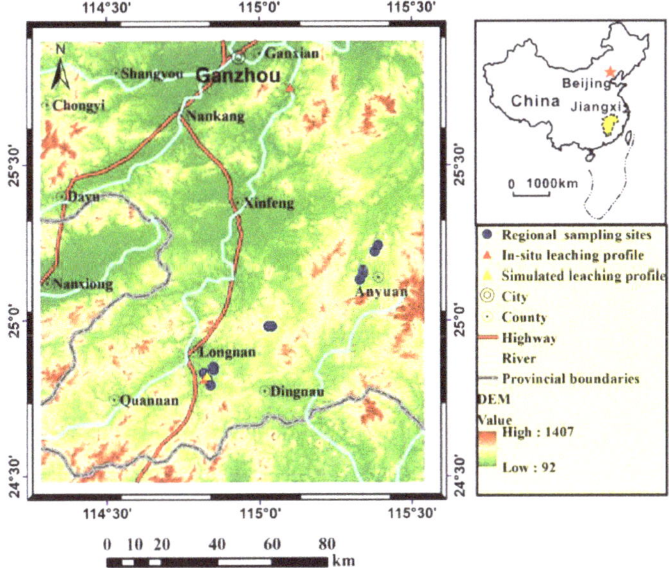

Figure 1. The location of sampling sites.

2.2. Regional Stratigraphy and Lithology

The study area mainly comprises sandstones, slates, phyllites, and carbonates of the Lower Paleozoic (Sinian) to Mesozoic ages, covered by Quaternary sediments, except for the Silurian, Ordovician, and Tertiary. Quaternary sediments consist of loose fluvial deposits in a river terrace with a high ratio [24].

Magmatic activity in the area can be resolved into four cycles: Caledonian, Variscan, Indosinian, and Yanshanian. The main lithology of igneous intrusions is an acidic to medium-acidic rock [25]. These three types of magmatic rock account for 99.2% of all magmatic rocks in the region. This corresponds to the diagenesis times of 461–384, 228–242, and 189–94 of REE ore-bearing rocks, respectively, as given in [26].

3. Materials and Methods

3.1. Sample Collection

A total of 49 surface soil samples were collected from REE mines; among these, 27 were sampled regionally, mainly from Anyuan, Longnan, and Dingnan. Ten profile samples were also collected from Jiangwozi, in Ganxian District [27] and from the Wenlong REE mine in Longnan County (longitude 114°49′31.99″E; latitude 24°49′21.68″N).

The in situ leaching profile of the Wenlong REE mine was sampled on three different occasions, October 2, 2016; January 13, 2017; and April 13, 2017, over 193 days (12 samples in total). These samples were collected at elevations from 255 to 249 m and at depths from 0.5 to 1.5 m from the surface; the location was mapped with GPS. The sampled weathering profile comprises, from the surface to the bedrock, purple clayey-sandy soil, red silty clay, gray yellow clayey- sand. The underlying bedrock lithology comprises siliceous slate with high organic content, fine sandstone, grayish-white medium-crystalline granodiorite, fine-medium biotite granite, and grayish-white medium-crystalline granodiorite.

3.2. Simulated Leaching Experiment

An unexploited weathering crust with soil horizons similar to those of the in situ leaching profile (i.e., completely weathered upper layer; the transitional middle layer; only partly weathered bottom layer) was selected as representative of the textural and structural characteristics of the typical ion-absorbed REE profile. In the laboratory, a custom column device consisting of a liquid injection barrel, soil column tube, and liquid collection basin was used for the leaching experiment.

The soil column tube used a PVC pipe with a height of 120 cm and inner diameter of 11 cm. An array of sampling holes located at 30, 50, 70, 90, and 110 cm from top to bottom were punched through the pipe wall. The tube bottom was sealed by a lid with a floor drain and plug. To prevent leaching of the sample with the liquid discharge, a paper filter of cotton fiber with 11 cm diameter, similar to the PVC pipe, was laid on the floor of the drain, and the bottom lid was perforated and connected to a plastic hose to receive the leaching ore concentrate. The field profile was sampled following its pedostratigraphy, and the samples were placed in the soil column tube to recreate this stratigraphy, from bottom to top layers. Deionized water spraying was used regularly to ensure that the water content and water holding rate in the soil column sample were approximately consistent with those in the field profile. A total of eight soil columns (T1–T8) were made, each tube was filled with 100 cm of soil, and filter paper with a diameter of 11 cm was placed on the top. The leaching solution was made from analytically pure $(NH_4)_2SO_4$ crystals dissolved in deionized water (from AK-RO-UP-500), to simulate leaching and rainfall. HCl was used to adjust the pH value of the solution. The leaching time was 40 days, with samples taken on the 5th, 15th, 23rd, and 40th day. The machine standard was RO < 0.7 mV and UP < 20 MΩ. The specific leaching parameters of each soil column are shown in Table 1.

Table 1. Experimental parameters of simulated leaching.

Soil Column No.	Leaching Solution	pH	Content of $(NH_4)_2SO_4$ (wt%)	Leaching Solution Flow Rate (mL/min)
T1	$(NH_4)_2SO_4$	5	8	3
T2	$(NH_4)_2SO_4$	3	8	3
T3	$(NH_4)_2SO_4$	4	8	3
T4	$(NH_4)_2SO_4$	5	12	3
T5	$(NH_4)_2SO_4$	5	4	3
T6	$(NH_4)_2SO_4$	5	8	1
T7	$(NH_4)_2SO_4$	5	8	5
T8 (Simulated rainfall)	Deionized water	6.8~7.2	0	3

3.3. Particle size Analysis

The particle size analysis was conducted using different analytical testing methods. Regionally collected samples (27) were processed by the wet sieving method [27]. Particles of less than 0.075 mm were tested on an LS908 (A) laser particle size analyzer (Henan Zhengzhou North-south Instrument Equipment Co. LTD, Zhengzhou, China) at the Jiangxi University of Science and Technology School of Resource and Environmental Engineering. The coarser (>0.075 mm) fraction of the in situ leaching profile was also analyzed by wet sieving, while the <0.075 mm fraction was analyzed with a Malvern MasterSizer 2000 laser particle size analyzer at the Peking University School of Urban and Environmental Sciences.

3.4. Clay Mineral Analysis

The XRD analysis of clay minerals followed standard procedures, as described in [28,29]. Bulk samples were pulverized to a fine powder using a planetary ball mill with agate elements. Specimens for XRD analysis were front-loaded using a blade; sieve rotation ensured random grain orientation. The clay fraction was separated in deionized water; the clay suspension was then deposited onto 0.45 μm Whatman filters in vacuum, and transferred to glass slides. Each concentrated clay sample was air-dried before XRD analysis, and then saturated with ethylene glycol for subsequent analysis. Occasionally, heat treatment was necessary; in this case, the slides were heated for one hour at 550 °C before further XRD analysis. Analyses were performed with Rigaku D/max 2550 XRD at the Oil and Gas Laboratory in ALS Houston, USA, and at the Key Laboratory of Nonferrous Metal Materials Science and Engineering of the Ministry of Education, in the Central South University, China. A Bruker Endeavor D4 XRD (Cu radiation, 40 kV, 40 mA) at 0.02564 degree/step/second was used for bulk analysis; clay analysis was performed at 0.02992 degree/step/second.

XRD patterns of separated clay fractions on glass were used for clay mineral identification. Mineral identification was facilitated by JADE (version 9.5). Quantitative analysis of minerals was performed by the Rietveld method [30], and amorphous phases were not accounted. The results were normalized to 100% based on the assumption that the complete mineral content of the sample was accounted for in the XRD patterns. Duplicate samples were analyzed at the two laboratories in the United States and China, and the error was less than 10%.

4. Results

4.1. Particle Size Analysis

A total of 27 regional particle size samples were analyzed. Particle size distribution curves from this sample set show both unimodal and bimodal patterns (Figure 2).

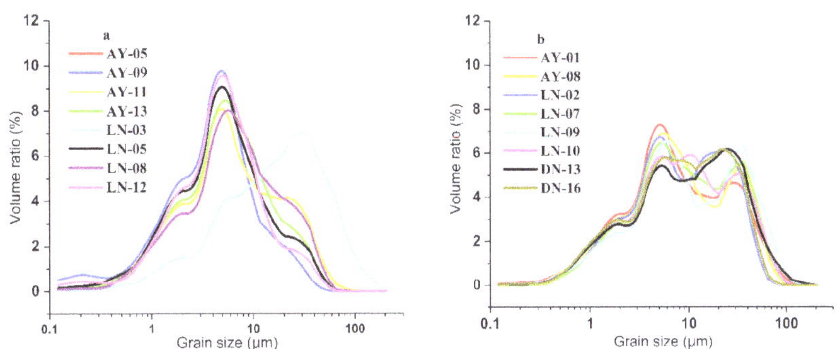

Figure 2. Grain size distribution of 27 regional samples: (**a**) shows unimodal patterns; and (**b**) shows bimodal patterns.

Particle size curves with unimodal patterns presented approximate normal distribution, with the exception of Sample LN-03 (Figure 2a). A significant peak was estimated in the particle size range of 2.1–17.2 μm, with volume ratio from 8.96 to 10.36. The range of grain size ratio was more than 70% and was controlled by fine particles. Significantly, Sample LN-03 showed a unimodal curve with obvious right-hand deviation (Figure 2a), with a grain size peak at 30.5 μm and volume ratio of 7.2%. Therefore, in Sample LN-03, the peak volume ratio was lower and the particle size was coarser than that in other unimodal samples. In samples of bimodal particle size distribution, the first significant peak appeared at 1.7 μm (Figure 2b), with an estimated particle size range of 1.4–2.6 μm and volume ratio of 1.5%–5%. The maximum peak interval was in the particle size range of 2.6–9.7 μm, whereas the volume ratio was in the range of 1.47%–10.07%. The third peak interval was in the particle size range of 14.2–36.9 μm, with a volume ratio of 4% to 7.5%.

Samples with unimodal particle size distribution (Figure 2a; e.g., AY-05 and AY-09) originated in weathering crust comprising red silty clay, indicative of advanced bedrock weathering and pedogenesis. This weathering crust overlies bedrock lithology with siliceous slate and fine sandstone interlayers. In contrast, samples with bimodal particle size distribution originated in soils over grayish-white medium-grained granodiorite with moderate weathering.

In the samples from the weathering crust profile, the grain size increased gradually up-profile, reflecting the degree of weathering in different profile horizons. Sample LN-03, with a coarser particle size overall (Figure 2a), came from a depth of 0.90 m, which is close to party weathering bedrock. The particle volume distribution curves indicate that particle size distribution in the REE mine does not follow a single normal distribution pattern, thus reflecting considerable differences in soil particle size gradation. These analyses show that sample location, degree of weathering, and bedrock lithology affect the distribution of soil particle size.

4.2. Regional Clay Mineral Analysis

The clay minerals in the regional samples were mainly kaolinite, followed by illite, chlorite, vermiculite (Table 2). Mineral residues from the parent rock included quartz and potassium feldspar with small quantities of plagioclase in some samples. The kaolinite content varied greatly in the regional samples (maximum: 62.1%; minimum: 8.8%; average: 31.91%; standard deviation of 11.90).

Table 2. Main clay minerals content of regional in soil.

Serial Number	Sample	Clay (%)				Other Minerals (%)		
		Vermiculite	Chlorite	Kaolinite	Illite-Mica	Quartz	Potassium Feldspar	Plagioclase
1	AY-01 *	-	1.8	37	9.4	42.8	8.3	-
2	AY-02	-	-	40.3	10.1	39.8	9.8	-
3	AY-03 *	-	-	37.5	13.5	40.8	6.7	-
4	AY-05 *	-	2.8	37.3	3.4	38	13.4	-
5	AY-06 *	-	2.7	38.1	6.3	30.8	15.9	-
6	AY-07	-	-	62.1	13.9	20.3	3.7	-
7	AY-08	-	-	56	16.4	22.8	4.7	-
8	AY-09 *	-	2.5	31.5	5.9	42.2	10.5	-
9	AY-10	-	-	36.9	9.5	42.2	11.3	-
10	AY-11	-	-	36.2	4.9	55.7	3.2	-
11	AY-12 *	-	3.4	29.2	0	43.8	6	-
12	AY-13 *	6.6	1	45.8	3.1	23.3	23.1	-
13	DN-13 *	-	2.6	28.6	1.8	31.4	31.8	-
14	DN-16 *	1.5	1.5	24.5	1.3	34.1	35.3	-
15	LN-01 *	1.9	0.9	32.8	-	48.7	6.8	-
16	LN-02	3.8	-	25.4	21	35	18.7	-
17	LN-03	-	-	8.8	28.6	54.4	8.2	-
18	LN-04 *	-	-	30.3	2	47.8	18.6	-
19	LN-05	0.1	-	23	19.1	44.1	13.8	-
20	LN-06 *		-	11.9	0.4	31.5	29.3	10

Table 2. *Cont.*

Serial Number	Sample	Clay (%)				Other Minerals (%)		
		Vermiculite	Chlorite	Kaolinite	Illite-Mica	Quartz	Potassium Feldspar	Plagioclase
21	LN-07	1.5	-	12.4	13	59.6	15	-
22	LN-08 *	-	-	30	0.3	47.8	13.2	0.4
23	LN-09 *	1.4	-	33.7	0.6	43.7	16.9	-
24	LN-10 *	0.6	-	37.9	1.3	39.3	15.4	1.2
25	LN-11	0.8	-	13.5	6.4	53.9	26.2	-
26	LN-12 *	-	-	30.1	0.3	38.3	25	1
27	LN-18 *	-	-	30.9	3.2	49.8	6.5	1.5

* Sample analyses were supported by Oil and Gas Laboratory in ALS Houston.

Since the chemical composition of kaolinite is the same as that of halloysite (except for weakly-bound interlayer water) [31], the kaolinite diffraction peak is the same as that of halloysite. Fang et al. [9] studied clay minerals in six REE mining areas in southern Jiangxi Province; the characteristic diffraction peaks of halloysite and kaolinite were 7.30–7.45 Å, 4.5–4.6 Å, 3.58–3.60 Å, and 3.32–3.37 Å. Chi et al. [3] found that the weathered crust leaching type REE ore was mainly composed of clay minerals, i.e., mainly halloysite, illite, kaolinite, and very small amounts of montmorillonite, alongside quartz sand and rock-forming minerals (feldspar). Halloysite is generally formed in the upper layer of a weathering crust from kaolinite interstratified minerals in noncrystalline stage, resulting from weathering and the dissolution of feldspar [13,15,32]. We infer that the kaolinite identified through XRD in our samples may contain kaolinite and kaolinite-interstratified minerals. The coexistence of both minerals in weathering profiles is frequently reported in studies with electron microscopy [32,33].

4.3. *Analysis of Clay Minerals in the in Situ Leaching Profile*

Samples from the Jiangwozi and Longnan profiles, Ganxian District, mainly comprised kaolinite and illite (i.e., mica), along with quartz and feldspar (Table 3). The kaolinite content ranged from 8.3% to 35.0% (average: 18.88%; standard deviation: 7.82).

Table 3. Main clay minerals content of soil profile (%).

Serial Number	Sample	Kaolinite	Illite-Mica	Quartz	Potassium Feldspar	Depth (cm)
1	GX-01	16.2	19.2	39.7	24.8	20
2	GX-02	18.7	13.5	37	30.8	45
3	GX-03-2	25.9	15.3	33.6	25.2	60
4	GX-03-1	20.3	7.3	39.7	32.7	80
5	GX-04-5	28.2	13.6	31	27.2	105
6	GX-04-4	14.9	15.9	34	35.2	130
7	GX-04-3	16.4	11.3	33.3	39	150
8	GX-04-2	16.5	22.5	32.3	28.6	170
9	GX-04-1	35	17.2	33.1	14.7	190
10	GX-05	9.8	12.4	39.9	37.9	215
11	P01-1-1	12.6	17.2	46.2	31.3	80
12	P01-2-1	12.3	16	40.5	34.1	150
13	P01-3-1	9.5	12.3	44	24.9	195
14	P01-4-1	11.8	10.1	53.2	16.4	280
15	P01-1-2	8.9	13.4	61.3	25.1	80
16	P01-2-2	8.3	13	53.6	18.3	150
17	P01-3-2	33	11.9	36.8	24.2	195
18	P01-4-2	17.7	15	43	20.9	280
19	P01-1-3	22.5	20.7	35.8	30.6	80
20	P01-2-3	22.8	19.9	26.7	21.4	150
21	P01-3-3	31.2	13.8	33.6	34.3	195
22	P01-4-3	22.8	15	27.9	31.3	280

In the Jiangwozi profile, the kaolinite content increases nonlinearly from the surface to the bottom, from 16.2% to 28.2% (Table 2), and then decreases gradually. The kaolinite content reached a maximum of 35.0% at 200 cm from the surface, where the soil texture is coarse and fine sand. At this depth, kaolinite is distributed like a network [27]. The REE ore of the Jiangwozi profile is mainly enriched in the weathering and leaching layer between 90 and 200 cm. The change of clay mineral content is closely related to the chemical weathering rate of the rock: the maximum kaolinite content (35.0%) is in GX-04-1 where potassium feldspar content is at its minimum (14.7%), suggesting that potassium feldspar is strongly weathered. At depths of 0–55 cm, the illite (mica) content showed the opposite trend to that of kaolinite. From 70 to 120 cm, illite (i.e., mica) follows the same trend as kaolinite, but its content is lower. At this depth, the content of coarse particles is relatively increased, probably due to enhancement of chemical weathering and vertical migration of particles. The kaolinite content peaks at about 200 cm, while the illite content peaks earlier, at around 190 cm (at 17.2%). Yang [34] researched the clay mineralogy of the REE weathering crust of Longnan granite, in Jiangxi Province, and found that the crystallization degree of kaolinite gradually increased down-profile, suggesting that the content of kaolinite also increased. This discovery was demonstrated in granite weathering crust profiles where kaolinite was dominant at the bottom of crust [13,35].

The analysis of soil clay minerals and rare earths in the Wenlong mine showed that at the onset of in situ leaching, kaolinite is absent; however, as leaching continues, the kaolinite content increases dramatically at depth of around 200 cm (Table 2). In comparison, with the significant change with depth in the Ganxian District profile, at the Wenlong mine profile, kaolinites formed in the course of weathering increase gradually as in situ leaching progresses. This is the result of the coupling of natural weathering and human activity (profile stripping in the course of REE mining).

4.4. Clay Minerals in the Simulated Leaching Profile

The clay (kaolinite and illite) and other mineral (quartz, potassium feldspar, and plagioclase) content of the eight soil profiles subjected to the simulated leaching experiment are presented in Table 4. Each soil column sample is denoted by a three-digit number: the first digit (T1 to T8) represents the number of the soil column; the second digit (1, 3, and 5) represents the depth of the sample in the soil profile (at 30, 70, and 110 cm, respectively); and the third digit (1, 2, 3, and 5) represents the time of mineral concentration measurement since the onset of the experiment (5th, 15th, 23rd, and 40th day, respectively).

Table 4. Clay and other mineral content of eight soil profiles (T1 to T8) at different times during simulated leaching (%).

Serial Number	Sample	Kaolinite	Illite-Mica	Quartz	Potassium Feldspar	Plagioclase
1	T1-1-1	22.3	7.4	45.7	24.6	-
2	T1-1-2	21.5	7.8	48.1	22.6	-
3	T1-1-3	20.2	5.1	45.2	29.6	-
4	T1-1-5	11.1	12.9	45.9	30.1	-
5	T1-3-1	18.9	8.8	55.6	16.7	-
6	T1-3-2	14.9	9.4	39.4	36.2	-
7	T1-3-3	13.8	13.2	51.7	21.3	-
8	T1-3-5	6.8	14.7	54.6	23.9	-
9	T1-5-1	15.2	14.6	36.6	33.6	-
10	T1-5-2	13.1	8.6	48.1	30.2	-
11	T1-5-3	16.1	12.8	37.2	33.9	-
12	T1-5-5	17.5	8.2	39.2	35.1	-
13	T2-1-1	20.5	11.4	42.5	25.5	-
14	T2-1-2	16.6	12.5	37.7	33.2	-
15	T2-1-3	21.7	7.2	45.3	25.8	-
16	T2-1-5	20.2	10.2	46.3	23.2	-

Table 4. *Cont.*

Serial Number	Sample	Kaolinite	Illite-Mica	Quartz	Potassium Feldspar	Plagioclase
17	T2-3-1	11.2	12.5	44.8	35.1	-
18	T2-3-2	9.2	15.0	43.0	32.8	-
19	T2-3-3	8.8	23.2	35.9	32.1	-
20	T2-3-5	10.5	23.0	43.1	23.5	-
21	T2-5-1	8.9	20.8	44.4	25.8	-
22	T2-5-2	8.5	16.0	42.2	33.4	-
23	T2-5-3	25.3	1.2	45.5	28.0	-
24	T2-5-5	31.5	5.3	31.0	32.2	-
25	T3-1-1	15.3	10.5	55.5	18.7	-
26	T3-1-2	16.2	12.8	40.2	30.8	-
27	T3-1-3	15.7	16.9	39.6	27.8	-
28	T3-1-5	18.4	9.7	48.4	23.5	-
29	T3-3-1	9.2	17.9	43.3	29.6	-
30	T3-3-2	11.2	14.2	45.7	28.9	-
31	T3-3-3	10.5	15.8	43.5	30.3	-
32	T3-3-5	18.6	20.2	38.0	23.2	-
33	T3-5-1	16.5	15.0	28.5	40.1	-
34	T3-5-2	9.2	11.4	44.8	34.6	-
35	T3-5-3	11.8	21.9	28.3	38.1	-
36	T3-5-5	16.2	17.5	25.9	40.5	-
37	T4-1-1	14.1	14.9	43.9	27.2	-
38	T4-1-2	18.1	12.3	48.5	21.1	-
39	T4-1-3	18.2	11.8	42.8	27.2	-
40	T4-1-5	18.5	10.6	34.0	36.9	-
41	T4-3-1	9.9	12.2	46.2	31.7	-
42	T4-3-2	6.9	16.1	41.8	35.2	-
43	T4-3-3	10.4	16.7	29.8	43.1	-
44	T4-3-5	15.7	20.4	37.3	26.6	-
45	T4-5-1	14.3	16.5	34.0	35.2	-
46	T4-5-2	9.1	14.0	42.5	34.4	-
47	T4-5-3	15.3	9.5	50.1	25.1	-
48	T4-5-5	11.2	6.8	51.4	27.4	3.1
49	T5-1-1	14.2	9.6	48.3	27.9	-
50	T5-1-2	16.9	14.0	43.6	25.5	-
51	T5-1-3	16.4	30.6	35.4	17.6	-
52	T5-1-5	26.7	9.6	34.2	29.5	-
53	T5-3-1	16.6	8.9	49.4	25.2	-
54	T5-3-2	12.4	12.1	47.9	27.5	-
55	T5-3-3	16.2	11.1	41.5	31.2	-
56	T5-3-5	12.1	17.1	39.8	31.0	-
57	T5-5-1	15.2	10.2	29.0	37.8	7.9
58	T5-5-2	9.4	8.8	31.8	32.4	17.6
59	T5-5-3	11.4	13.2	36.7	29.7	9.1
60	T5-5-5	9.6	11.5	54.2	24.6	-
61	T6-1-1	10.9	6.9	52.9	23.9	5.4
62	T6-1-2	10.5	8.4	57.7	19	4.4
63	T6-1-3	9.8	8.6	57.7	23.9	-
64	T6-1-5	14.0	12.5	46.9	26.6	-
65	T6-3-1	13.3	13.2	40.6	30.4	2.4
66	T6-3-2	15.1	12.9	49.7	22.3	-
67	T6-3-3	13.7	10.3	47.0	29.0	-
68	T6-3-5	13.3	12.8	60.6	13.2	-
69	T6-5-1	9.8	13.9	47.8	20.8	7.7
70	T6-5-2	7.0	11.6	59.1	22.3	-
71	T6-5-3	15.4	12.9	40.8	26.1	4.8
72	T6-5-5	16.4	12.3	38.0	29.8	-
73	T7-1-1	16.8	10.2	43.0	22.5	7.4
74	T7-1-2	15.7	13.2	40.1	31.0	-
75	T7-1-3	20.6	13.3	39.9	26.2	-
76	T7-1-5	19.1	21.3	38.3	21.3	-

Table 4. *Cont.*

Serial Number	Sample	Kaolinite	Illite-Mica	Quartz	Potassium Feldspar	Plagioclase
77	T7-3-1	7.5	15.5	40.5	30.5	6.1
78	T7-3-2	10.3	13.1	51.3	22.4	2.9
79	T7-3-3	15.3	4.5	38.5	31.1	10.6
80	T7-3-5	11.3	10.9	53.3	19.6	4.9
81	T7-5-1	13.5	8.0	34.1	38	6.4
82	T7-5-2	11.0	14.6	32.8	41.6	-
83	T7-5-3	18.3	12.4	28.0	41.3	-
84	T7-5-5	25.3	17.7	41.5	15.5	-
85	T8-1-1	16.0	9.6	46.4	28.0	-
86	T8-1-5	20.6	19.3	30.7	29.4	-
87	T8-3-1	8.5	5.4	34.7	51.4	-
88	T8-3-5	15.2	11.5	35.7	37.6	-
89	T8-5-1	14.6	11.3	49.9	24.2	-
90	T8-5-5	19.6	17.7	33.8	28.9	-

-: represented that its content was below the detection limit.

The lowest kaolinite content is 6.8%, at T1-3-5, i.e., at sampling port number 3 in the T1 soil column after 40 days of simulated leaching. The highest kaolinite content is 31.5%, at T2-5-5, i.e., sampling port number 5 in the T2 soil column after 40 days of simulated leaching. The average content of kaolinite in the REE ore is 14.67% and the standard deviation is 4.72. In all columns except T1 and T2, the first sampling port (depth: 30 cm) shows that the variation in kaolinite content increases with the longer leaching. This is most obvious in T5, where the kaolinite content after 40 days of simulated leaching (sample No: T5-1-5) is 1.88 times the initial value. In soil columns T1 and T2, the kaolinite content changed little in the first 23 days, but after 40 days, it declined significantly in T1. The highest illite content is 30.6%, at T5-1-3, i.e., sampling port number 1 in the T5 soil column after 23 days of simulated leaching. Overall, the average illite content in the soil column sample set is 12.78%, with a standard deviation of 4.63, slightly less than that of kaolinite. Similarly, the illite-mica content of the first sampling port (depth of 5 cm) increased gradually in T1 and T6–T8 as leaching progressed, while other soil columns showed no obvious variation. However, in the T5 soil column, the illite-mica peaked at 30.6% after 23 days of leaching, and then fell back to the initial level after 40 days of leaching. The second and third sampling ports show that the content of kaolinite, illite and other clay minerals in the RE ore is complex under different leaching conditions (Table 3), which may be controlled by many factors.

The minimum content of potassium feldspar is 13.2%, after 40 days of leaching (sample No: T6-3-5), which is 56.6% lower than the potassium feldspar content at the sampling port at the initial stage of leaching (sample No: T6-3-1: 30.4%). This result indicates that potassium feldspar at the bottom of the soil column may have been weathered and mobilized with the leaching solution after prolonged leaching. The maximum potassium feldspar content is 51.4% (T8-3-1); average value is 28.81%, and the standard deviation is 6.60. Quartz fluctuates from 25.9 to 60.6% in the course of ore leaching, with an average value of 42.6%, and a standard deviation of 7.67, which is greater than that of potassium feldspar. In addition, plagioclase was detected in 15 samples, and its content fluctuated between 2.4% and 17.6%, with an average value of 6.71% and a standard deviation of 3.68. In the other samples, the plagioclase content was below the detection limit.

5. Discussion

5.1. Soil Particle Size and Distribution of Clay Minerals in REE Mining Areas

Ion-absorbed REE ore is mainly formed by advanced weathering of granite. It is a loose, earthy substance comprising quartz, feldspar, and clay minerals [36]. Therefore, this loose soil mantle, formed by surface weathering is closely related to mineral grain size. The cumulative curves of regional particle size distribution are S-shaped (Figure 3), which is consistent with earlier report [36].

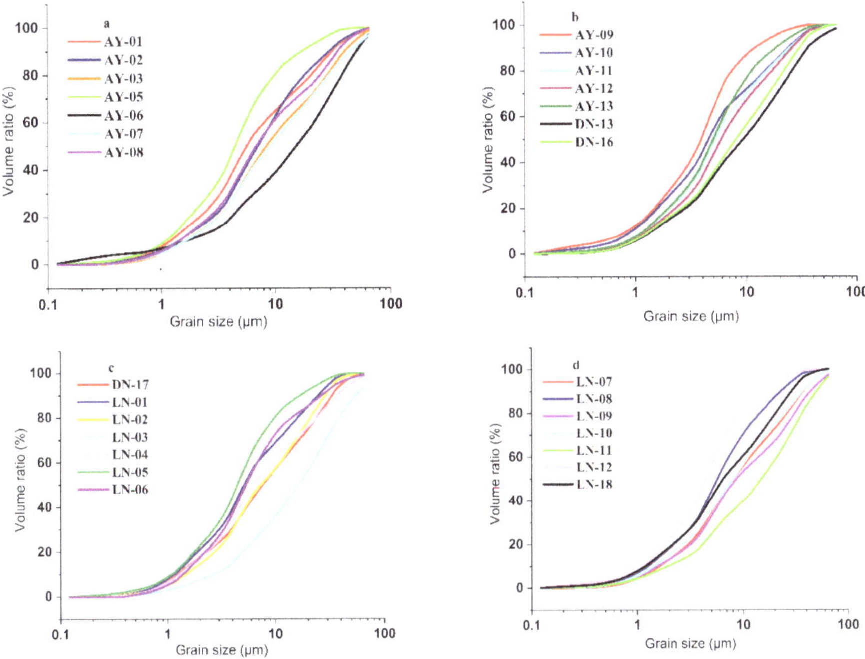

Figure 3. Cumulative particle size distribution curves from ion-absorbed RE in (regional samples). (**a**) shows the curves with AY-01, AY-02, AY-03, AY-05, AY-06, AY-07 and AY-08; (**b**) shows the curves with AY-09, AY-10, AY-11, AY-12, AY-13, DN-13 and DN-16; (**c**) shows the curves with DN-17, LN-01, LN-02,LN-03, LN-04, LN-05 and LN-06; (**d**) shows the curves with LN-07, LN-08, LN-09, LN-10, LN-11, LN 12 and LN 18.

Additionally, the particle distribution curves in Figure 3a–d show that particle size ranges between 3.7 and 30.5 μm. According to Aberg's classification of granular materials [36,37], some of the cumulative curves of the regional sample set are A-shaped (i.e., the left end of the cumulative particle distribution curve is relatively steep, with a concave side downward), indicating that the particle gradation changes significantly with an increase in coarse particles in the soil. Yan et al. [38] used a wet sieving method to classify REE ores into eight types of particle size distribution. In this paper, particle size analysis concentrated on the <0.075 mm size fraction, and the analytical method was different (Malvern-2000 laser particle size analyzer); therefore, our particle size distribution curves are different.

The cumulative particle size distribution curves of AY-05 (Figure 3a), AY-09 (Figure 3b), LN-05 (Figure 3c), and LN-12(Figure 3d), nevertheless, are inclined steeply to left in their upper part (a B-type structure in Arberg's terms [37]), indicating that grain size changes abruptly from fine to coarse. A possible reason for this is that simulated leaching continues to promote rapid decomposition of feldspar (Table 3). Dissolution of feldspar due to leaching releases SiO_2, which migrates downward, where it recrystallizes resulting in particle thickening [36].

Eigenvalue analysis shows that the 10 particle size (the particle size at which the cumulative particle size distribution curve reaches 10% of the volume) was in the range of 0.82–5.03 μm (Table 5). The minimum D10 value (0.82), corresponds to Sample AY-09, a red sand sampled from 1.5 m below- surface, above Sinian feldspar quartz and slate bedrock in the age of Sinian period (Z). Field investigation in this area revealed that the weathering crust is approximately 3.5–3.8-m thick, and that the bedrock is strongly deformed and fractured, and thus particularly susceptible to physical and chemical weathering. D90 (the particle size at which the cumulative particle size distribution curve reaches 90% of the volume) ranges from is 11.99 μm (minimum) to 60.99 μm (maximum), with a standard deviation of 11.4, respectively, thus suggesting that discreteness increases with particle size. The average particle size, Dav and volumetric average particle size, D [4,3] have similar maximum, minimum, and standard deviation values. The standard deviation of the median particle diameter, D50 (the particle size at which the cumulative particle size distribution curve reaches 50% of the volume) is more significant than that of D [3,2] (surface area average particle size), indicating that the median particle diameter is more discrete than the surface area average particle size.

Table 5. Characteristic parameters of particle size distribution; regional weathering crust samples.

Sample No.	D10 (μm)	D50 (μm)	D90 (μm)	Dav (μm)	D [3,2] (μm)	D [4,3] (μm)
AY-01	1.32	6.25	33.88	12.88	3.09	12.88
AY-02	1.43	7.28	29.54	11.65	3.36	11.65
AY-03	1.64	8.47	40.28	15.64	4.09	15.64
AY-05	1.10	4.50	17.10	7.02	2.30	7.02
AY-06	1.79	15.51	50.28	21.57	3.67	21.57
AY-07	2.21	12.76	40.04	17.22	4.75	17.22
AY-08	1.44	6.93	35.50	13.21	3.48	13.21
AY-09	0.82	3.88	11.99	5.40	1.77	5.40
AY-10	0.96	4.57	23.70	8.76	2.21	8.76
AY-11	1.34	5.38	20.86	8.42	2.89	8.42
AY-12	1.13	6.39	28.22	11.06	2.61	11.06
AY-13	1.20	5.02	18.49	7.70	2.58	7.70
DN-13	1.64	12.75	44.85	18.90	4.37	18.90
DN-16	1.37	8.07	30.07	12.42	3.50	12.42
LN-01	1.16	5.14	24.69	9.41	2.84	6.10
LN-02	1.50	7.45	28.71	11.85	3.59	11.85
LN-03	5.03	20.86	61.00	27.80	8.19	27.80
LN-04	1.49	8.39	38.06	14.92	3.67	14.92
LN-05	1.37	6.42	33.75	13.00	3.53	13.00
LN-06	1.41	5.47	26.90	10.20	3.31	10.20
LN-07	1.62	8.17	37.10	14.42	4.07	14.42
LN-08	1.26	5.68	22.95	9.08	2.89	9.08
LN-09	1.55	8.63	43.42	17.11	3.74	17.11
LN-10	1.41	8.08	36.38	13.98	3.58	13.98
LN-11	1.91	15.00	49.19	21.02	4.51	21.02
LN-12	1.12	4.41	13.85	6.45	2.08	6.45
LN-18	1.21	6.68	28.74	11.42	2.79	11.42

With Dav as an independent variable and the other characteristic parameters as dependent variables (Figure 4), the regression coefficient is D90 > D [4,3] > D50 > D [3,2] > D10. This finding illustrates that the increase in surface average particle size has a more significant impact on coarse particles than on fine particles (D10). D [4,3] has the highest correlation with Dav (correlation coefficient: 0.99102), followed by D90, D50, D [3,2], and the lowest correlation with D10 (correlation coefficient: 0.727).

D10 residual analysis (Figure 5a) shows that when the average particle size Dav increases, other residuals decrease (with very few exceptions). This shows that for D10, as the average particle size increases, the volume of particle grain size decreases by less than 10%. Residuals are normally distributed (Figure 5b), and the regression analysis of the dependent variable is similar to that of the independent variable. Residuals present a linear shape at a 99.5% confidence interval (Figure 5c,d).

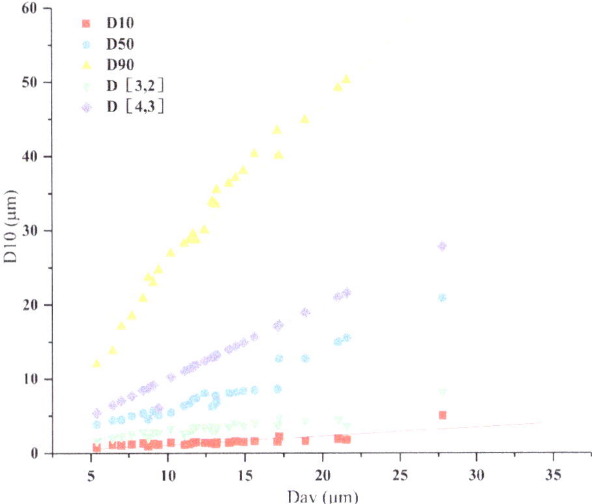

Figure 4. Scatterplot of characteristic parameters of particle size distribution; regional weathering crust samples.

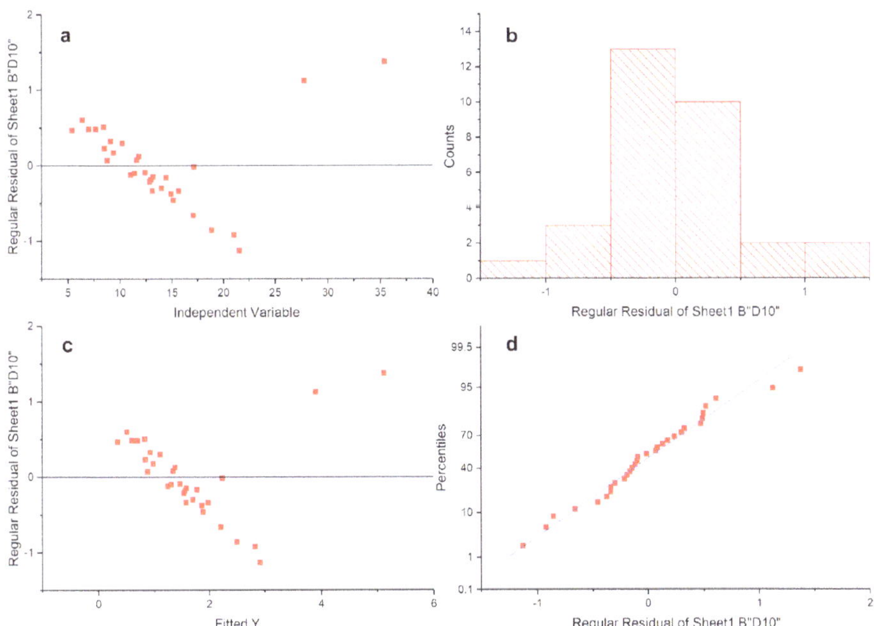

Figure 5. Residuals characteristic parameter D10; regional samples: (**a**) shows scatter plot of residuals; (**b**) shows histogram of regular residual; (**c**) shows scatter plot of residuals with fitted Y; and (**d**) shows scatter plot of residual flatten and significance testing.

The dominant clay minerals in the regional soil samples were kaolinite, followed by illite, and some vermiculite, and chlorite (Table 2). Kaolinite content has a weak correlation with rock-forming minerals, such as potassium feldspar and quartz. This finding indicates a nonlinear process of potassium feldspar alteration into kaolinite during granite weathering. Moreover, clay minerals and quartz were present in the weathering crust over metamorphic sandstone and slate in the Anyuan County area, and their correlation reached 0.68 (Figure 6). This indicates that the conversion ratio of feldspar to clay minerals, after weathering of this metamorphic bedrock, is higher than that of granite in the Longnan County granite.

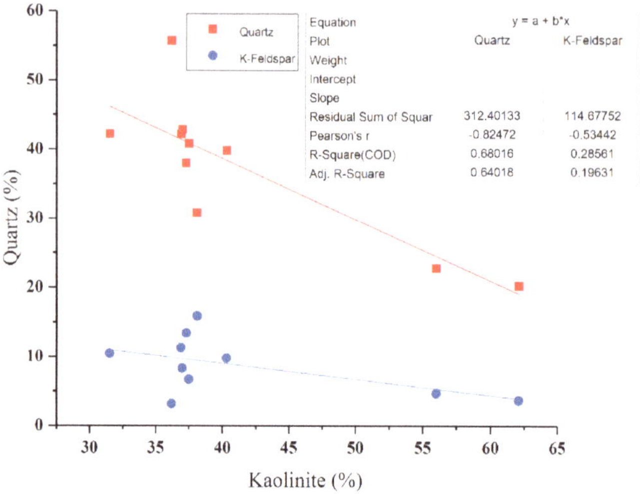

Figure 6. Correlation trend of kaolinite, Potassium feldspar, and quartz in samples from Anyuan County.

The relationship between average particle size (Dav) and kaolinite, quartz, and potassium feldspar indicates that the mineral particle size had only a minor effect on the clay mineral content. Quartz and potassium feldspar are mostly distributed on both sides of kaolinite. Quartz is further away from the *X*-axis, while potassium feldspar is closer to the *X*-axis. This distribution, therefore, further confirms that weathering of potassium feldspar has a significant impact on the formation of kaolinite.

Wang et al. (2018) [39] found that the main minerals of the low-grade metamorphic rocks (e.g., meta-sandstone, meta- siltstone, and slate) in the Anyuan County area of southern Jiangxi are 30–70% quartz, 5–30% feldspar, 3–10% biotite, and 3–12% muscovite. The CIA index is in the range of 68–75%. The Yanshanian granite in Longnan County is mainly compounded of 25–32.7% quartz, 31.1–42.4% potassium feldspar, 17–28.9% plagioclase, 3.4–6% biotite, and 1–3.4% muscovite [40]. According to the analysis of major elements in granite in the Zudong mining area, Longnan County [8,41] (Table 6), CIA is in the range of 61–65%, with an average of 63%, which is significantly lower than that of Anyuan County. This suggests that a higher CIA value reflects more extensive loss of Na^+, K^+, and Ca^{+2} during leaching, enrichment in Al and Si, and more advanced conversion of feldspar to clay minerals [42,43]. This study confirms that the weathering of the parent rock has a significant effect on the formation of soil clay minerals. Weathered feldspar minerals are converted to kaolinites, which are then converted into kaolin minerals under moderate silica and salt-based ion conditions [13]. Furthermore, layered silicate minerals, such as muscovite, biotite, and chlorite, are weathered at varying degrees to form kaolinite minerals [44]. The original rocks of Anyuan County are predominantly metamorphic sandstone, siltstone, and slate, which were found to be relatively broken in the field, and provides favorable conditions for further weathering. As shown by the analysis of kaolinite in Table 1, the average content of kaolinite, illite, and potassium feldspar in Anyuan County (samples AY-01 to AY-12) is 40.66%, 8.03%, and 9.72%, respectively. This indicates that most of the potassium feldspar in the Anyuan County metamorphic bedrock was converted into kaolinite minerals. In Longnan County (samples LN-01 to LN-12), the bedrock is medium-grained granodiorite, with an average kaolinite mineral ratio of 24.67%, i.e., much less than that in Anyuan County.

Table 6. Chemical composition (wt %) of Longnan granites (Zudong mining area).

Sample No.	1	2	3	4	5	6
SiO_2	70.34	75.55	76.14	74.88	72.48	74.58
TiO_2	0.44	0.25	0.03	0.05	0.13	0.15
Al_2O_3	14.59	12.13	12.97	13.44	13.4	13.34
Fe_2O_3	2.03	1.18	0.09	0.21	0.51	0.56
FeO	0.64	1.01	1.07	1.42	1.28	1.25
MnO	0.25	0.07	0.03	0.08	0.11	0.05
MgO	0.57	0.34	0.07	0.19	0.17	0.18
CaO	0.55	0.42	0.62	0.58	1.51	0.73
Na_2O	3.37	2.74	4.25	3.97	3.62	3.65
K_2O	5.7	5.36	4.52	4.61	5.37	4.8
P_2O_5	0.05	0.04	0.02	0.06	0.04	0.05
LOI	0.92	0.7	0.96	0.57	1.62	0.6
Total	100.33	99.85	100.77	100.28	100.24	100.03
CIA	64	63	63	65	61	64

CIA = Chemical Index of Alteration, Data from Bao et al., 2008 [8] and Zhang, 1990 [41].

Differences in kaolinite and illite content between Anyuan and Longnan are not only related to the composition of the parent rock. The sampling depth and topography also have a significant impact on the clay mineral and soil formation [45]. Samples from Anyuan County came from average elevations of between 311 and 468 m, with relatively gentle relief (gradient: 24°–25°) and dense vegetation cover. Owing to the influence of bedrock lithology, geological structure and surface erosion, a concave slope formed in this area, and the weathering profile was deep. Samples from Longnan County, on the other hand, came from average elevations of between 278 and 321 m, from a relief of lightly-weathered residual hills with linear or convex slopes controlled by granite lithology. These nuances of landforms have a crucial influence on parent rock weathering and soil formation.

5.2. Vertical Variation of Clay Minerals in REE Ores

An analysis of the Ganxian District soil profile [27] showed that the main clay minerals are kaolinite and illite (Figure 6). Kaolinite content fluctuates from top to bottom, with the lowest content (9.8%) at 115 cm from the bottom of the section. In this section, the layers are mostly located at the bottom of the semiweathered layer, where the granite structure is visible and the weathering degree of the rock is weakened. The content of rock-forming minerals (i.e., quartz, 39.9%; potassium feldspar, 37.9%) in (Table 2) shows that conversion of potassium feldspar to clay minerals was minor, which is consistent with this deeper level of the weathering profile. The peak of kaolinite (35.0%) appears at a depth of 190 cm; the kaolinite content is relatively low in the 130–190 cm interval (14.9-16.5: Table 3). From 20 to 105 cm below surface, the kaolinite content increases irregularly, reaching 28.2% at 105 cm. This increase is probably due to the rapid conversion of feldspar and mica minerals into kaolinite and other clay minerals.

Furthermore, illite content shows variation similar to that of kaolinite, but with sharper changes. Potassium feldspar and quartz also exhibit different variation characteristics. Two distinct horizons are thus resolved in the Ganxian weathering profile can be divided on the basis of clay mineral distribution, as follows (Figure 7).

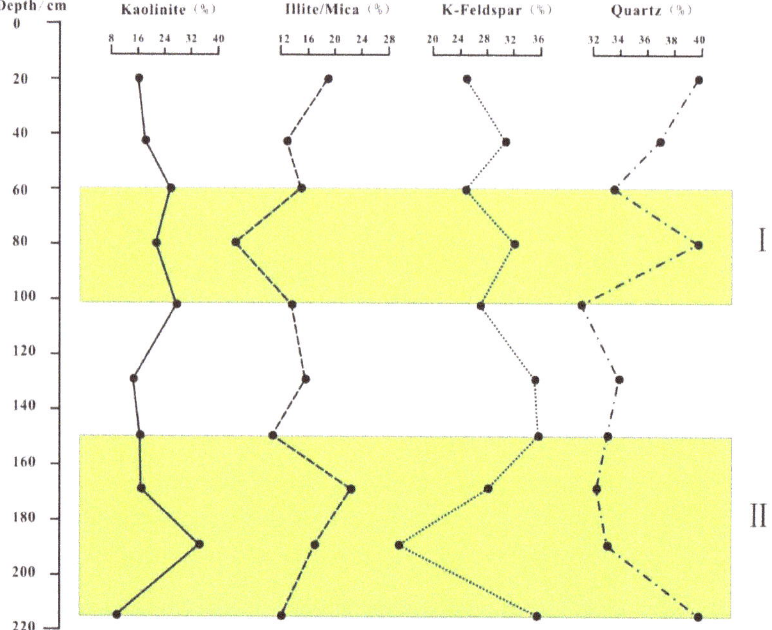

Figure 7. Clay and other mineral content in the Ganxian weathering profile.

Band I (60–105 cm): clay mineral content shows a high-low-high variation trend, while the illite content shows a broader range of variation than that of potassium feldspar. The content of quartz, and feldspar shows the opposite trend to that of clay minerals, and the variation range of the quartz content was broader than that of potassium feldspar. Over the course of weathering, clay minerals are converted into kaolinite minerals due to physical and chemical weathering processes. According to Uzarowicz et al. (2011) [46], the composition of clay minerals in the soil surface follows the acidic soil formation process, i.e., it is strictly controlled by the content of chlorite and mica debris, with subsequent conversion of chlorite and mica to montmorillonite and vermiculite. Our analysis shows that kaolinite and illite were reduced concurrently in the profile, while the quartz and potassium feldspar content increased. It is suggested that large quantities of mica and chlorite minerals were formed after weathering and alteration of the parent rock (i.e., granite), thereby controlling the formation and transformation of kaolinite.

An XRD analysis of Sample GX-04-4, showed seven illite-mica diffraction peaks, between d = 4.47 Å and d = 2.50 Å, with a cumulative peak height of 117.8% (Figure 8). The GX-04-1 sample showed a total of four distinct illite-mica diffraction peaks, with a cumulative peak height of 17.4%. XRD analysis further confirmed that, as a result of surface weathering of the granite bedrock, feldspars have altered to chlorite and mica; other clay minerals are due to the weathering and alteration of mica in the original rock. The formation and conversion of kaolinite were limited.

Band II (150–215 cm): clay mineral shows a low-high-low variation trend, opposite to that of band I. Potassium feldspar and quartz show greater changes in an opposite trend. The content of kaolinite and potassium feldspar shows a particularly evident reversal at 190 cm (increase of kaolinite; decrease of potassium feldspar), indicating that the weathering of potassium feldspar contributes to the formation of kaolinite in the soil. In the supergene weathering realm, the conversion of potassium feldspar into kaolinite can be expressed using the following chemical Equation (1) [47]:

$$4KAlSi_3O_8 + 4H^+ + 2H_2O \rightarrow 4K^+ + Al_4Si_4O_{10}(OH)_8 + 8SiO_2 \quad (1)$$

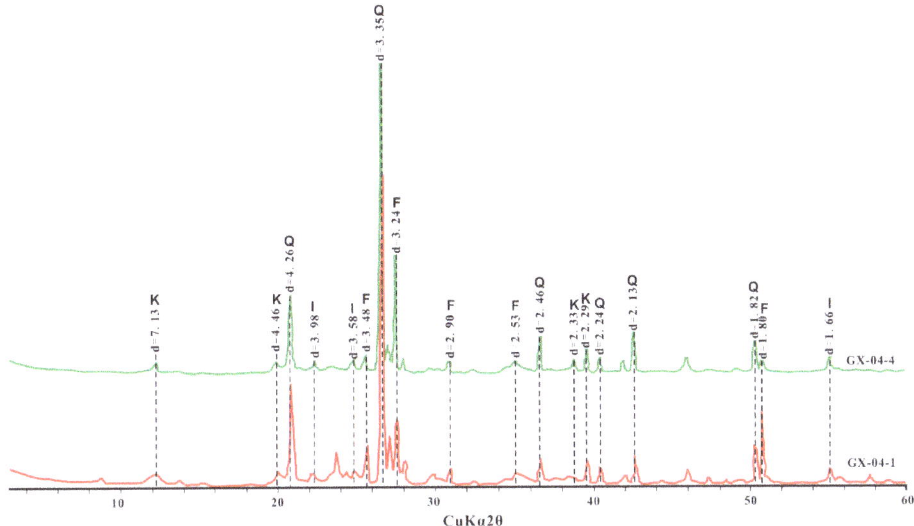

Figure 8. XRD spectra of randomly oriented bulk sample from the Ganxian Profile, I: illite-mica, K: kaolinite, Q: quartz, F: feldspars.

Equation (1) shows that decomposition of potassium feldspar releases a large quantity of free metal cations in the soil. This promotes further enrichment in kaolinite. According to Dixon (1989), the formation of kaolinite in the soil requires an adequate amount of silica and a small number of metal cations [48], while the chemical decomposition of potassium feldspar needs to consume HCO_3^- produced from H^+ and CO_2 [49]. This provides a favorable environment for the formation of kaolinite, forming the distinct volatile characteristic of band II. Moreover, the potassium feldspar diffraction peak in GX-04-1 was significantly reduced as compared to that in GX-04-4 (Figure 9). For quartz, the cumulative diffraction peak height is 195% in Sample GX-04-1, and 79.1% in Sample GX-04-4. This suggests that formation of kaolinite was favored in the SiO_2-rich environment.

An analysis of samples collected at different stages of the in situ leaching profile from Longnan County showed that the clay mineral content fluctuated regularly as leaching progressed (Figure 8). In the early stage of leaching, the kaolinite content in the soil was less than 15%, while the illite-mica content was slightly higher than that of kaolinite, fluctuating from 10.1% to 17.2% (Table 3). The potassium feldspar content fluctuated between 16.4%–34.1%, and the quartz content was relatively high. The total kaolinite content increased, and the total potassium feldspar content decreased as leaching progressed (Figure 8). In the later stage of leaching, from the surface to the bottom of soil column, the kaolinite content increased rapidly from 8.9% to 33.0%, then gradually decreased. The illite-mica content also decreased slightly compared to the previous period, with the exception of the bottom of the Longnan soil column, where it increased weakly (from 10.1% to 15%). The content of quartz increased significantly in the upper layer between 46.2%–61.3% and 40.5%–53.65% (Table 3), with an average increase of 32.5%. In the final stage of leaching, the kaolinite content increased significantly: from 22.5% to 31.2% (average growth: 46.2%; Table 3). The illite-mica content increased by between 15% and 20.7%. The potassium feldspar content increased significantly in the final stage of leaching (average growth: 32.9%), while the content of quartz decreased (27.9% to 35.8%).

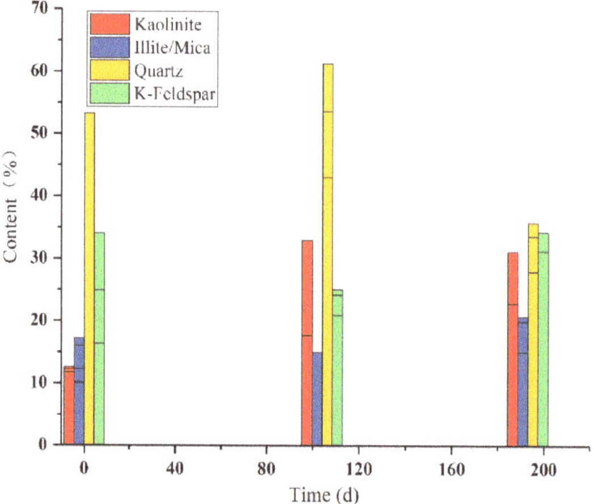

Figure 9. Clay content variation during in situ leaching, Longnan Profile.

Parfitt et al. (1983) [20] found that the Si concentration in the soil solution decreased as rainfall increased, reflecting increasing leaching of the soil. Wu et al. (2016) [36] suggested that granite with compact structure has higher strength, a low degree of weathering, and a higher content of residual feldspar. In conditions of sustained weathering by acid rain, cations increased in the leached upper soil, and residual feldspar decomposed rapidly, demonstrating a process of feldspar decomposition, as described by Equation (1). Fine particles of SiO_2 gradually migrated from the upper to the lower soil horizons, leading to silica enrichment in the latter.

An analysis of the major clay minerals in different stages of in situ leaching (Figure 10) showed that the vertical migration of clay minerals was significant. Kaolinite did not change significantly in the initial stage of leaching; its content fluctuated between 9.5% and 12.6%, with a standard deviation of 1.22. In the course of leaching, potassium feldspar was consistently weathered and converted to kaolinite [36]. The kaolinite content peaked at a depth of 200 cm, reaching a maximum value of 33%, which is 3.47 times the initial value. During the late leaching stage, the kaolinite content fluctuated slightly. Although it increased slightly from 200 cm, it rapidly reduced afterward. Field investigation in the Longnan profile revealed that about 50 m from the northeast end of the profile, there were rows of 150–180 cm-deep injecting holes along the hill slope (i.e., along 330°–150° direction).

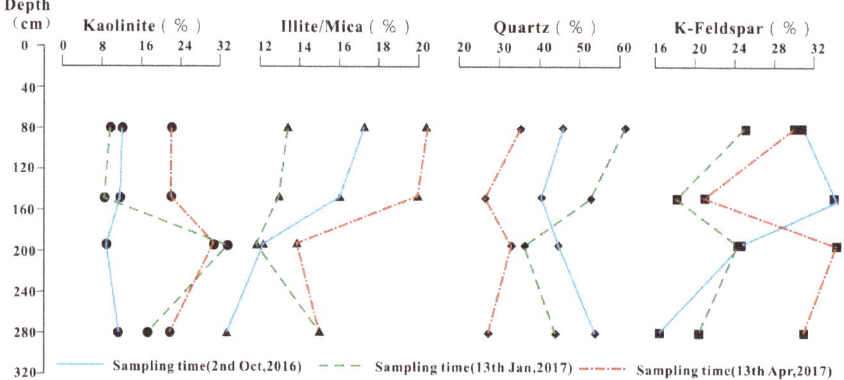

Figure 10. Clay mineral, quartz, and potassium feldspar content on the Longnan soil column at different stages of the leaching experiment.

These injection holes received $(NH_4)_2SO_4$ electrolyte solutions over a long period [50]. Due to this, soil in the lower part of the profile was saturated, and its acidity was enhanced, which further enhanced kaolinization of potassium feldspars. In the initial stage of simulated leaching, the illite-mica content was high near the surface of the sediment column (17.2%), and decreased downwards to only 10.1% at the bottom of column. As leaching progressed, the illite-mica content continued to decrease with increasing depth in the 0–200 cm interval; from 200 cm downwards, however, it increased significantly in the later stage of leaching. The range of illite-mica contents was similar to that of the leaching process, but higher than that in the leaching process. Different clay minerals have different geochemical behavior, and may have different physical and chemical responses to factors such as pH, salinity, and blocking cations [51]. Previous studies have shown that illite forms from potassium feldspar alteration in two different types of microsystems [14]: a) in the early stages of the weathering process, along crystal joints of orthoclase with muscovite or biotite; and b) in the final stages of weathering, where the original structure of the parent rock is destroyed. In both cases, illite forms in association with other clay minerals, i.e., smectite in the early weathering stage, and kaolinite in the late weathering stage. With the exception of its interlayer charge and consequent potassium content, illite is, in many ways, similar to phengite mica [14]. As indicated above, this depth presented an increase in the illite-mica contents due to potassium feldspar weathering and dissolution, silicon release, and hydrated layers mineral formation.

Comparing kaolinite with illite-mica, no correlation between the two was evident in the initial stage of in situ leaching, while a significant negative correlation was present in the later stage of leaching. This trend indicates that, besides the decomposition of potassium feldspar into kaolinite, a large quantity of interlayer silicates such as illite-mica are converted to kaolinite as leaching progresses.

The change in quartz and potassium feldspar content became complex with increasing depth in the soil column (Figure 10). In the early leaching stage, the quartz content initially decreased, then increased, while potassium feldspar content showed the opposite trend. For both minerals, the inflection points were at a depth of 150 cm. Between 80 and 150 cm in the soil column, the content of both quartz and potassium feldspar decreased as leaching progressed. However, quartz inherited the characteristics of the initially decreasing trend, i.e., downwards from 195 cm, it turned into an increasing trend. In contrast, potassium feldspar showed the opposite trend downwards from 150 cm. In the later leaching stage, both quartz and potassium feldspar showed a decreasing-increasing-decreasing trend with depth, but the variation range of potassium feldspar was broader than that of quartz.

Although the content of vermiculite was not measured in the simulated leaching experiment, regional sampling of REE ore revealed low vermiculite content (Table 2). Vermiculite formation occurs in two stages: a) in the early stage, the common mica weathering products are dioctahedral vermiculites whose layer charge is lower than that of the parent mica; b) in the second stage, mica dissolution advances further, and corroded zones of polyphase assemblage of dioctahedral hydroxy-vermiculite appear within mica crystals [14]. Vermiculite has good ion-adsorption properties; its adsorption capacity of REE ions is nearly 0.2 mmol/g. Vermiculites adsorbing REE ions can be regenerated by cation ion-exchange reagents according to the following reaction [52]:

$$\begin{aligned}&\left\{(\mathrm{Mg,Fe,Al})_6[(\mathrm{Si,Al})_8\mathrm{O}_{20}](\mathrm{OH})_4\right\}_m\cdot n\mathrm{RE}^{3+}\cdot e\mathrm{H_2O}+3n\mathrm{M}^+\\&\quad=\left\{(\mathrm{Mg,Fe,Al_6})[(\mathrm{Si,Al})_8\mathrm{O}_{20}](\mathrm{OH})_4\right\}_m\cdot 3n\mathrm{M}^+\cdot e\mathrm{H_2O}+n\mathrm{RE}^{3+}\end{aligned} \quad (2)$$

Chemical Equation (2) shows that the decomposition of potassium feldspar releases large quantities of free Al, Fe and Mg cations, and Si in the soil. This favors the formation of vermiculite (Figure 10). However, in the in situ leaching profile, the vermiculite content was low, probably due to the flow of leaching liquid flow and surface water elution.

Previous studies indicated that ion-absorbed REE ores mainly contain halloysite, illite, and kaolinite, and less smectite [1–3,9]. The factors that strongly favor the formation of smectite include low-lying topography, poor drainage, and base-rich parent material, leading to chemical conditions of high pH, high silica activity, and an abundance of basic cations [13]. In leaching conditions with lower pH, as in our leaching experiment, it was impossible to form abundant smectite. With the exception of vermiculite transformed to smectite, the original REE ores in our area of study contained less than 1% smectite [3].

5.3. Simulating Migration of Clay Minerals during Leaching

A total of eight simulated soil columns were subjected to different experimental conditions of pH, immersion concentration, and leaching rate. A high-acidity leaching mining solution was used to further decompose the remaining feldspar in the soil to clay minerals [17]. The simulated leaching experiment showed that kaolinite was further enriched in the soil column. The leaching solution concentration and leaching rate also had an effect the rate of decomposition of silicate minerals in the soil columns [53].

In the T1 soil column (Figure 11), the kaolinite content was initially high; subsequently, it decreased and then increased slowly as leaching progressed. The most prominent diffraction peak of kaolinite (d = 7.20 Å in Figure 10) is relatively weak in the middle part of the soil column, compared with the upper and lower parts. Similar results have been reported from the weathering profiles of other Mesozoic granites [54]. Other relatively prominent kaolinite diffraction peaks were at d = 2.33 and 1.99 Å. As shown in Table 3, the variation was also weaker in the middle of the soil column compared with the upper and lower parts. Kaolinite and illite-mica had similar diffraction peak characteristics across the soil column. For illite-mica, the most prominent initial diffraction peak corresponded to d = 10.01 Å, and the (002) crystal planes showed significantly high diffraction. Other evident peaks were at d = 5.0 Å and d = 4.46 Å. As the diffraction angle increased, strong diffraction peaks appeared at d = 2.44 Å, and d = 1.99 Å. Quartz showed a prominent diffraction peak for d = 4.26 Å (peak height: 2615; diffraction intensity: 21.9%). High quartz diffraction peaks at d = 3.34 Å were present in all samples. Potassium feldspar presented the first evident diffraction peak for d between 6.6 and 6.45 Å (corresponding to a diffraction angle (2-Theta) at between 13.4° and 13.7°. This finding reflects the different diffraction intensities of different crystal faces. Significant diffraction peaks were also present at d = 3.24, 2.28, and 1.98 Å.

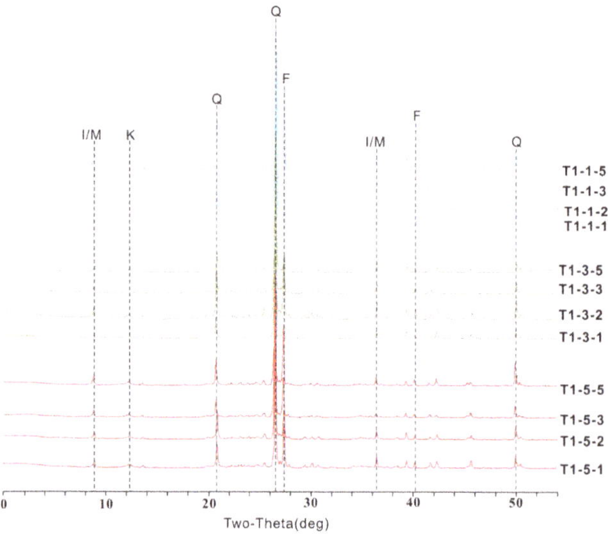

Figure 11. XRD spectra of randomly oriented bulk samples from Soil Column T1 (simulated leaching). I/M: illite-mica; K: kaolinite; Q: quartz; F: potassium feldspars.

Here, we discuss clay mineral content under three key conditions of simulated leaching.

① Same concentration of leaching solution and leaching rate; different pH values (soil columns T1–T3 in Table 3, Figure 12):

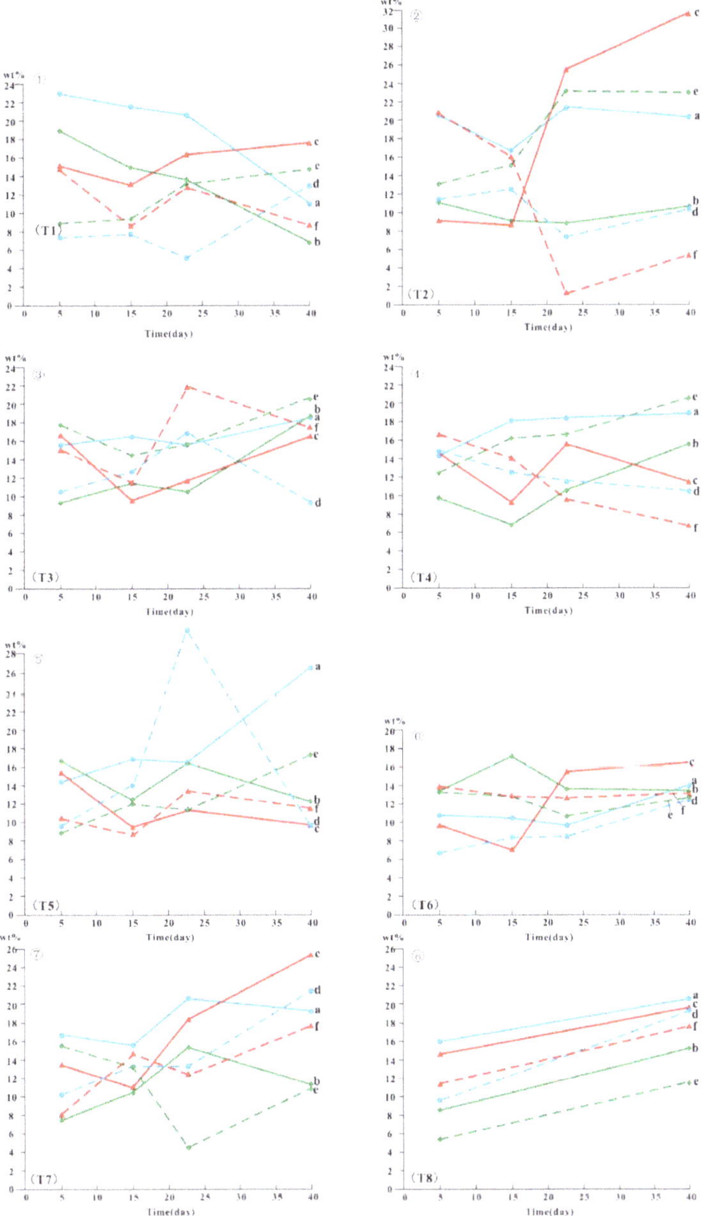

Figure 12. Changing clay mineral content in different conditions of simulated leaching. (**a**–**c**) kaolinite content sample at sample locations of 30, 70, and 110 cm, respectively; (**d**–**f**) illite-mica content at sample locations of 30, 70, and 110 cm, respectively.

With the concentration of leaching solution and rate of leaching being stable, the migration and enrichment of clay minerals is controlled by pH. Soil column T1 shows that as leaching progresses, the kaolinite content decreases at depths of 30 and 70 cm in the soil column. After two weeks of leaching, the soil column was gradually enriched in the clay minerals, and the content of kaolinite content in the bottom layer increased (① in Figure 12); the lower the pH, the more favorable the soil conditions for the decomposition of parent rock, particularly for the hydrolysis of feldspars and the formation of clay minerals. As indicated above, the soil column T2 was leached with solution of pH = 3, the lowest pH value in this group. After 15 days of leaching, the kaolinite content in soil Column T2 increased markedly, reaching a maximum of 31.5% (② in Figure 12). In the final leaching stage, the kaolinite content increased significantly at a depth of 110 cm, indicating that REEs were adsorbed and released due to the high recovery rate. Other studies have demonstrated that the higher the pH of the leaching solution, the higher the adsorption capacity of clay minerals for rare earth ions. In a weak acid environment (pH = 4), the kaolinite content in different layers of soil Column T3 soil increased slowly (③ in Figure 12).

② Same pH and leaching rate; different concentration of the leaching solution (soil Columns T4, T5):

In soil Columns T4 and T5, the kaolinite content increased gradually in the near-surface layer (30 cm) as leaching progressed (③ in Figure 12 and ④ in Figure 12. The higher the concentration of the leaching solution, the flatter the change in clay mineral composition. After 15 days of leaching, the concentration of the leaching solution in soil Column T5 decreased to 4% $(NH_4)_2SO_4$, while the clay mineral content near the surface increased rapidly, eventually reaching 26.7%. At a depth of 70 cm, Soil Columns T4 and T5 showed different responses: with the lower-concentration leaching solution of T5, the kaolinite content in this layer decreased slowly, with an average decrease of about 25%; with the more concentrated leaching solution, i.e., T4, the kaolinite content showed increasing volatility. At a depth of 110 cm, different leaching concentrations produced similar effects on the kaolinite content.

At the same time, illite-mica minerals exhibited different responses under different leaching conditions. The high-concentration leaching solution in soil Column T4 gradually reduced the illite-mica content. On the 23rd day, with the 4% $(NH_4)_2SO_4$ solution, the illite-mica content increased abruptly, further verifying the enhanced adsorption capacity of illite in a weakly acidic environment. As indicated above, at the upper level of the soil columns, clay minerals were concentrated by the high-concentration leaching solution, but at the bottom of the soil columns, this was not observed (⑤ in Figure 12). Our experiment suggested that the effective concentration of leaching solution was 8%.

③ Same pH and leaching concentration; different leaching rates:

Soil Columns T6 and T7 showed that the leaching rate had an impact on the content of kaolinite (⑥ in Figure 12, ⑦ in Figure 12). At 30 cm, the content of kaolinite in soil Column T6 increased slowly as leaching progressed. After 15 days, the kaolinite content increased significantly in soil column T7; in soil column T6, at 70 cm, it increased slightly, and gradually decreased as leaching progressed. In soil column T7, after 20 days of leaching, the total kaolinite content increased, as it did in the upper layer (30 cm), demonstrating that clay minerals have similar structures in REE ore. At a depth of 110 cm, the kaolinite content decreased after 15 days in columns T6 and T7, and then increased again. This change probably reflects the translocation of fine-grained kaolin minerals from the upper and middle to the lower parts of the profile as leaching progresses. The longer the leaching time, the higher the content of fine-grained clay minerals at the bottom of profile; some of these even clog the porosity, which reduces the flushing out of rare-earth ions and hampers leaching [36]. In the leaching conditions of this sample group, with a leaching rate of 5 mL/min, enrichment in clay minerals was at a rate of over 1 mL/min. This indicated that a high recovery rate is not possible at a slower rate of leaching.

A comparison between different simulated leaching conditions showed that the content of kaolinite and other clay minerals tended to increase from the initial stage until the completion of leaching. The fluctuation of clay mineral content is the result of the combination of different pHs, leaching concentrations, and leaching rates (⑧ in Figure 12). However, it cannot be assumed that a certain leaching condition determines the outcome of the leaching mining process. Leaching mining is a complex chemical process, and the variation of clay mineral content only reflects one aspect of it. It is impossible to adequately simulate ore leaching conditions in sediment columns, in view of the boundary restrictions of a soil column, the horizontal flow of ore leaching solution, and the ore texture and structure. For these reasons, our experimental results were not as expected, although the leaching conditions in our soil column experiment were controlled. However, comparing the results experimental leaching of soil Columns T2, T4, T6, and T7 at 110 cm after 23 days of leaching (Figure 11) with the results of in situ leaching in the Longnan section (Figure 9), in both cases, the kaolinite content increased while the illite-mica content decreased. We expect that simulated leaching experiments applying many different leaching conditions will permit us to explore how various factors influence clay mineral fluctuation during leaching.

The REE distribution on the kaolinite-water interface is considered to be the result of the adsorption of REE ions by kaolinite, and is strongly controlled by pH [55]. Tian et al. (date) [52] found that the REE recovery was up to over 96% with a NH_4^+ concentration in the raffinate solution of 0.2 g/L and a pH of 2. Other research suggested if the pH of the leaching agent is either too high or too low, recovery of REE is reduced. The optimal pH values were between 4 and 8. The maximum leaching efficiency of REE was 91% [56]. In our experimental study, the pH values were between 3 and 5. Disregarding other external factors, we suggest that REE recovery of over 90% can be achieved through leach mining with a leaching solution of 8 wt% concentration and a pH of 5, at a leaching rate of 5 mL/min.

6. Conclusions

Analyses of REE ore clay mineral properties and grain size, and the distribution and variation of clay minerals, quartz, and feldspars in in situ leaching and simulated leach mining led to the following conclusions:

(1) In the surveyed REE mine areas, the soil particle size (i.e., volume frequency) curve showed unimodal and bimodal distribution. Many cumulative particle size distribution curves had a "B" shape, with particle sizes of 3.74–30.46 μm. Other curves showed an "A" shape, which indicates that the increase of coarse particles in the soil affects particle gradation. An analysis of particle size eigenvalues showed that D10 was 0.82–5.03 μm, while D90 ranged from 11.99 to 60.99 μm, with a standard deviation of 11.44. These values reflect discreteness among the coarse particle fraction. Taking the Dav parameter as the independent variable and other characteristic parameters as the dependent variables of the regression analysis, the regression coefficient order was D90 > D [3,4] > D50 > D [2,3] > D10. This finding revealed that the increase of surface average particle size in soils has a greater effect on coarse particles. A D10 residual analysis showed that with an increase of the average particle size of Dav content, the volume fraction below 10% was reduced. The residual error showed a linear relationship with a 99.5% confidence interval.

(2) A regional clay mineral analysis showed that ion-absorbed REE ores formed on different bedrock lithologies have similar clay mineral contents. The main clay minerals were kaolinite, illite, chlorite, and vermiculite. The kaolinite content ranged from a minimum of 8.8% to a maximum of 62.1%. In the granite weathering area of Longnan County, the kaolinite content had weak correlation with the content of potassium feldspar, quartz, and other rock-forming minerals. The chemical weathering index (CIA) was in the range of 61–65, and the average kaolinite content was low. In the metamorphic terrain of Anyuan County, the kaolinite content was strongly correlated with quartz, and the CIA ranged from 68 to 75. This reveals a higher degree of feldspar weathering and conversion to clay minerals in metamorphic bedrock.

(3) Studying the natural weathering profile of granite in the Ganxian District showed that the content of kaolinite was relatively low at 60–105 cm below surface. There, conversion of feldspars to chlorite and mica in the course of weathering, and the increased content of mica in the parent rock, limited the formation of kaolinite. At 150–215 cm below the surface, the kaolinite content increased.

This was attributed to the desiliconization of overlying feldspar minerals, which increased the SiO_2 content in the weathering mantle and promoted the decomposition of feldspars into kaolinite.

(4) An analysis of in situ leaching mining profile showed that in the early stage of leaching, the content of kaolinite in the soil is relatively low, i.e., lower than that of illite-mica. As the leaching progressed and potassium feldspar continued to convert to kaolinite, significant kaolinite peaks appeared in the profile, and the illite-mica content became slightly reduced. This indicates that interlayer silicate minerals such as illite-mica were progressively converted to kaolinite. In a later stage of leaching, the kaolinite content became slightly reduced in comparison with that of the earlier leaching stage, and the content of illite-mica and potassium feldspar increased.

(5) Simulated leaching mining reveals that under the same conditions of leaching solution concentration and leaching rate, the migration and enrichment of clay minerals were controlled by pH; the lower the pH value, the more favorable the conditions for clay mineral formation. With the same pH and leaching rate, but different concentrations of the leaching solution, a high-concentration leaching solution (i.e., 12% $(NH_4)_2SO_4$)) resulted in a slow increase in the kaolinite content in the upper part of the soil profile (30 cm). Under stable pH and immersion concentration conditions, the variation of the leaching rate influences soil formation and clay mineralization processes in different horizons within the soil profile. Based on these results, and disregarding other external factors, we suggest that REE recovery of over 90% can be achieved through leach mining with a leaching solution of 8% concentration and a pH of 5, at a leaching rate of 5 mL/min.

Author Contributions: Conceptualization and methodology, L.C., and H.C.; investigation, L.C., X.J., L.Q., and H.C.; experimental analysis, X.J.; writing-review and editing, L.C. and Z.H.; partial writing, X.J., and L.Q.; plotting, H.C.; software, H.D. All authors have read and agreed to the published version of the manuscript.

Funding: This project was funded by the National Science Foundation of China (Grant no. 41967038), China Postdoctoral Science Foundation (Grant no. 2015M582530), and the Science Foundation of Jiangxi Provincial Department of Education (Grant no. GJJ150657).

Acknowledgments: We are indebted to Kaixing Wu and Tao Sun for their helpful discussions and suggestions. We are also sincerely thanks for Houston–Oil and gas laboratory and the Key Laboratory of Nonferrous Metal Materials Science and Engineering of the Ministry of Education. The authors are grateful to two anonymous reviewers for their helpful suggestions and comments. We want to thank Editage for the English language editing.

Conflicts of Interest: The authors declare no conflict of interest.

References

1. Yang, Y.Q.; Hu, C.S.; Luo, Z.M. Geological characteristic of mineralization of rare earth deposit of the ion-absorption type and their prospecting direction. *Bull. Chin. Acad. Geol. Sci.* **1981**, *2*, 102–118. (In Chinese with English abstract)
2. Chen, D.Q.; Wu, J.S. The mineralization mechanism of ion-adsorbed REE deposit. *J. Chin. Rare Earth Soc.* **1990**, *8*, 175–179. (In Chinese with English abstract)
3. Chi, R.A.; Tian, J. Review of weathered crust rare earth ore. *J. Chin. Rare Earth Soc.* **2007**, *25*, 641–650. (In Chinese with English abstract)
4. Harlavan, Y.; Erel, Y. The release of Pb and REE from granitoids by the dissolution of accessory phases. *Geochim. Cosmochim. Acta.* **2002**, *66*, 837–848. [CrossRef]
5. Galan, E.; Fernandez-Caliani, J.C.; Miras, A.; Aparicio, P.; Marquez, M.G. Residence and fractionation of rare earth elements during kaolinization of alkaline peraluminous granites in NW Spain. *Clay Miner.* **2007**, *42*, 341–352. [CrossRef]
6. Sanematsu, K.; Murakami, H.; Watanabe, Y.; Duangsurigna, S.; Vilayhack, S. Enrichment of rare earth elements (REE) in granitic rocksand their weathered crusts in central and southern Laos. *Bull. Geol. Surv. Jpn.* **2009**, *60*, 527–558. [CrossRef]
7. Zhao, Z.; Wang, D.H.; Chen, Z.H.; Chen, Z.Y. Progress of Research on Metallogenic Regularity of Ion-adsorption Type REE Deposit in the Nanling Range. *Acta Geol. Sin.* **2017**, *91*, 2814–2827.
8. Bao, Z.W.; Zhao, Z.H. Geochemistry of Mineralization with Exchangeable REY in the Weathering Crusts of Granitic Rocks in South China. *Ore Geol. Rev.* **2008**, *33*, 519–535. [CrossRef]

9. Fan, C.Z.; Zhang, Y.; Chen, Z.H.; Zhu, Y.; Fan, X.T. The study of clay minerals from weathered crust rare earth ores in southern Jiangxi Province. *Acta Petrol Mineral.* **2015**, *34*, 803–810. (In Chinese with English abstract)
10. Coppin, F.; Berger, G.; Bauer, A.; Castet, S.; Loubet, M. Sorption of lanthanides on smectite and kaolinite. *Chem. Geol.* **2002**, *182*, 57–68. [CrossRef]
11. Brasbury, M.H.; Baeyens, B. Sorption of Eu on Na- and Ca-montmorillonites: Experimental investigations and modelling with cation exchange and surface complexation. *Geochim. Cosmochim. Acta* **2002**, *66*, 2325–2334. [CrossRef]
12. Chi, R.A.; Tian, J.; Luo, X.P.; Xu, Z.G.; He, Z.Y. The basic research on the weathered crust elution-deposited rare earth ores. *Nonferrous Met. Sci. Eng.* **2012**, *3*, 1–13. (In Chinese with English abstract)
13. Wilson, M.J. The origin and formation of clay minerals in soils pat, present and future perspectives. *Clay Miner.* **1999**, *34*, 7–25. [CrossRef]
14. Meunier, A.; Velde, B. *The Geology of Illite*; Verlag Berlin/Heidelberg, Springer: Berlin/Heidelberg, Germany, 2004; pp. 63–143.
15. Zhang, Z.Y.; Huang, L.; Liu, F.; Wang, M.K.; Fu, Q.L.; Zhu, J. Characteristics of clay minerals in soil particles of two Alfisols in China. *Appl. Clay Sci.* **2016**, *120*, 51–60. [CrossRef]
16. Cudahy, T.; Caccett, M.; Thomas, M.; Hewson, R.; Abrams, M.; Kato, M.; Kashimura, O.; Ninomiya, Y.; Yamaguchi, Y.; Collings, S.; et al. Satellite-derived mineral mapping and monitoring of weathering, deposition and erosion. *Sci. Rep.* **2016**, *6*, 23702. [CrossRef] [PubMed]
17. Moldoveanu, G.A.; Papangelakis, V.G. An overview of rare-earth recovery by ion-exchange leaching from ion-adsorption clays of various origins. *Mineral. Mag.* **2016**, *80*, 63–76. [CrossRef]
18. Ishihara, S.; Hua, R.; Hoshino, M.; Murakami, H. REE abundance and REE minerals in granitic rocks in the Nanling range, Jiangxi province, Southern China, and generation of the REE-rich weathered crust deposits. *Resour. Geol.* **2008**, *58*, 355–372. [CrossRef]
19. Wang, X.; Lei, Y.; Ge, J.; Wu, S. Production forecast of China's rare earths based on the generalized Weng model and policy recommendations. *Resour. Pol.* **2015**, *43*, 11–18. [CrossRef]
20. Parfitt, R.L.; Russell, M.; Orbell, G.E. Weathering sequence of soils from volcanic ash involving allophane and halloysite, New Zealand. *Geoderma* **1983**, *29*, 41–57. [CrossRef]
21. Botero, Y.L.; López-Rendón, J.E.; Ramírez, D.; Zapata, D.M.; Jaramillo, F. From Clay Minerals to Al_2O_3 Nanoparticles: Synthesis and Colloidal Stabilization for Optoelectronic Applications. *Minerals* **2020**, *10*, 118. [CrossRef]
22. Tian, J.; Yin, J.Q.; Chi, R.A.; Rao, G.H.; Jiang, M.T.; Ouyang, K.X. Kinetics on leaching rare earth from the weathered crust elution-deposited rare earth ores with ammonium sulfate solution. *Hydrometallurgy* **2010**, *101*, 166–170.
23. Wu, D.R.; Wu, A.Q.; He, X.Y.; Li, M. Trends of Extreme Climate in South Jiangxi from 1956 to 2013. *J. Yangtze River Sci. Res. Inst.* **2017**, *34*, 24–29.
24. Jiangxi Bureau of Geological and Mineral Resources. *Regional Geology of Jiangxi Province*; Geological Publishing House: Beijing, China, 1984; pp. 1–308.
25. Jiangxi Bureau of Geological and Mineral Exploration and Development. *Drawing Instructions of Digital Geological Map of Jiangxi Province*; Geological Publishing House: Beijing, China, 1998; pp. 1–80.
26. Zhao, Z.; Wang, D.H.; Chen, Z.Y.; Chen, Z.H.; Zheng, G.D.; Liu, X.X. Zircon U-Pb Age, Endogenic Mineralization and Petrogenesis of Rare Earth Ore-bearing Granite in Longnan, Jiangxi Province. *Acta Geosci. Sin.* **2014**, *35*, 719–725. (In Chinese with English abstract)
27. Liu, C.Y.; Liu, J.H.; Zhong, L.X.; Xu, S.; Chen, L.K. Soil particle size distribution characteristics of ionic rare earth: A case study in rare earth mine of Jiangwozi in Gan County. *Nonferrous Met. Sci. Eng.* **2017**, *8*, 125–130. (In Chinese with English abstract)
28. Hu, X.P.; Shi, Y.X.; Dai, X.R.; Wang, J.T.; Liu, B.; Wang, N.Y. Characteristics and source of city wall's earth in the Neolithic Liangzhu City based on XRD analyses of clay minerals. *Acta. Petrol. Mineral.* **2013**, *32*, 373–382.
29. Mukasa-Tebandeke, I.Z.; Ssebuwufu, P.J.M.; Nyanzi, S.A.; Schumann, A.; Nyakairu, G.W.A.; Ntale, M.; Lugolobi, F. The Elemental, Mineralogical, IR, DTA and XRD Analyses Characterized Clays and Clay Minerals of Central and Eastern Uganda. *Adv. Mater. Phys. Chem.* **2015**, *5*, 67–86. [CrossRef]
30. Rietveld, H.M. A Profile Refinement Method for Nuclear and Magnetic Structures. *J. Appl. Cryst.* **1969**, *2*, 65–71. [CrossRef]

31. Gi, Y.J. Formation of Vermicular Kaolinite from Halloysite Aggregates in the Weathering of Plagioclase. *Clays Clay Miner.* **1998**, *46*, 270–279.
32. Banfield, J.F. The mineralogy and chemistry of granite weathering. Master's Thesis, Australian National University, Canberra, Australia, 1985; 130p.
33. Banfield, J.E.; Eggleton, R.A. Analytical transmission electron microscope studies of plagiodase, muscovite, and Potassium feldspar weathering. *Clays Clay Miner.* **1990**, *38*, 77–89. [CrossRef]
34. Yang, Z.M. A study on clay minerals from the REE-rich weathered crust developed on the Longnan granite in Jiangxi. *Sci. Geo. Sin.* **1987**, *1*, 70–81. (In Chinese with English abstract)
35. Righi, D.; Meunier, A. *Origin of Clays by Rock Weathering and Soil Formation*; Springer: Berlin/Heidelberg, Germany, 1995; pp. 1–119.
36. Wu, K.X.; Zang, L.; Zhu, P.; Chen, L.K.; Tian, Z.F.; Xin, X.Y. Research on Particle Size Distribution and Its Variation of Ion-adsorption Type Rare Earth Ore. *Chin. Rare Earths.* **2016**, *37*, 67–74. (In Chinese with English abstract)
37. Aberg, B. Void ratio of noncohesive soils and similar materials. *J. Geotech Eng.* **1992**, *118*, 1315–1334. [CrossRef]
38. Yan, J.B.; Wu, K.X.; Liu, H.; Liao, C.L.; Jing, X.W.; Zhu, P.; Ouyang, H. Patterns of particle size distribution and genesis of Ion-adsorption REE Ore-exemplified by Dabu REE Ore Deposit, Ganxian County, Southern Jiangxi Province. *Chin. Rare Earths* **2018**, *36*, 372–384. (In Chinese with English abstract)
39. Wang, Z.; Zhao, Z.; Zou, X.Y.; Chen, Z.Y.; Tu, X.J. Petrogeochemical Characteristics and Metallogenetic Potential of Epimetamorphic Rocks in South Jiangxi Province. *Rock Miner. Anal.* **2018**, *37*, 96–107. (In Chinese with English abstract)
40. Zhang, L.; Wu, K.X.; Chen, L.K.; Zhu, P.; Ouyang, H. Overview of Metallogenic Features of Ion-adsorption Type REE Deposits in Southern Jiangxi Province. *J. Chin. Soc. Rare Earth.* **2015**, *33*, 10–17. (In Chinese with English abstract)
41. Zhang, Z.H. A Study on weathering crust ion adsorption type REE deposits, south China. *Contrib. Geol. Miner. Resour. Res.* **1990**, *5*, 57–71. (In Chinese with English abstract)
42. Goldberg, K.; Humayun, M. The applicability of the Chemical Index of Alteration as a paleoclimatic indicator: An example from the Permian of the Paraná Basin, Brazil. *Palaeogeogr. Palaeoclimatol. Palaeoecol.* **2010**, *293*, 175–183. [CrossRef]
43. Nesbitt, H.W.; Young, G.M. Early Proterozoic climates and plate motions inferred from major element chemistry of lutites. *Nature* **1982**, *299*, 715–717. [CrossRef]
44. Samotoin, N.D.; Bortnikov, N.S. Formation of Kaolinite Nanoand Microcrystals by Weathering of Phyllosilicates. *Geol. Ore Depos.* **2011**, *53*, 340–352. [CrossRef]
45. Oliveira, D.P.; Sartor, L.R.; Souza Júnior, V.S.; Corrêa, M.M.; Romero, R.E.; Andrade, G.R.P.; Ferreira, T.O. Weathering and clay formation in semi-arid calcareous soils from Northeastern Brazil. *Catena* **2018**, *162*, 325–332. [CrossRef]
46. Uzarowicz, Ł.; Skiba, S.; Skiba, M.; Segvic, B. Clay-mineral Formation in Soils Developed in the Weathering Zone of Pyrite-Bearing Schists: A Case Study from the Abandoned Pyrite Mine in Wieściszowice, Lower Silesia, SW Poland. *Clays Clay Miner.* **2011**, *6*, 581–594. [CrossRef]
47. Liu, S.K.; Han, C.; Liu, J.M.; Li, H. Hydrothermal decomposition of potassium feldspar under alkaline conditions. *RSC Adv.* **2015**, *5*, 93301–93309. [CrossRef]
48. Dixon, J.B.; Weed, S.B.; Parpitt, R.L. Minerals in Soil Environments. *Soil Sci.* **1989**, *150*, 283–304.
49. Yuan, G.H.; Cao, Y.C.; Schulz, H.M.; Hao, F.; Gluyas, J.; Liu, K.; Yang, T.; Wang, Y.Z.; Xi, K.L.; Li, F.L. A review of feldspar alteration and its geological significance in sedimentary basins: From shallow aquifers to deep hydrocarbon reservoirs. *Earth-Sci. Rev.* **2019**, *191*, 114–140. [CrossRef]
50. Liu, J.H.; Chen, L.K.; Liu, C.Y.; Qiu, L.R.; He, S. Pb speciation in rare earth minerals and use of entropy and fuzzy clustering methods to assess the migration capacity of Pb during mining activities. *Ecotox. Environ. Safe.* **2018**, *165*, 334–342.
51. Cuevas, J.; Leguey, S.; Garralón, A.; Rastrero, M.R.; Procopio, J.R.; Sevilla, M.T.; Jiménez, N.S.; Abad, R.R.; Garrido, A. Behavior of kaolinite and illite-based clays as landfill barriers. *Appl. Clay Sci.* **2009**, *42*, 497–509. [CrossRef]

52. Tian, J.; Yin, J.Q.; Chen, K.H.; Rao, G.H.; Jiang, M.T.; Chi, R.A. Extraction of rare earths from the leach liquor of the weathered crust elution-deposited rare earth ore with non-precipitation. *Int. J. Miner. Process.* **2011**, *98*, 125–131.
53. Xiao, Y.; Feng, Z.; Huang, X.; Huang, L.; Chen, Y.; Wang, L.; Long, Z. Recovery of rare earths from weathered crust elution-deposited rare earth ore without ammonia-nitrogen pollution: I. Leaching with magnesium sulfate. *Hydrometallurgy* **2015**, *153*, 58–65.
54. Tan, P.; Oberhardt, N.; Dypvik, H.; Riber, L. Weathering profiles and clay mineralogical developments, Bornholm. Denmark. *Mar. Pet. Geol.* **2017**, *80*, 32–48. [CrossRef]
55. Wan, Y.X.; Liu, C.Q. Study on Adsorption of Rare Earth Elements by Kaolinite. *J. Rare Earth* **2005**, *23*, 337–341.
56. He, Z.Y.; Zhang, Z.Y.; Yu, J.X.; Zhou, F.; Xu, Y.L.; Xu, Z.G.; Chen, Z.; Chi, R.A. Kinetics of column leaching of rare earth and aluminum from weathered crust elution-deposited rare earth ore with ammonium salt solutions. *Hydrometallurgy* **2016**, *163*, 33–39.

Article

Characteristics of Precipitation of Rare Earth Elements with Various Precipitants

Kenneth N. Han

South Dakota School of Mines and Technology, Rapid City, SD 57701-3995, USA; kennethhydro@gmail.com

Received: 7 January 2020; Accepted: 11 February 2020; Published: 17 February 2020

Abstract: The effective and selective leaching of rare earth elements (REEs) from various sources is frequently possible in practice by adopting a carefully coordinated strategy incorporating a selective precipitation of these elements from undesired ones in solution. In this study, the behavior of chemical precipitation of REEs with commonly used precipitants such as sulfate, carbonate, fluoride, phosphate, and oxalate was examined using thermodynamic principles and calculations. It was found that the pH of the system has a profound effect on determining particular chemical species of precipitants, which are subsequently responsible for the precipitation of REEs. The role of various anions such as Cl^-, NO_3^-, and SO_4^{2-} derived from the acid used in the leaching process on the precipitation behavior of REEs was examined. These anions form complexes with REEs and display a very positive effect on the precipitation behavior. The nitrate environment exhibits most conducive to precipitation followed by sulfate and then chloride.

Keywords: precipitation; leaching; complexation; anion effect; equilibrium calculation; rare earth elements

---

1. Introduction

Rare earth elements (REEs) are becoming very important in modern society. REEs are essential ingredients in the transition to green technology and there have been numerous recent articles describing their extraction processes from various sources [1,2].

Most rare earth bearing minerals are refractory in nature and therefore, a pretreatment or strong acid treatment of ores bearing these minerals is usually necessary [3–6] before leaching is affected. Such pretreated ores are then subjected to the leaching process of REEs. The leach liquor is frequently subjected to the removal of impurities such as iron, calcium, uranium, and thorium by going through stage-wise precipitation using various compounds such as hydroxide, carbonate, sulfate, and oxalate [7–13]. After the removal of these impurities by chemical precipitation, REEs in solution are then subjected to precipitation using various precipitants, which in general takes place at pH above 3–4, since most of these impurities are precipitated at around these pHs. However, during the precipitation of impurities, some REEs are also removed from the solution by coprecipitation or adsorption. Therefore, it is desirable to precipitate REEs before removal of the impurities. It is the hope of this study to identify the precipitants and chemical conditions to achieve this objective.

The chemical complexity of the aqueous medium introduced by various chemical treatment of leach liquor has introduced a complex nature to the system making the analysis of the subsequent chemical precipitation a challenge. In addition to metal ions in the leach liquor extracted from ores, many anions such as Cl^-, NO_3^-, and SO_4^{2-} mostly derived from acids used in the leaching operation also play an important role in leaching as well as precipitation.

The effect of various anions on the leaching behavior has been discussed in earlier studies [14,15]. For example, the leaching of Re-phosphate in the presence of sulfuric acid can be simply described by

Equation (1). (It is noted that Re is the elemental symbol for rhenium. However, Re(III) refers to only REE(III) in this paper.)

$$<RePO_4> + \{H_2SO_4\} = \{Re^{3+}\} + \{PO_4{}^{3-}\} + \{SO_4{}^{2-}\} + 2\,\{H^+\} \quad (1)$$

Here the symbols, < > and { } stand for solid and liquid form, respectively. The oxidation state of REEs in solution and solids is assumed to be Re(III) throughout this presentation. It should be noted however, that some REEs can be oxidized to Re(IV) or reduced to Re(II) in aqueous media in very rare cases [16]. The concentration of the free REEs at equilibrium can be estimated if the concentration of all species involved in Equation (1) is known. However, such calculated values are far from real values since the chemicals produced from the leaching process undergo hydrolysis and complexation [14,15]. As a result, the correct evaluation of the product concentration requires a very complicated process involving calculations of equilibrium equations simultaneously in some cases.

The chemical precipitation of REEs is similarly complicated. For example, a simple chemical precipitation of a free Re-ion, Re^{3+} with phosphate ion, $PO_4{}^{3-}$ can be given by Equation (2).

$$\{Re^{3+}\} + \{PO_4{}^{3-}\} = <RePO_4> \quad (2)$$

It should be noted that REEs precipitate with the phosphate ion as a free Re-ion, Re^{3+} at relatively high pHs such as 4 and 6. However, when the pH of the solution changes to a lower value, for example 1 or 2, by adding acids of HCl, HNO_3, or H_2SO_4, the anions such as Cl^-, $NO_3{}^-$, and $SO_4{}^{2-}$ will form various complexes with REEs [3,14,15], and then these complexes are subjected to precipitation. For example, in the presence of sulfate ions, there will be at least Re^{3+}, $Re(SO_4)^+$, $Re(SO_4)_2{}^-$, and $Re_2(SO_4)_3$, present in solution, and for most of the sulfate concentration, $Re(SO_4)_2{}^-$ is the predominant species when the concentration of sulfate ions is especially more than 0.1 M [14,15]. As a result, in such cases, $Re(SO_4)_2{}^-$ should be used instead of the free ion and the resulting concentration of rare earth elements after precipitation is an order of magnitude different from when the free ion is used.

In addition, the phosphate ion also may not exist depending upon the pH of the system. It should be noted that the values of pKa are 2.0 for $H_3PO_4/H_2PO_4{}^-$, 7.2 for $H_2PO_4{}^-/HPO_4{}^{2-}$, and 12.0 for $HPO_4{}^{2-}/PO_4{}^{3-}$ (see Table 1) [17]. This means that H_3PO_4 should be used instead of the phosphate ion in Equation (2) for the system whose pH lies below 2 and likewise, $H_2PO_4{}^-$ for pH between 2 and 7.2 and so on.

Table 1. The pKa values for various precipitants [17].

Precipitants	pKa
$H_3PO_4/H_2PO_4{}^-$	2.1
$H_2PO_4{}^-/HPO_4{}^{2-}$	7.2
$HPO_4{}^{2-}/PO_4{}^{3-}$	12.0
$H_2CO_3/HCO_3{}^-$	6.38
$HCO_3{}^-/CO_3{}^{2-}$	10.34
HF/F^-	3.0
$H_2C_2O_4/HC_2O_4{}^-$	1.0
$HC_2O_4{}^-/C_2O_4{}^{2-}$	4.2
$HSO_4{}^-/SO_4{}^{2-}$	2.0

In this paper, factors such as pH, complexation, and hydration of precipitants affecting the chemical precipitation with various precipitants will be examined and their practical implication in the analysis of such precipitation processes will be discussed. The precipitation behavior of REEs with carbonate, fluoride, phosphate, sulfate, and oxalate was considered in this study, since these precipitants are most widely used in precipitation of REEs in practice.

2. Acquisition of Thermodynamic Information

Table 1 presents the values of pKa for the precipitants considered in this study. It should be noted that there are some variations in the numerical values of pKa based on various literature sources. For example, the values for oxalic acid given by Pourbaix [17] and Dean [18] are different from those by HSC [19] and Wagman [20]. However, the effect of such differences on the final outcome of the behavior of chemical precipitation is not very significant.

The thermodynamic data used in this study are listed in Table 2. Most data given in the table were taken from the literature information [21–26] and those unavailable were estimated using the parallel plot method as described earlier [14].

Table 2. The Gibbs standard free energy formation of various compounds in kCal/mol. Information taken from [14,15] with some modifications.

	Carbonate	Phosphate	Sulfate	Fluoride	Oxalate	Oct-Sulfate
La	−768.5	−455.3	−892.6	−399.8	−872.2	−1399.9
Ce	−761.4	−455.2	−884.7	−396.8	−868.3	−1340.2
Pr	−764.5	−451.0	−879.4	−394.2	−860.6	−1343.5
Nd	−761.6	−448.7	−877.2	−391.5	−861.8	−1335.0
Sm	−759.6	−447.6	−878.8	−392.3	−860.1	−1330.5
Eu	−759.4	−446.7	−875.3	−391.0	−859.1	−1275.3
Gd	−758.6	−445.5	−874.2	−390.5	−857.7	−1328.1
Tb	−757.0	−444.7	−873.7	−388.7	−856.9	−1313.2
Dy	−750.6	−442.2	−868.9	−385.9	−856.0	−1323.3
Ho	−744.2	−440.4	−861.3	−383.0	−849.6	−1314.3
Er	−743.9	−439.1	−855.9	−380.9	−842.0	−1309.5
Tm	−736.9	−438.2	−862.9	−379.6	−846.6	−1308.1
Yb	−735.5	−437.6	−861.5	−378.4	−839.5	−1293.4
Lu	−734.5	−435.7	−860.1	−376.8	−841.3	−1319.1
Y	−749.9	−443.0	−866.8	−389.4	−851.5	−1320.6
Sc	−714.9	−418.2	−845.1	−361.5	−828.4	−1300.0

Oct-sulfate: $Re_2(SO_4)_3{\cdot}8H_2O$ (here Re represents REEs).

Some of the standard Gibbs free energy formation values, such as Y- and Sc-oxalates were not available in the literature and, therefore, were estimated using the parallel method as described by earlier studies [14,15]. It should be noted that there were typographical errors in the values for Re-octa-hydrated sulfates in earlier publications [14,15]. It should also be noted that subsequent relevant calculations in the same articles were carried out using correct values, however.

3. Carbonate System

It is well known that rare earth elements are readily precipitated with carbonate at high pHs. Rare earth elements are readily subjected to precipitation with carbonate [27,28] especially at high pHs. Re-carbonates are ideal precipitates, since carbonates are readily available, relatively less expensive than other precipitants, and precipitates can be easily dissolved in a mild acid for further treatments of REEs in subsequent processes of purification. However, it is not easily precipitated in acidic media. As a result, the leach liquor is usually subjected to a neutralization step before the precipitation of REEs with carbonate is affected. During the neutralization process, however, impurities such as Fe^{3+}, Ca^{2+} are precipitated first resulting in loss of REEs due to adsorption and coprecipitation of REEs with oxides or sulfates of impurities.

A simple stoichiometric equation describing the precipitation of REEs with carbonate can be written in Equation (3), which is analogous to Equation (2).

$$2\,\{Re^{3+}\} + 3\,\{CO_3^{2-}\} = \langle Re_2(CO_3)_3 \rangle \quad (3)$$

As discussed earlier, the concentration of CO_3^{2-} at pH 1–6 that was considered in this study, which is consistent with the range of pH practiced in industry, is practically zero, since the pK value for HCO_3^-/CO_3^{2-} is 10.34. The stable species at pH less than 6, is H_2CO_3, which is the dominant species as seen in Table 1 and, therefore, H_2CO_3 should be used in Equation (3). Furthermore, when pH of the solution is less than 6, Re^{3+} is subjected to complexation to become either $ReCl^{2+}$, $Re(NO_3)^{2+}$, or $Re(SO_4)_2^-$ depending upon the kind of acid used [14,15]. Therefore, for the working pH range (pH 1–6) in this study, Equations (4)–(6) should be used for chloride, nitrate, and sulfate systems instead of Equation (3).

$$2\,\{ReCl^{2+}\} + 3\,\{H_2CO_3\} = \langle Re_2(CO_3)_3\rangle + 2\,\{Cl^-\} + 6\,\{H^+\} \tag{4}$$

$$2\,\{Re(NO_3)^{2+}\} + 3\,\{H_2CO_3\} = \langle Re_2(CO_3)_3\rangle + 2\,\{NO_3^-\} + 6\,\{H^+\} \tag{5}$$

$$2\,\{Re(SO_4)_2^-\} + 3\,\{H_2CO_3\} = \langle Re_2(CO_3)_3\rangle + 4\,\{SO_4^{2-}\} + 6\,\{H^+\} \tag{6}$$

In order to illustrate the steps involved in calculating the equilibrium concentration of REEs in solution, let us take Equation (4) as an example. The equilibrium constant for Equation (4), K_{eq} can be found knowing the Gibbs standard free energy formation of each component involved in the reaction. Such found equilibrium constant is given in Equation (7) below:

$$K_{eq} = \frac{\{Cl^-\}^2\{H^+\}^6}{\{ReCl^{2+}\}^2\{H_2CO_3\}^3} \tag{7}$$

Here $\{H^+\}$ is known if the pH of the system is known, $\{Cl^-\}$ is known since the pH of the system is adjusted by HCl in this reaction, (see Table 3) and $\{H_2CO_3\}$ was set as 10^{-4} M. Therefore, $\{ReCl^{2+}\}$ can be determined.

It should be noted that Equation (6) is valid above pH 2 but SO_4^{2-} should be replaced by HSO_4^- when the pH of the system is less than 2 (See Table 3). When Equations (4)–(6) are solved to calculate the equilibrium concentrations of REEs in solution, it was assumed that the final concentrations of Cl^-, NO_3^-, and SO_4^{2-} are those set by respective acid concentration to adjust the pH of the solution as discussed above. This assumption is reasonable for low pHs but the accurate estimation of the final equilibrium concentrations of REEs at pH 6 should be carried out taking into account all of practical conditions of the system.

It should also be noted that most of precipitants considered here are mostly weak acids and, therefore, they contribute in setting the pH of the solution (see Table 4). This can be a cause of inaccuracy in the final results especially at relatively high pHs, say pH greater than 4. The correction of such effects can be made but no attempt on such corrections was made in this study. This is because the objective of the study is to observe the precipitation behavior at low pHs.

Table 3. Concentrations of various anions used in the study at different pH's [17].

	HCl	HNO_3	H_2SO_4	
pH	Cl^- (M)	NO_3^- (M)	SO_4^{2-} (M)	HSO_4^- (M)
6	1.0×10^{-6}	1.0×10^{-6}	1.0×10^{-6}	9.9×10^{-11}
5	1.0×10^{-5}	1.0×10^{-5}	1.0×10^{-5}	9.9×10^{-9}
4	1.0×10^{-4}	1.0×10^{-4}	9.9×10^{-5}	9.9×10^{-7}
3	1.0×10^{-3}	1.0×10^{-3}	9.1×10^{-4}	9.0×10^{-5}
2	1.0×10^{-2}	1.0×10^{-2}	5.0×10^{-3}	5.0×10^{-3}
1	1.0×10^{-1}	1.0×10^{-1}	9.0×10^{-3}	9.1×10^{-2}
0	1.0	1.0	1.0×10^{-2}	9.9×10^{-1}

Table 4. The pH values of some weak acids for various concentrations.

Concentration, M.	H_3PO_4	HF	H_2CO_3	$H_2C_2O_4$
1.0×10^{-6}	5.97	6	6.31	6
1.0×10^{-4}	4	4.05	5.2	4
1.0×10^{-3}	3	3.26	4.69	3
1.0×10^{-2}	2.3	2.64	4.18	2.09
1.0×10^{-1}	1.63	2.1	3.68	1.31

It is important to note that any calculated value exceeding the highest concentration of REEs in practice should be treated with caution. The highest concentration of REEs possible in the aqueous system is based on the grade of REEs in the ore being leached. Assuming the highest grade of REEs of the ore being treated is 50%, the highest concentration of REEs possible in the aqueous media could be about 1.5 M [15]. It is safe to assume that when the calculated value is more than 1–2 M, the precipitation is impossible under the conditions considered. On the other hand, when the calculated value is less than the target value, 1 ppm, then the precipitation could be favorable.

Figure 1 presents the equilibrium concentration of dissolved REEs in solution when REEs are subjected to precipitation with carbonate. In this study, the concentration of the precipitant, was kept constant at 10^{-4} M throughout the paper as stated earlier. The choice of the precipitant concentration being 10^{-4} M, which is the final concentration after the precipitation is completed, is arbitrary and deemed reasonable in view of practical application. The horizontal dashed line across the figure represents the concentration of REEs in solution being 3×10^{-6} M, which is about 1 ppm of REEs. This is an arbitrary concentration representing an acceptable target level of precipitation of REEs after subjecting to the carbonate precipitation.

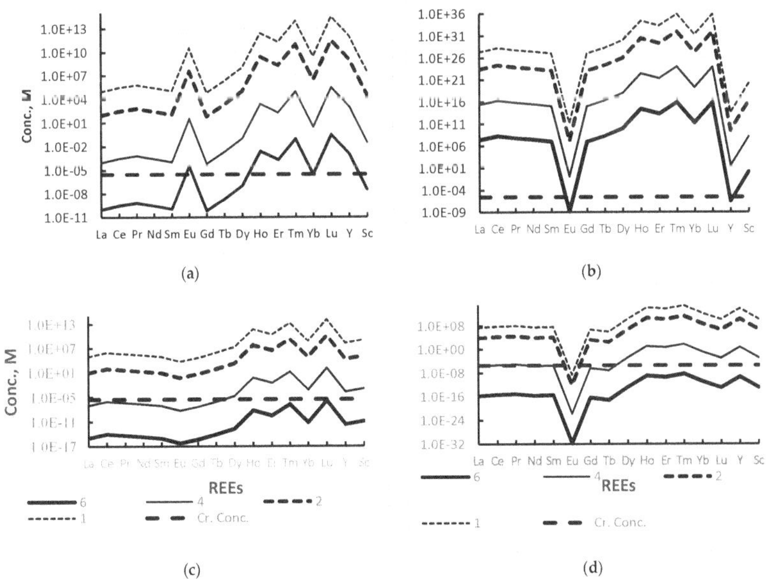

Figure 1. Concentrations of (**a**) Re^{3+}, (**b**) $ReCl_2^{+}$, (**c**) $Re(NO_3)_2^{+}$, and (**d**) $Re(SO_4)_2^{-}$ in equilibrium with carbonate precipitate with 10^{-4} M of total concentration of carbonate in solution at pH 6, 4, 2, and 1. Cr.concentration: 1 ppm line as a target concentration.

It is a reasonable assumption that the initial concentration of REEs in the leach liquor would be in the range of 3×10^{-1} to 3×10^{-3} M [15]. It is a usual practice that the initial concentration of the precipitant is a little more than the stoichiometric amount. It is obvious that carbonate is incapable of precipitating most of REEs, except a few light REEs at pH 6. It is interesting to note that the precipitation of nitrate and sulfate complexes is significantly better than free REE ions. The anomaly behavior by Eu seems to be due to inaccuracy in the standard Gibbs free energy data given in the literature.

4. Fluoride System

It has been well understood that fluoride readily complexes with rare earth elements [6,29]. The pKa value of HF/F^- is 3 as seen in Table 1 and, therefore, the relevant equations governing the precipitation of REEs with fluoride should be formulated in such a way that HF is the precipitant at pH less than 3 and F^- should be used at higher pHs. For example, the precipitation of $ReCl_2^+$ with fluoride is described by Equation (8) when the pH of the solution is 6 or 4, while Equation (9) is for pH 2 and 1. Here too, the concentration of the precipitant is assumed to be 10^{-4} M. The precipitation of the complexed REEs is remarkable as in the case with the carbonate system. Most of REEs is easily complexed with chloride, nitrate, and sulfate as long as the concentration of these anions is more than 0.1 M, below which they remain as Re^{3+} [14,15].

$$\{Re^{3+}\} + 3\,\{F^-\} = <ReF_3> \qquad (8)$$

$$\{Re^{3+}\} + 3\,\{HF\} = <ReF_3> + 3\,\{H^+\}9 \qquad (9)$$

Figure 2a–d present the precipitation of Re^{3+}, $ReCl_2^+$, $Re(NO_3)_2^-$, and $Re(SO_4)_2^-$ with fluoride, respectively. With regard to Figure 2a, there appears to be only one line representing pH 6 and 4. This is because when pH is above 3, F^- is responsible for the precipitation reaction as seen in Equation (7), in which the precipitation becomes independent of pH. As seen in these figures, fluoride is a more powerful precipitant for REEs than carbonate in the pH range considered in this study.

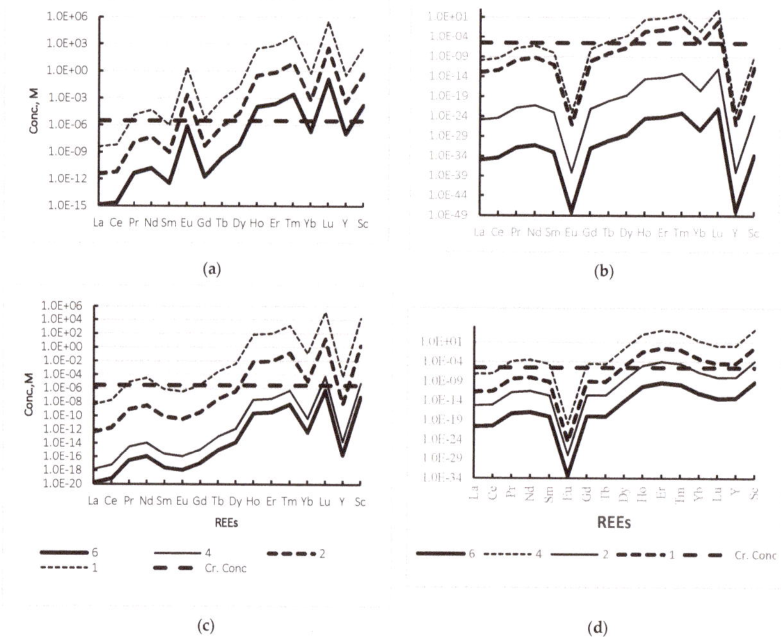

Figure 2. Concentrations of (**a**) Re^{3+}, (**b**) $ReCl_2^+$, (**c**) $Re(NO_3)_2^+$, and (**d**) $Re(SO_4)_2^-$ in equilibrium with fluoride precipitate with 10^{-4} M of total concentration of fluoride in solution at pH 6, 4, 2, and 1. Cr. Concentration: 1 ppm line as a target concentration.

5. Phosphate System

When phosphoric acid, H_3PO_4 is placed in water, it dissociates into dihydrogen phosphate, $H_2(PO_4)^-$, hydrogen phosphate, $H(PO_4)^{2-}$, or phosphate ion, $PO_4{}^{3-}$ depending upon the pH of the solution. The pKa values of these conjugated species are 2.1 for $H_3PO_4/H_2PO_4{}^-$, 7.2 for $H_2PO_4{}^-/HPO_4{}^{2-}$, and 12.0 for $HPO_4{}^{2-}/PO_4{}^{3-}$ as noted in Table 1. Therefore, the relevant equilibrium equations for the precipitation of REEs with phosphate should involve phosphoric acid for pH 1 and 2, and dihydrogen phosphate for pH 4 and 6. These are shown in Equations (10) and (11), respectively, for precipitation of $ReCl_2{}^+$.

$$\{ReCl_2{}^+\} + H_2PO_4{}^- = <RePO_4> + 2\{Cl^-\} + 2\{H^+\} \quad (10)$$

$$\{ReCl_2{}^+\} + \{H_3PO_4\} = <RePO_4> + 2\{Cl^-\} + 3\{H^+\} \quad (11)$$

Figure 3 illustrates the precipitation of Re^{3+}, $ReCl_2{}^+$, $Re(NO_3)_2{}^-$ and $Re(SO_4)_2{}^-$ with phosphate. Unlike the carbonate but similar to fluoride systems, the phosphate system shows a powerful precipitation behavior even for free REEs. REEs are readily precipitated for all forms of rare earth element ions in solution except Re-chloride complexes. Here again, the unusual behavior shown by europium is most likely due to erratic thermodynamic data given in the literature. It should be noted that industrially important REE-bearing minerals are monazite and xenotime, both of which are phosphate compounds. Chemical interaction between REEs and phosphate has been investigated in numerous studies [30,31].

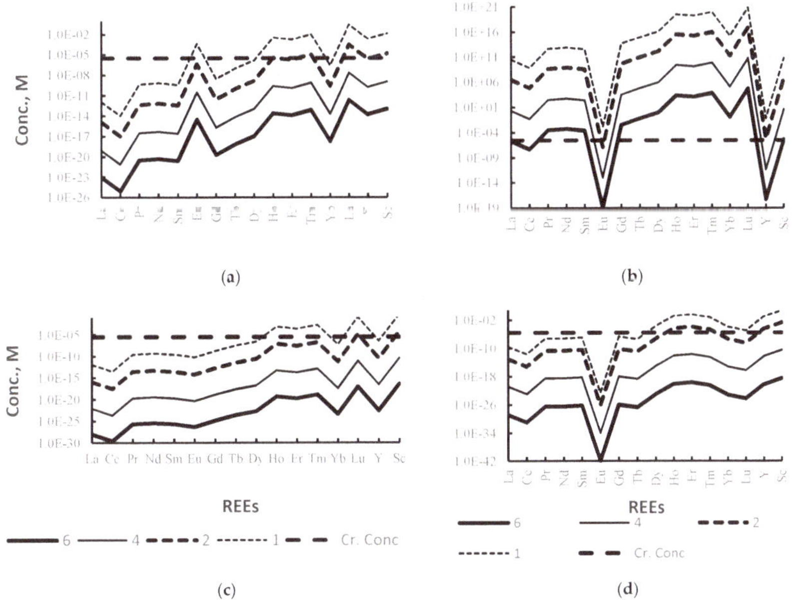

Figure 3. Concentrations of (**a**) Re^{3+}, (**b**) $ReCl_2{}^+$, (**c**) $Re(NO_3)_2{}^+$, and (**d**) $Re(SO_4)_2{}^-$ in equilibrium with phosphate precipitate with 10^{-4} M of total concentration of phosphate in solution at pH 6, 4, 2, and 1. Cr. Concentration: 1 ppm line as a target concentration.

6. Sulfate System

6.1. Sulfate (SO_4^{2-})

The precipitation of REEs with sulfate is given in Figure 4. In Figure 4a, the concentration of free REEs in solution in equilibrium with Re-sulfate precipitates is shown. In the construction of this figure, Equations (12) and (13) were used. Equation (12) was used for pH 6 and 4, while Equation (13) was used for pH 2 and 1. As mentioned earlier, the concentration of precipitants was kept constant at 10^{-4} M. At pH 6, the concentration of sulfate ion is 10^{-4} M plus 10^{-6} M, in which the first was added by the rule and the latter was from sulfuric acid to adjust the pH. In addition, at pH 4, on the other hand, the concentration of sulfate ion is 2×10^{-4} M, because of the added sulfate plus sulfate provided by sulfuric acid to adjust the pH of the solution. The same is also applicable for Figure 4d, in which case, Equations (14) and (15) were used instead of Equations (12) and (13).

It is noted that the precipitation is better at low pHs, namely pH 2 and 1, compared to those at pH 6 and 4. The reason for this is because there is more sulfate present at low pHs due to the supply of sulfate via sulfuric acid to adjust the pH of the solution.

$$2\,\{Re^{3+}\} + 3\,\{SO_4^{2-}\} = <Re_2(SO_4)_3> \quad (12)$$

$$2\,\{Re(SO_4)_2^-\} = <Re_2(SO_4)_3> + \{SO_4^{2-}\} \quad (13)$$

$$2\,\{Re(SO_4)_2^-\} = <Re_2(SO_4)_3> + \{SO_4^{2-}\} \quad (14)$$

$$2\,\{Re(SO_4)_2^-\} + \{H^+\} = <Re_2(SO_4)_3> + \{HSO_4^-\} \quad (15)$$

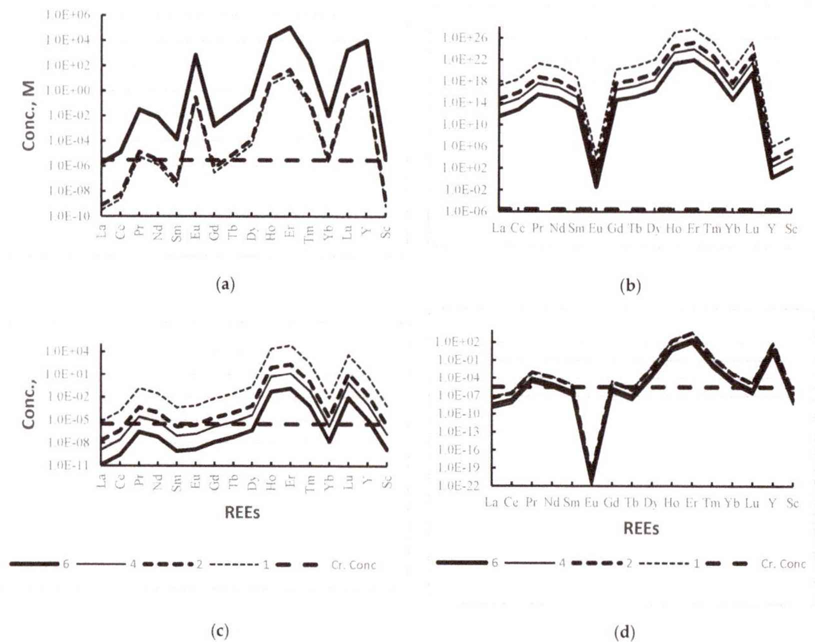

Figure 4. Concentrations of (**a**) Re^{3+}, (**b**) $ReCl_2^+$, (**c**) $Re(NO_3)_2^+$, and (**d**) $Re(SO_4)_2^-$ in equilibrium with sulfate precipitate with 10^{-4} M of total concentration of sulfate in solution at pH 6, 4, 2, and 1. Cr. Concentration: 1 ppm line as a target concentration. Additional sulfate is provided by sulfuric acid in (**a**) and (**d**).

For the precipitation of $Re(SO_4)_2^-$ using sulfate, the precipitation calculations were done using Equations (14) and (15). The precipitation of $Re(SO_4)_2^-$ with sulfate is unusual in that the species does not need any precipitant but it has to release one sulfate or bisulfate to form the precipitate $<Re_2(SO_4)_3>$ as seen in Equations (14) and (15). Such precipitation can be referred to as a decomposition precipitation.

It should be noted that in practice, REEs are usually precipitated with double salt sulfate [5,32], since the precipitation is more effective with double sodium sulfate. However, in this study, precipitation of REEs with only sulfate was carried out. Although double salt sulfate precipitation is more powerful than simple sulfate precipitation, the precipitation behavior of the two systems is expected to be very similar.

6.2. Octa-Hydrated Sulfate ($SO_4{\cdot}8H_2O$)

The chemical precipitation of REEs with sulfate to form octa-hydrated sulfate is very similar to the precipitation behavior of sulfate. The relevant equations used in this analysis are given in Equations (16)–(19).

$$2\,\{Re^{3+}\} + 3\,\{SO_4^{2-}\} + 8\,\{H_2O\} = <Re_2(SO_4)_3{\cdot}8H_2O> \tag{16}$$

$$2\,\{Re^{3+}\} + 3\,\{HSO_4^-\} + 8\,\{H_2O\} = <Re_2(SO_4)_3{\cdot}8H_2O> + 3\,\{H^+\} \tag{17}$$

$$2\,\{Re(SO_4)_2^-\} + 8\{H_2O\} = <Re_2(SO_4)_3{\cdot}8H_2O> + \{SO_4^{2-}\} \tag{18}$$

$$2\,\{Re(SO_4)_2^-\} + \{H^+\} + 8\,\{H_2O\} = <Re_2(SO_4)_3{\cdot}8H_2O> + \{HSO_4^-\} \tag{19}$$

As in the case with the sulfate system, the appearance of sulfate or bisulfate is based on the pH of the solution. In other words, when the pH of the system is greater than 2, sulfate is the predominant species and at low pHs, bisulfate should be used as shown in Table 3. It was expected that the octa-hydrated system is favored in the chemical precipitation compared to simple sulfate precipitation [14,15]. However, the results shown in Figure 4; Figure 5 show that the two systems behave very similarly.

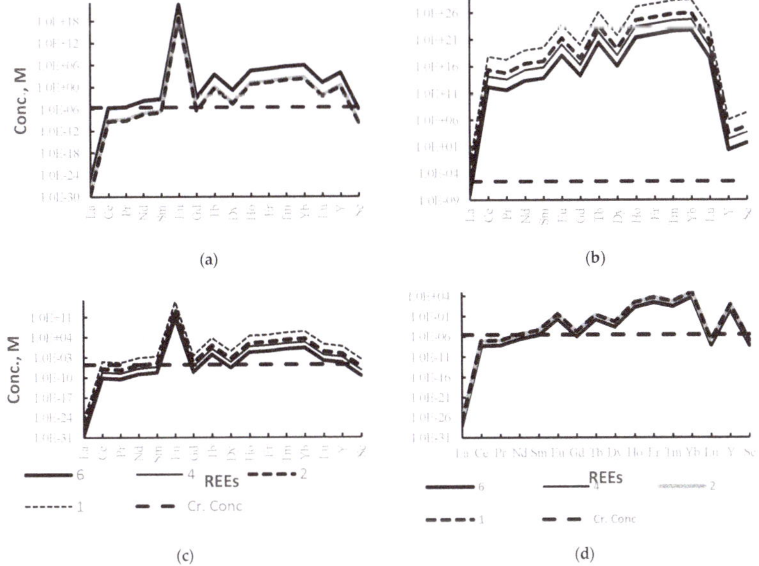

Figure 5. Concentrations of (**a**) Re^{3+}, (**b**) $ReCl_2^+$, (**c**) $Re(NO_3)_2^+$, and (**d**) $Re(SO_4)_2^-$ in equilibrium with octa-hydrated sulfate precipitate with 10^{-4} M of total concentration of sulfate in solution at pH 6, 4, 2, and 1. Cr. Concentration: 1 ppm line as a target concentration. Additional sulfate is provided by sulfuric acid in (**a**) and (**d**).

7. Oxalic Acid

Oxalic acid is the most commonly used precipitant for REEs from solution and the majority of investigators have used this precipitant to separate REEs in the solution from other dissolved ions. This is because oxalic acid has in general a very favorable affinity with REEs in general [4,6,9,13]. As seen in Table 1, the pKa values are 1 and 4.2 for $H_2C_2O_4/HC_2O_4^-$ and $HC_2O_4^-/C_2O_4^{2-}$, respectively. As a result, the relevant equations to be used to solve the equilibrium concentrations of REEs in solution after precipitation are Equations (20) and (21) for pH 6 and 4, and Equation (22) for pH 2 and 1, respectively.

$$2\,\{Re(SO_4)_2\} + 3\,\{C_2O_4{}^{2-}\} = <Re_2(C_2O_4)_3> + 4\,\{(SO_4)^{2-}\} \quad (20)$$

$$2\,\{Re(SO_4)_2{}^-\} + 3\,\{HC_2O_4{}^-\} = <Re_2(C_2O_4)_3> + 4\,\{(SO_4)^{2-}\} + 3\,\{H^+\} \quad (21)$$

$$2\,\{Re(SO_4)_2{}^-\} + 3\,(HC_2O_4{}^-) + \{H^+\} = <Re_2(C_2O_4)_3> + 4\,\{(HSO_4)^-\} \quad (22)$$

The concentrations of (a) Re^{3+}, (b) $ReCl_2{}^+$, (c) $Re(NO_3)_2{}^+$, and (d) $Re(SO_4)_2{}^-$ in equilibrium with oxalate precipitate with 10^{-4} M of total concentration of oxalate in solution at pH 6, 4, 2, and 1 is plotted in Figure 6.

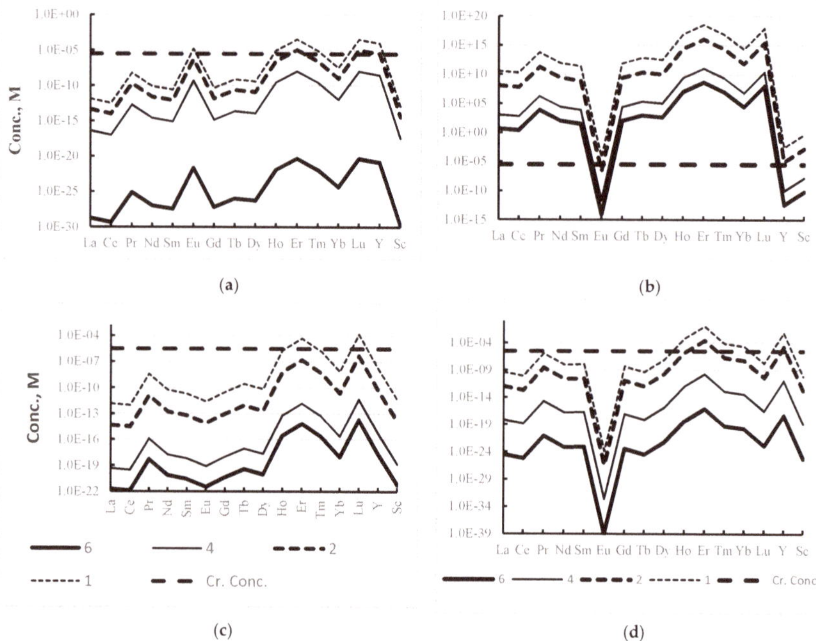

Figure 6. Concentrations of (**a**) Re^{3+}, (**b**) $ReCl_2{}^+$, (**c**) $Re(NO_3)_2{}^+$, and (**d**) $Re(SO_4)_2{}^-$ in equilibrium with oxalate precipitate with 10^{-4} M of total concentration of oxalate in solution at pH 6, 4, 2, and 1. Cr. Concentration: 1 ppm line as a target concentration.

As seen in this figure, oxalic acid is proven to be the most effective precipitant for REEs in solution. The degree of precipitation is very favorable for all pH ranges.

8. Comparison Among Precipitants

As expected, all of the precipitants considered in this study have shown good precipitation characteristics. In general, the most effective precipitant is oxalate, which has shown an outstanding precipitation behavior for REEs even at low pHs. On the other hand, most of precipitants are not effective unless the pH of the system is relatively high. As seen in Figure 7, the precipitation of free Re-ions, Re^{3+} is relatively poor even at pH 6 with the exception of oxalate and phosphate. When the pH of the system is low, such as pH 1 (Figure 7b), even oxalate loses its effectiveness. As seen in Figure 7, it is in general that the order of precipitation power follows: oxalate > phosphate > fluoride > sulfate > carbonate.

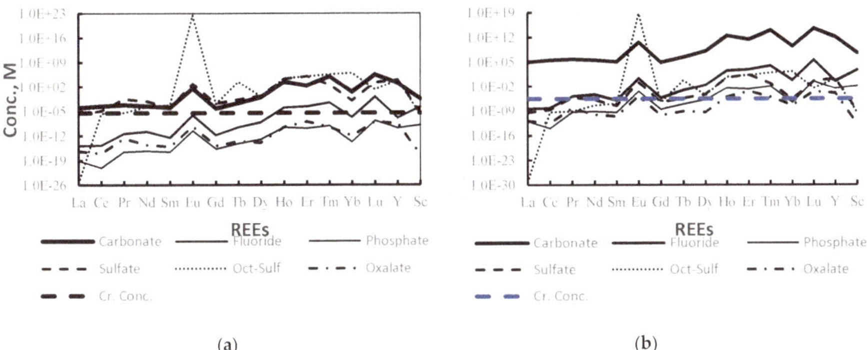

Figure 7. Concentration of Re^{3+} in equilibrium with various precipitants at 10^{-4} M and at (**a**) pH 4 and (**b**) pH 1.

However, when REEs complex with the anions such as Cl^-, NO_3^-, and SO_4^{2-} or HSO_4^-, which almost always exist, especially at low pHs, since acids frequently used in dissolving Re-bearing minerals in the leaching operation are either HCl, HNO_3, or H_2SO_4. As discussed in the earlier studies [8,9], when such acids are present in the aqueous system, REEs immediately form complexes with these anions. In the presence of chloride, $ReCl_2^+$ is the most dominating species, $Re(NO_3)_2^+$ for the nitrate system and $Re(SO_4)_2^-$ is for the sulfate system. As a result, the calculations involved to find the equilibrium concentrations of REEs should be carried out using these chemical moieties in place of Re^{3+}. Consequently, the precipitation behavior of REEs will also change and therefore, Figure 8 will be the relevant figure in place of Figure 7 for the nitric system as an example.

It is very notable in the case of the nitrate system, the precipitation power of carbonate has overtaken that of fluoride and sulfate. Similar plots were made for the chloride system as well as the sulfate system and the results were quite similar but the intensity of precipitation was not as strong as the nitric system.

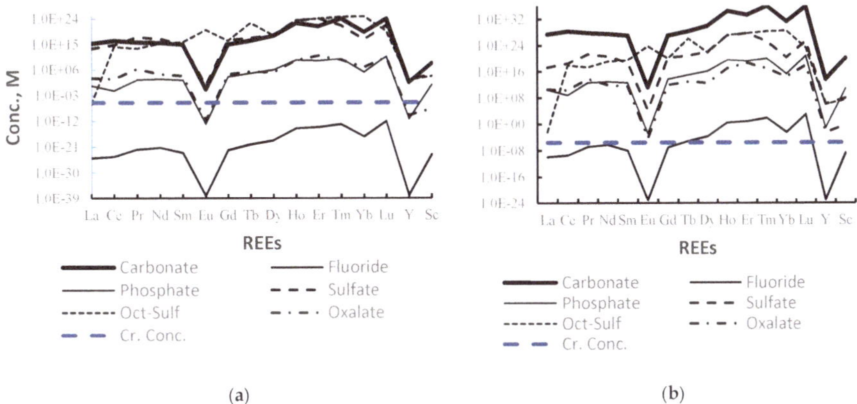

Figure 8. Concentration of $ReCl_2^+$ in equilibrium with various precipitants at 10^{-4} M and at (**a**) pH 4 and (**b**) pH 1.

9. Conclusions

The chemical precipitation of REEs with precipitants including carbonate, fluoride, phosphate, sulfate, and oxalate was studied over a wide range of pH such as 1, 2, 4, and 6. The concentration of precipitants was fixed at 10^{-4} M chosen as the standard concentration for comparison. Such precipitation occurs in practice after primary leaching as a part of the purification process.

With all of the five precipitants studied, Ligjt REEs tend to precipitate much easier than Heavy REEs. The effect of anions associated with acids used in the leaching process was found to have had a significant impact on the overall precipitation process. In general, the complexed REEs with these anions tend to provide a conducive environment in precipitation resulting in an order of magnitude difference in the precipitation extent. The nitrate system was found to be most effective followed by sulfate and chloride systems. The pH of the system has another significant factor in determining the precipitation in that the difference in pKa values of the precipitants introduces precipitant moiety of the precipitants, which has a pronounced effect on the overall precipitation.

Funding: This research received no external funding.

Acknowledgments: This paper was written in preparation of a lecture to be given at the 7th International Conference on Hydrometallurgy to be held in Ganzhou, China in May of 2020.

Conflicts of Interest: This author declares no conflict of interest in relation to this paper.

References

1. Amato, A.; Becci, I.; De Michelis, B.; Innocenzi, F.V.; Ippolito, N.M.; Gomez, P.J.; Beolchini, V.F. Sustainability analysis of innovative technologies for the rare earth elements recovery. *Renew. Sustain. Energy Rev.* **2019**, *106*, 41–53. [CrossRef]
2. Binnemans, K.; Jones, P.T.; Blanpain, B.; Gerven, T.V.; Yang, Y.; Walton, A.; Buchert, M. Recycling of rare earths: A critical review. *J. Clean. Prod.* **2013**, *51*, 1–22. [CrossRef]
3. Han, K.N.; Kellar, J.J.; Cross, W.M.; Safarzadeh, S. Opportunities and challenges for treating rare-earth elements. *Geosyst. Eng.* **2014**, *17*, 178–194. [CrossRef]
4. Gupta, C.K. Krishnamurthy. Extractive metallurgy of rare earth. *Int. Mater. Rev.* **1992**, *37*, 197–248. [CrossRef]
5. Vijayalakshmi, R.; Mishra, S.L.; Singh, H.; Gupta, C. Processing of xenotime concentrate by sulphuric acid digestion and selective thorium precipitation for separation of rare earths. *Hydrometallurgy* **2001**, *61*, 75–80. [CrossRef]
6. Yan, C.; Jia, J.; Liao, C.; Wu, S. Rare earth separation in China. *Tsinghua Sci. Technol.* **2006**, *11*, 241–247. [CrossRef]
7. Ahmed, S.H.; Helaly, O.S.; Ghany, E. Evaluation of rare earth double sulphate precipitation from monazite leach solutions. *Int. J Inorg. Bioinorg. Chem.* **2015**, *5*, 1–8.
8. Helaly, O.S.; Abd, M.S.; Ghany, E.; Borai, E.H.; Aly, H.F.; Abded Fattan, T.M. Separation of cerium, light and heavy rare earth concentrates from Egyptian crude monazite. *Chem. Technol.* **2015**, *10*, 184–192.
9. Chi, R.; Xu, Z. A solution chemistry approach to the study of rare earth element precipitation by oxalic acid. *Metall. Mater. Trans. B* **1999**, *30*, 189–195. [CrossRef]
10. Konishi, Y.; Noda, Y. Precipitation stripping of rare-earth carbonate powders from rare-earth-loaded carboxylate solutions using carbon dioxide and water. *Ind. Eng. Chem. Res.* **2001**, *40*, 1793–1797. [CrossRef]
11. Cantrell, K.J.; Byrne, R.H. Rare earth element complexation by carbonate and oxalate ions. *Geochim. Cosmochim. Acta* **1987**, *51*, 597–605. [CrossRef]
12. Vasconcellos, M.E.; da Queiroz, C.A.; Abrao, A. Sequential separation of the yttrium—Heavy rare earths by fractional hydroxide precipitation. *J. Alloys Compd.* **2004**, *374*, 405–407. [CrossRef]
13. Khawassek, Y.M.; Eliwa, A.A.; Gawad, E.A.; Abdo, S.M. Recovery of rare earth elements from El-Sela effluent solutions. *J. Radiat. Res. Appl. Sci.* **2015**, *8*, 583–589. [CrossRef]
14. Han, K.N. Effect of anions on the solubility of rare earth element-bearing minerals in acids. In *Mining, Metallurgy & Exploration*; Springer: Berlin/Heidelberg, Germany, 2019; Volume 36, pp. 215–225. [CrossRef]

15. Han, K.N. Effect of metal complexation on the solubility of rare earth compounds. In *Critical and Rare Earth Elements/Recovery from Secondary Resources*; Abhilash Akcil, A., Ed.; CRC Press: Boca Raton, FL, USA, 2019; pp. 59–84.
16. Aide, M.T.; Aide, C. Rare Earth Elements: Their Importance in Understanding Soil Genesis. *Int. Sch. Res. Not.* **2012**. [CrossRef]
17. Pourbaix, M. *Atlas of Electrochemical Equilibria*; National Association of Corrosion Eng: Houston, TX, USA, 1974; p. 644.
18. Dean, J.A. *Lange's Handbook of Chemistry*, 15th ed.; McGraw-Hill: New York, NY, USA, 1999; p. 51.
19. HSC Chemistry 5.11. *Chemical Reaction and Equilibrium Software with Extensive Thermochemical Database*; Version 5.0; Outokumpu Research Oy: Piori, Finland, 2002.
20. Wagman, D.D. The NBS tables of chemical thermodynamic properties: Selected values for inorganic and C_1 and C_2 organic substances in SI units. American chemical society and the American institute of physics for the national bureau of standards. *J. Phys. Chem. Ref. Data* **1982**, *11*, 407.
21. Kim, E.; Osseo-Asare, K. Aqueous stability of thorium and rare earth metals in monazite hydrometallurgy:Eh-pH diagrams for the systems Th-, La-, Nd-, (PO_4)-(SO_4)-H_2O at 25 °C. *Hydrometallurgy* **2012**, *113*, 67–78. [CrossRef]
22. Firsching, F.H.; Mohammadzadel, J. Solubillty Products of the Rare-Earth Carbonates. *Am. Chem. Soc.* **1986**, *31*, 40–42.
23. Firsching, F.H.; Brune, S.N. Solubility Products of the Trivalent Rare-Earth Phosphates. *J. Chem. Eng. Data* **1991**, *36*, 93–95. [CrossRef]
24. Speddin, F.H.; Jaffe, G.S. Conductances, Solubilities and Ionization Constants of Some Rare Earth Sulfates in Aqueous Solutions at 25°. *Inst At. Res. Dep. Chem.* **1953**, *76*, 882–884.
25. Migdiscov, A.; Williams-Jones, A.E.; Wagner, T. An experimental study of the solubility and speciation of the rare earth elements (III) in fluoride- and chloride-bearing aqueous solutions at temperatures up to 300 °C. *Geochim. Cosmochim. Acta* **2009**, *73*, 7087–7109. [CrossRef]
26. Chung, D.Y.; Kim, E.H.; Lee, E.H.; Yoo, J.Y. Solubility of rare earth oxalate in oxalic and nitric acid media. *J. Ind. Eng. Chem.* **1998**, *4*, 277–284.
27. De Vasconcellos, M.E.; da Rocha, S.M.R.; Pedreira, W.R.; da Queiroz, C.A.; Abrao, A. Solubility behavior of rare earths with ammonium carbonate and ammonium carbonate plus ammonium hydroxide: Precipitation of their peroxicarbonates. *J. Alloys Compd.* **2008**, *451*, 426–428. [CrossRef]
28. Kim, P.; Anderko, A.; Navrotsky, A.; Riman, R.E. Trends in structure and thermodynamic properties of normal rare earth carbonates and rare earth hydroxycarbonates. *Minerals* **2018**, *8*, 106. [CrossRef]
29. Tokunaga, S.; Haron, M.J.; Wasay, S.A.; Wong, K.F.; Laosangthum, K.; Uchiumi, A. Removal of fluoride ions from aqueous solutions by multivalent metal compounds. *Int. J. Environ. Stud.* **1995**, *48*, 17–28. [CrossRef]
30. Wu, S.; Zhao, L.; Wang, X.; Huang, X.; Zhang, Y.; Feng, Z.; Cui, D. Simultaneous recovery of rare earth elements and phosphorus from phosphate rock by phosphoric acid leaching and selective precipitation: Towards green process. *J. Rare Earths* **2019**, *37*, 652–658. [CrossRef]
31. Zakharova, B.; Komissarova, L.; Trasking, V.; Naumov, S.; Melnikov, P. Precipitation of Rare Earth Phosphates from H_3PO_4 Solutions. *Phosphorus* **1996**, *111*, 1–4. [CrossRef]
32. Moeller, T.; Kremers, H.E. Observations on Rare Earths Double Sodium Sulfate Precipitation for Separation of the Terbium and Yttrium Earths. *Ind. Eng. Chem. Anal.* **1945**, *17*, 44–45. [CrossRef]

Review

Magnetic Nanohydrometallurgy Applied to Lanthanide Separation

Fernando M. de Melo, Sabrina N. Almeida and Henrique E. Toma *

Instituto de Química, Universidade de São Paulo, São Paulo 05508-000, Brazil; fernando.menegatti.melo@usp.br (F.M.d.M.); sabrina_nobrega@hotmail.com (S.N.A.)
* Correspondence: henetoma@iq.usp.br

Received: 24 April 2020; Accepted: 8 June 2020; Published: 11 June 2020

Abstract: Lanthanides play an important role in modern technology because of their outstanding optical, electronic, and magnetic properties. Their current hydrometallurgical processing involves lixiviation, leading to concentrates of elements whose separation requires exhaustive procedures because of their similar chemical properties. In this sense, a new nanotechnological approach is here discussed, involving the use of iron oxide nanoparticles functionalized with complexing agents, such as diethylenetriaminepentaacetic acid (DTPA), for carrying out the magnetic extraction and separation of the lanthanide ions in aqueous solution. This strategy, also known as magnetic nanohydrometallurgy (MNHM), was first introduced in 2011 for dealing with transition metal recovery in the laboratory, and has been recently extended to the lanthanide series. This technology is based on lanthanide complexation and depends on the chemical equilibrium involved. It has been better described in terms of Langmuir isotherms, considering a uniform distribution of the metal ions over the nanoparticles surface, as evidenced by high angle annular dark field microscopy. The observed affinity parameters correlate with the lanthanide ion contraction series, and the process dynamics have been studied by monitoring the nanoparticles migration under an applied magnetic field (magnetophoresis). The elements can be reversibly captured and released from the magnetically confined nanoparticles, allowing their separation by a simple acid-base treatment. It can operate in a circular scheme, facilitated by the easy magnetic recovery of the extracting agents, without using organic solvents and ionic exchange columns. MNHM has been successfully tested for the separation of the lanthanide elements from monazite mineral, and seems a promising green nanotechnology, particularly suitable for urban mining.

Keywords: lanthanide separation; magnetic nanohydrometallurgy; complexing nanoparticles; magnetic nanoparticles; magnetophoresis; monazite processing; urban mining

---

1. Introduction

As the major group of the rare earth (REE) elements, lanthanides are increasingly important in modern technology because of their remarkable performance in optical, electronic and magnetic materials [1]. They are relevant in catalysis [1], metallurgy, ceramics, medicine [2–5], electronics, and energy [6,7], being also widely used in domestic apparatus such as TV (Eu, Ce), computer monitors (Ce, Nd), cell phones (La, Y, Nd, Gd, Dy, Eu, Tb), fluorescent lamps (Ce, La, Eu, Y, Tb), as well as in solar panels (La), aircrafts, automobile parts (La, Ce, Y, Dy, Eu), and wind turbines (Nd, Dy) [7].

The most important lanthanide minerals are represented by monazite, bastnäsite and xenotime. Monazite is a phosphate mineral of REE, rich in light elements such as Ce (46%), La (23%), Nd (19%), Pr (5%), Sm (3%), and Gd (1.7%), with less than 1% of the remaining lanthanides, in addition to Th. Similarly, bastnäsite is a fluorocarbonate mineral containing the light elements Ce (50%), La (32%), Nd (13%), and Pr (4%). It should be noticed that such light elements are the most relevant ones in modern technology, accounting for about 95% of global world consumption. In particular, Nd demand is becoming increasingly important, because of its use in commercial supermagnets based on $Nd_2Fe_{14}B$ alloys. Xenotime is an yttrium phosphate mineral encompassing the major heavy elements Dy (7%), Er (6.2%), Yb (6%), Gd (3.6%), Ho (2%), Tm (1.2%), Sm (1.2%), and Tb (1%).

China is currently the main world producer, possessing the largest known reserves (55 Mt). Recent discoveries are placing Brazil in the second world position [7,8], with confirmed reserves of 22 Mt. Although Brazil has pioneered the rare earth production in the mid-20th century, such competence has been lost and the country is no longer a REE producer [9].

Lanthanide processing involves the crushing of the ore into a gravel size, followed by milling, and flotation. The collected materials are then treated with acids or alkalis. In this process, the existing thorium and cerium elements are removed as thorium phosphate and CeO_2 products, leaving aqueous concentrates of dissolved light lanthanide elements. Such concentrates are then chemically treated to undergo solvent extraction and ionic exchange processes [10–16]. However, since the lanthanide elements are chemically very similar, their separation becomes rather difficult. The complexation equilibrium constants involved span only a narrow range, where the minor differences are dictated by trends in the so-called lanthanide contraction series accompanying the increase of the atomic number.

In the last years, we have been working on a new nanotechnological approach to hydrometallurgy, based on iron oxide nanoparticles functionalized with suitable complexing agents for capturing and processing the metal ions in solution. It has been denominated magnetic nanohydrometallurgy (MNHM) [17–21] and can be performed under mild conditions in aqueous solution, similarly to conventional hydrometallurgy. The difference, however, is that after sequestering the metal ions, the complexing superparamagnetic nanoparticles can be magnetically confined and isolated, allowing to selectively extract and collect the elements according to their stability constants. In the case of transition metal elements, the nanoparticles have been collected directly onto the electrode surfaces, using external magnets, allowing to perform electrowinning and obtaining the pure elements. After electrodeposition, the active nanoparticles are released just by removing the magnet, returning to the process in a green and sustainable fashion. MNHM has already been successfully tested for copper(II), silver(I) and mercury(II) ions [19]. The technology can be completely automated [18], using small reactors, and is particularly suitable for urban mining.

In the case of the lanthanides, after their selective sequestering by the complexing nanoparticles, the elements can be released in acidic media, e.g., below pH 2, generating a solution enriched in the heavy elements. By adjusting the pH to 6, a new equilibrium takes place, increasing even more the amount of heavy elements on the complexing nanoparticles, according to their respective affinity constants. In this way, by using MNHM, lanthanide separation can be feasible, involving sequential complexation, confinement and releasing cycles. It can be carried out in aqueous solution and at room temperature, without using organic solvents, ionic exchange resins and exhaustive chemical processing.

2. Complexing Magnetic Iron Oxide Nanoparticles

Nanoparticles design is the key point in magnetic nanohydrometallurgy. In order to be employed as a magnetic carrier, one should use a suitable core material of nanometric size, exhibiting a large response to magnetic fields, in a superparamagnetic behavior. One of the best materials for this purpose is magnetite, Fe_3O_4. It is widespread in nature and represents an environmentally compatible material. The black solid exhibits a mixed valence composition $Fe^{II}Fe^{III}{}_2O_4$ displaying the iron(II) ions in octahedral sites and the iron(III) ions equally distributed in octahedral and tetrahedral sites [22]. The nanoparticles exhibit nanocrystalline structure as shown in Figure 1. In the crystals, only the iron(II) ions contribute to the magnetization behavior of magnetite, since the iron(III) ions are antiferromagnetically coupled.

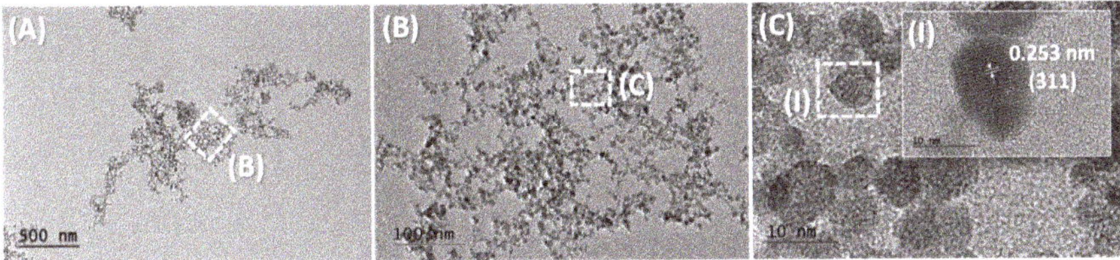

Figure 1. Transmission electron microscopy (TEM) images of nanocrystalline magnetite nanoparticles, in different magnifications (**A–C**). (I) Inset: Overfocused high resolution transmission electron microscopy (HRTEM) image of magnetite nanoparticle highlighting the interplanar distance of 311 plane (Scale bar: 10 nm).

The nanometric crystals exhibit typical magnetization values around 55–65 emu g^{-1} at room temperature with little hysteresis (Hc < 13 Oe), reflecting the existence of single magnetic domains. Magnetite is a stable material that can be easily prepared in large amounts, with low cost, by several methods, including the controlled oxidation of iron(II) polymeric complex suspensions [23], coprecipitation of iron(II) and iron(III) hydroxides [24], and the thermal decomposition of iron(III) metal organic complexes in organic media [25]. A similar alternative is maghemite, γ-Fe_2O_3, a brown solid composed exclusively by iron(III) ions. It can be obtained by the air oxidation under stirring and reflux for 3 h, of magnetite suspensions [26]. Its typical saturation magnetization, around 40–50 emu g^{-1}, is slightly smaller in relation to magnetite, but it exhibits greater stability in air. For this reason, most magnetite suspensions are slowly converted into the maghemite form after ageing under normal conditions.

Both magnetite and maghemite exhibit a superparamagnetic behavior characterized by the lack of hysteresis, or magnetic memory. This behavior can be seen in their corresponding magnetization plots, by scanning the magnetic field in a cyclic way, leading to superimposed curves in the two directions, as shown in Figure 2. The lack of hysteresis implies the existence of very weak interconversion barriers between the spin states, leading to a single magnetic domain. At very low temperatures, a very small hysteresis is often observed, reflecting the particularities of the nanoparticle structures. The superparamagnetic behavior is particularly relevant for magnetic nanohydrometalurgical applications, because it allows to readily disperse the nanoparticles after the capture and release of the metal ions, just by removing the magnetic [27]. In this way, the nanoparticles can be reused in a new cycle, without forming permanent agglomerates.

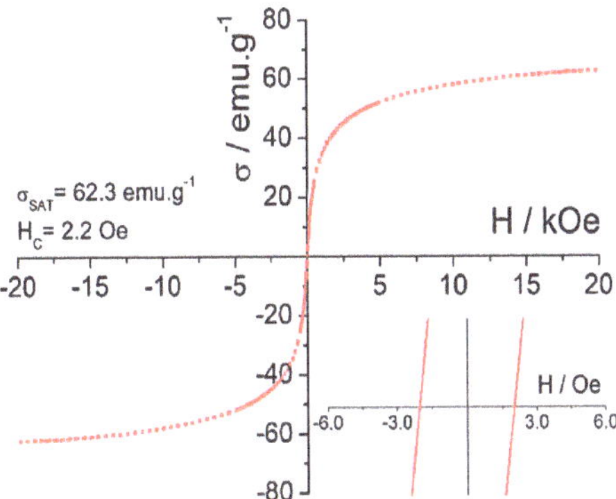

Figure 2. Magnetization curve for Fe_3O_4@SiO_2 nanoparticles, at room temperature, showing a negligible hysteresis effect (Hc = 2.2 Oe) in the expanded scale (inset).

In order to extend their lifetime and protect from the chemical environment, the nanoparticles can be coated with a thin silicate layer, by treating with tetraethoxysilane (TEOS) according to the Stöber method [28]. This step can be followed by the treatment with ethylenediaminopropyltrimethoxysilane (EAPTMS) in order to introduce the chemical linkers required for binding the DTPA complexing agent via the amino/anhydride reaction. The overall procedure is shown in Scheme 1.

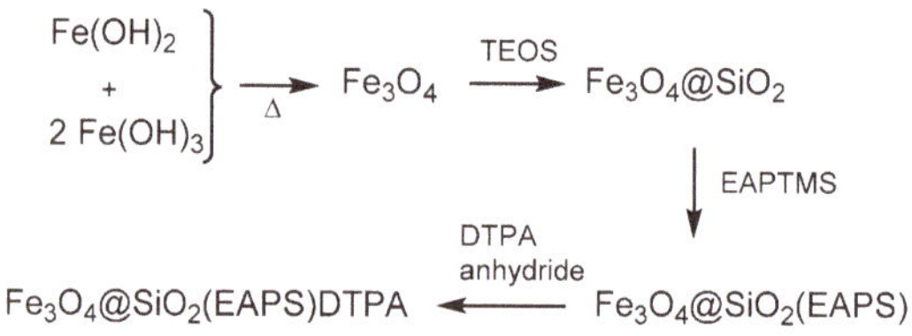

$$\left.\begin{array}{c} Fe(OH)_2 \\ + \\ 2\,Fe(OH)_3 \end{array}\right\} \xrightarrow[\Delta]{} Fe_3O_4 \xrightarrow{TEOS} Fe_3O_4@SiO_2 \xrightarrow{EAPTMS} Fe_3O_4@SiO_2(EAPS) \xrightarrow{\text{DTPA anhydride}} Fe_3O_4@SiO_2(EAPS)DTPA$$

Scheme 1. Synthetic routes for obtaining the complexing nanoparticles.

The coating of the magnetic core with silica helps in preventing the exposure of the surface metal ions, thus precluding the interaction with the solvent and electrolytes, including the lanthanide ions and buffers. The silica coating also improves the protection against air, decreasing the conversion rates of magnetite into maghemite, and preventing the decay of the magnetization properties. In addition, the protecting coating is also important for anchoring aminoalkylsilane linkers, such as ethylenediaminepropyltrimethoxisilane, EAPTMS, required for attaching the final complexing agents. In the case of the lanthanide ions, because of their hard-base characteristics, the diethylenetriaminepentaacetic acid (DTPA) ligand is one of the best choices as a sequestering agent. This complexing agent can be covalently linked to the anchored (ethylenediamine)propylsilane (EAPS) group, via the reaction with DTPA anhydride, through the formation of amide bonds, as shown in Figure 3. The final, engineered nanoparticles $Fe_3O_4@SiO_2$/EAPS/DTPA will be simply referred to as NP@DTPA, except when their specific composition becomes relevant.

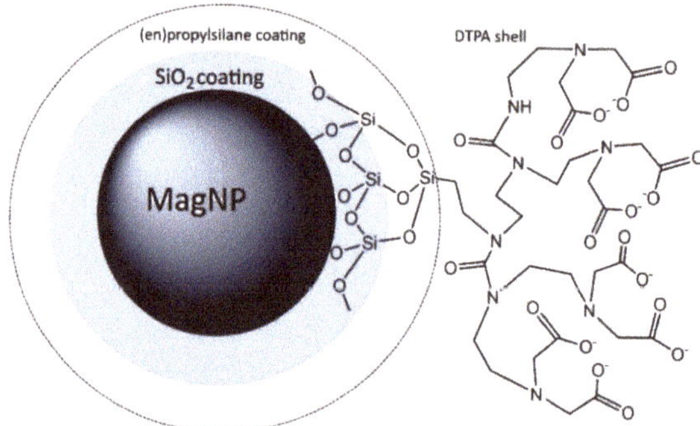

Figure 3. Magnetic nanoparticles protected by a SiO_2 coating, containing an (ethylenediamine)propylsilane shell for supporting the covalently bound DTPA molecules, $Fe_3O_4@SiO_2$/EAPS/DTPA, or simply NP@DTPA (adapted with permission from [29]).

The stepwise functionalization monitored by means of elemental analysis leads to a systematic increase of C and N content, accompanying the binding of the organosilane and DTPA agents, in the $Fe_3O_4@SiO_2$/EAPS and $Fe_3O_4@SiO_2$/EAPS/DTPA nanoparticles. The presence of the organic coating can also be se$_2$en in the FTIR spectra, as illustrated in Figure 4.

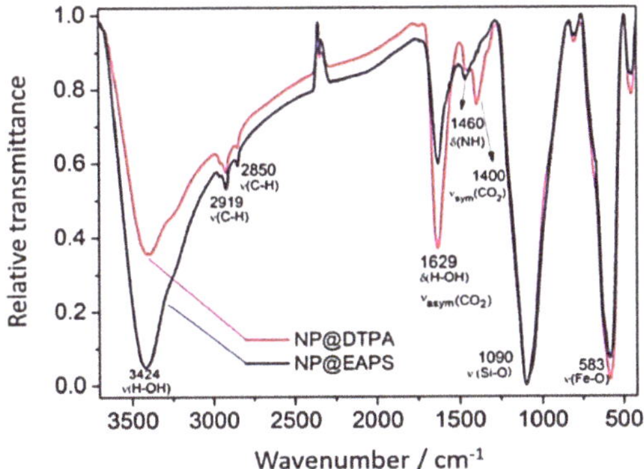

Figure 4. Typical FTIR spectra of the (—) $Fe_3O_4@SiO_2$/EAPS and (—) $Fe_3O_4@SiO_2$/EAPS/DTPA nanoparticles (adapted with permission from [29]).

The vibrational spectra of these two particles exhibit characteristic broad peaks at 1460 cm^{-1} ascribed to ν(N–H) stretching, and the sharp peaks at 2850 and 2919 cm^{-1} ascribed to ν(C–H) vibrations. Strong peaks observed at 1090 cm^{-1} correspond to ν(Si–OSi) vibrations, while the ν(Fe–O) peaks are shown at 583 cm^{-1}, in agreement with the magnetite composition. Differential peaks can be observed in the 1600–1400 cm^{-1} region, corresponding to the asymmetric and symmetric stretching vibrations of the carboxylate groups. The peak at 1630 cm^{-1} corresponds to the $\nu_{asym}(CO_2)$ vibration, but it seems superimposed to the δ(O–H) vibrations of the water molecules present in the system. The relative intensity peak is stronger for the DTPA particles, as expected for the presence of the large number of carboxylate groups in relation to $Fe_3O_4@SiO_2$/EAPS. The weak peak at 1465 cm^{-1} is associated with ν(N–H) vibrations in both cases, but there is a differential peak at 1400 cm^{-1} for the $Fe_3O_4@SiO_2$/EAPS/DTPA particle, corresponding to the $\nu_{sym}(CO_2)$ vibration. The $Fe_3O_4@SiO_2$/EAPS and Fe_3O_4 $@SiO_2$/EAPS/DTPA nanoparticles were also monitored by dynamic light scattering and zeta potential measurements, as shown in Figure 5.

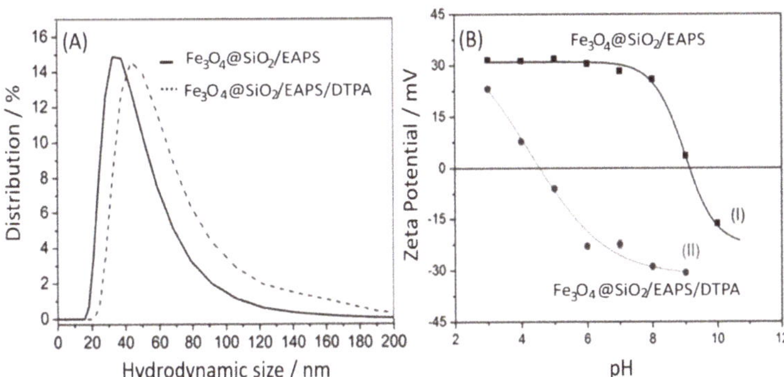

Figure 5. (**A**) Hydrodynamic sizes of the $Fe_3O_4@SiO_2$/EAPS and of the $Fe_3O_4@SiO_2$/EAPS/DTPA nanoparticles and (**B**) variation of their corresponding zeta potentials as a function of pH (adapted with permission from [29]).

In Figure 5A, the comparison between $Fe_3O_4@SiO_2$/EAPS and the $Fe_3O_4@SiO_2$/EAPS/DTPA nanoparticles shows that the external DTPA shell increases slightly the nanoparticles hydrodynamic size, as expected theoretically. Its presence is also reflected in the nanoparticles' zeta potentials, as shown in Figure 5B. The zeta potentials refer to electrokinetic potentials measured at the interfacial double layer also denoted as a slipping plane relative to the bulk fluid away from the interface [30]. It is particularly related to the surface plane where the nanoparticles interact with the metal ions ion solution. In addition, the zeta potentials are important for promoting the nanoparticles stability in solution by means of electrostatic repulsion forces [31].

As shown in Figure 5B, the Fe_3O_4@SiO_2/EAPS nanoparticles containing ethylenediamine groups are protonated below pH 8, leading to positive zeta potentials around 30 mV. Such high positive charges stabilize the nanoparticles in solution. The isoelectric point is observed at pH 9.12, where the electrostatic stability is lost. In the case of the Fe_3O_4@SiO_2/EAPS/DTPA nanoparticles, the zeta potentials at low pH are a little smaller, starting at 20 mV at pH 3, indicating a positive charge from the protonation of the amino groups, in the presence of the neutral carboxylic acid groups. The isoelectric point occurs at pH 4.4, where the partial deprotonation of the carboxylic groups compensates the positive charges of the protonated amino groups. Above pH 5 most carboxylic groups are deprotonated, and the nanoparticles become electrostatically stabilized by a negative charge, corresponding to −30 mV potential.

3. Lanthanide Capture by the Iron Oxide Nanoparticles

DTPA is a well-known complexing agent for metal ions, and perhaps the best one in the market, for lanthanide ions. Its interaction with metal ions in aqueous solution can be readily probed by separating the superparamagnetic nanoparticles using an external magnet and analyzing its metal content, as well as of the remaining solution, by means of chemical, spectroscopic or X-ray fluorescence techniques.

For a systematic work, it is important to find the best experimental conditions, considering the pKa and the zeta potentials involved, using controlled conditions and the same batch of iron oxide nanoparticles in order to keep constant the chemical and physical parameters, such as composition, size, mobility, and complexation activity. This is a very important precaution, because the nanoparticle characteristics depend upon the synthetic procedures employed, turning the direct comparison with related nanoparticles obtained from different batches or methods, which is not always feasible. For comparison purposes, only a few related systems have been reported in the literature since our first work on MNHM in 2011 [21], or on the lanthanum/neodymium separation in 2016 [32]. In particular, one should mention those based on metallic iron [33], or using polymeric coating [34], which, in spite of their different scopes and methodologies, there is a good agreement with the lanthanide sequestering properties reported in our work.

It should be noted that the experiments involving the capture of lanthanide ions by the NP@DTPA nanoparticles have been carried out at pH 6 [32]. At this pH, most of the carboxylic groups are in the anionic form, and the nanoparticles are stabilized by the electrostatic repulsion from the surface carboxylate charges, as described by the DLVO theory [35]. In addition, at this pH, there is no risk of precipitating the lanthanide hydroxides before performing the complexation reaction. Another point observation is that for technological applications, MNHM requires suitably engineered iron oxide nanoparticles, exhibiting, for instance: (i) strong magnetic cores, with crystalline structures, displaying little hysteresis at room temperature; (ii) great stability in aqueous solution, in order to perform under normal conditions, e.g., in the presence of oxygen (air), high stirring rates and variable pH; (iii) large surface area and great mobility in solution; (iv) a protecting coating for preserving the magnetic core from the chemical environment; (v) a functional external coating encompassing the complexing agents; (vi) a large concentration of the complexing agents at the surface, in order to improve the yield of sequestered metal ions; (vii) a good chemical reversibility for exchanging the complexed metal ions with protons in the processing steps and performing the separation process; (viii) high magnetic recovery rates for effectively recycling the functional nanoparticles; (ix) low toxicity and safe manipulation, offering minor risks for the environment; (x) low cost and allowing production in large scale for industrial applications.

In the complexation studies, it is important to start from stable nanoparticles in solution because of their mobility, thus improving the interaction of the DTPA groups with the metal ions in order to form strong chelate complexes. The analysis is facilitated by the easy removal of the nanoparticles, magnetically, in order to perform energy dispersive X-ray fluorescence (EDXRF) measurements and evaluate the amount of captured metal ions [29]. Typical EDXRF data for Fe_3O_4@SiO_2/EAPS/DTPA nanoparticles containing Nd^{3+} ions, separately, and their residual content in solution, as well as of the bare Fe_3O_4 nanoparticles, are shown in Figure 6.

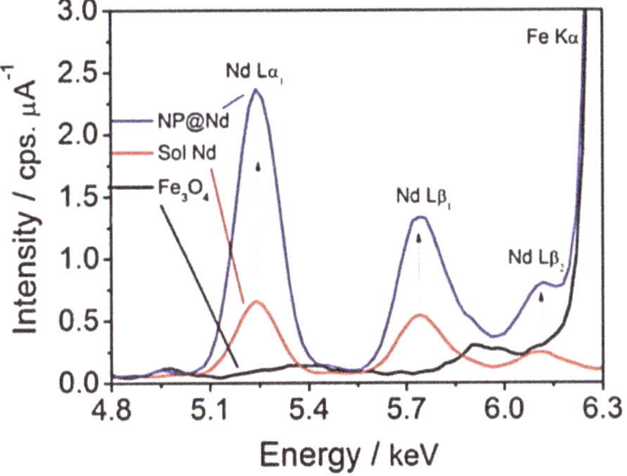

Figure 6. Typical X-ray fluorescence spectra of the Fe_3O_4 nanoparticles, of the neodymium nitrate solutions at 25 mmol L^{-1}, and of the nanoparticles after the capturing of Nd^{3+} ions (adapted with permission from [29]).

The complexation studies have been carried out for the several lanthanide ions, independently, in order to estimate their affinity or equilibrium constants. The process involves mobile nanoparticles and free ions in solution and can be described in terms of the Langmuir isotherms proposed for adsorption of molecules onto surfaces displaying similar binding sites, according to a monolayer distribution. There are also many other models available, such as the Freundlich isotherms, which can accommodate more complicated systems, including the formation of adsorption multilayers. In our case, the Langmuir isotherms have been preferred because their theoretical formalism keeps a close similarity with the chemical equilibrium in solution.

Considering the complexation equilibrium for the lanthanide ions (Ln^{3+}) and the superparamagnetic nanoparticles, one can write:

$$\text{NP@DTPA} + \text{Ln}^{3} \rightleftharpoons \text{NP@DTPA-Ln} \tag{1}$$

where the equilibrium constant can be expressed by

$$K = \frac{[\text{NP@DTPA} - \text{Ln}]}{\left[\text{Fe}^{3+}\right][\text{NP@DTPA}]} \tag{2}$$

If θ is the fraction of the adsorption sites to which the Ln^{3+} ion has become attached, the fraction of the unattached sites is (1 − θ). Therefore,

$$\frac{[\text{NP@DTPA} - \text{Ln}]}{[\text{NP@DTPA}]} = \frac{\theta}{(1-\theta)} \tag{3}$$

Calling the concentration of the lanthanide ions at equilibrium $[Ln^{3+}] = C_{eq}$ the adsorption constant can be written as

$$\text{K} = \frac{\theta}{\text{C}_{\text{eq}}(1-\theta)} \tag{4}$$

or rearranging,

$$\theta = \frac{KC_{eq}}{\left(1 + KC_{eq}\right)} \tag{5}$$

The adsorbed mass of lanthanide ions at equilibrium concentration is equal to the difference of the initial and equilibrium Ln^{3+} concentrations, C_o and C_{eq}, respectively. Defining q_{eq} as the adsorbed mass of lanthanide ions at equilibrium concentration per mass of adsorbent (m), NP, in a given volume V,

$$q_{eq} = \frac{\left(C_o - C_{eq}\right)V}{m} \tag{6}$$

Considering that

$$\theta = \frac{q_{eq}}{q_{max}} \tag{7}$$

it follows that

$$q_{eq} = \frac{KC_{eq}q_{max}}{\left(1 + KC_{eq}\right)} \tag{8}$$

Therefore, the plot of q_e versus C_{eq} should exhibit a characteristic curvature, approaching a saturation point at higher concentrations of the lanthanides, corresponding to q_{max}. The corresponding plots for La^{3+} and Nd^{3+} ions can be seen in Figure 7.

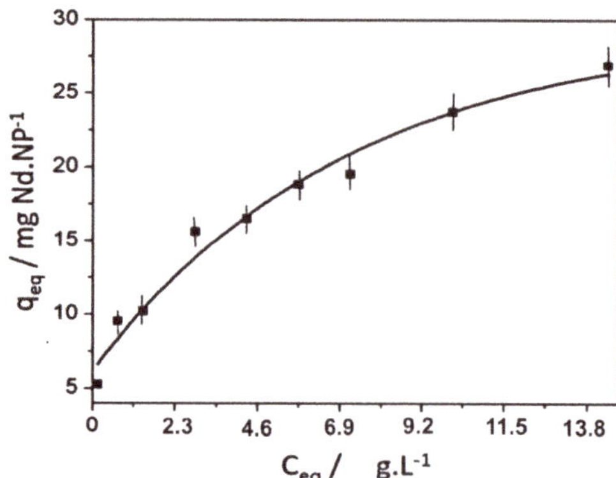

Figure 7. Adsorption isotherm of NP@DTPA in Nd^{3+} solutions at 298 K (adapted with permission from [29]).

A more convenient way is to express Equation (8) in the linearized form as shown in Figure 8.

$$\frac{C_{eq}}{q_{eq}} = \frac{1}{K\,q_{max}} + \frac{C_{eq}}{q_{max}} \tag{9}$$

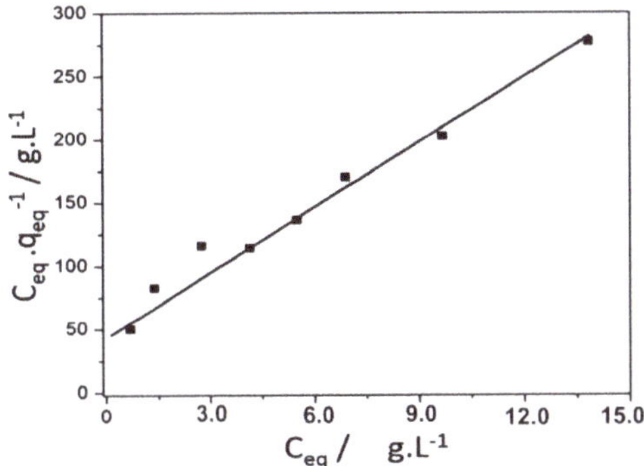

Figure 8. Linear plot of the adsorption isotherm of NP@DTPA in Nd^{3+} solutions based on the Langmuir Equation (9).

In this way, from the analysis of the linear data in Figure 8, the maximum coverage for the Nd^{3+} ions was evaluated as 28 mg(Nd) g(NP)$^{-1}$ and the equilibrium constant as 4.22×10^{-4} g(NP) L^{-1}.

4. Desorption Equilibrium

Desorption of the sequestered lanthanide ions in the NP@DTPA-Ln nanoparticles can be carried out by lowering the pH below 2 in order to protonate the carboxylate groups (Equation (10)).

$$\text{NP@DTPA-Ln} + H^+ \rightleftharpoons \text{NP@DTPAH} + Ln^{3+} \quad (10)$$

The process can be monitored by X-ray fluorescence as illustrated in Figure 9. In this example, the NP@DTPA-Nd nanoparticles were magnetically confined, and the Nd content determined. After this, the particles were treated with an acidic solution, keeping the pH around 2. Then, after equilibration, the Nd content of the magnetic nanoparticle was analyzed again. As one can see, there is a complete release of the Nd^{3+} ions to the solution.

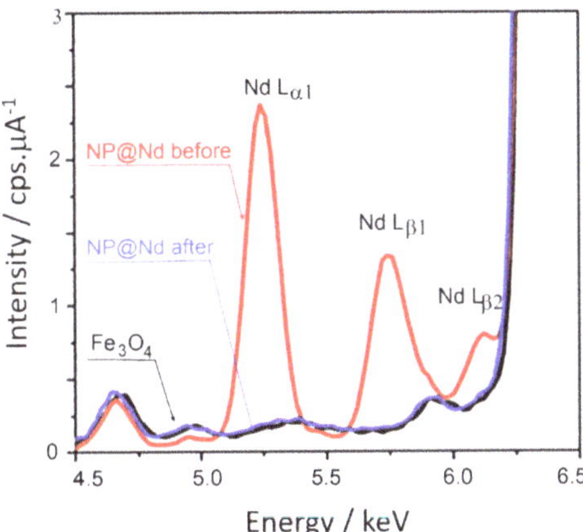

Figure 9. X-ray fluorescence analysis of the magnetic nanoparticles containing lanthanide ions, and after the treatment in acidic solutions at pH 2, showing the complete release of the metal elements (adapted with permission from [29]).

The reversible process can be controlled by the pH, allowing the use of the superparamagnetic nanoparticles in the separation of the lanthanide elements.

5. Magnetic Behavior, Kinetics and Magnetophoresis

Dy^{3+} ions have been specially employed for investigating the magnetic behavior, kinetics and magnetophoresis of the iron oxide nanoparticles because of their large magnetic moment. In addition, dysprosium is the major lanthanide component of the xenotime ore, being strategic in the fabrication of especially strong neodymium supermagnets. The capture of Dy^{3+} ions by the NP@DTPA nanoparticles can be represented as:

$$\text{NP@DTPA} + Dy^{3+} \rightleftharpoons \text{NP@DTPA-Dy} \tag{11}$$

The process has been probed by the X-ray fluorescence spectra, by monitoring the corresponding $L\beta_2$ band, as shown in Figure 10.

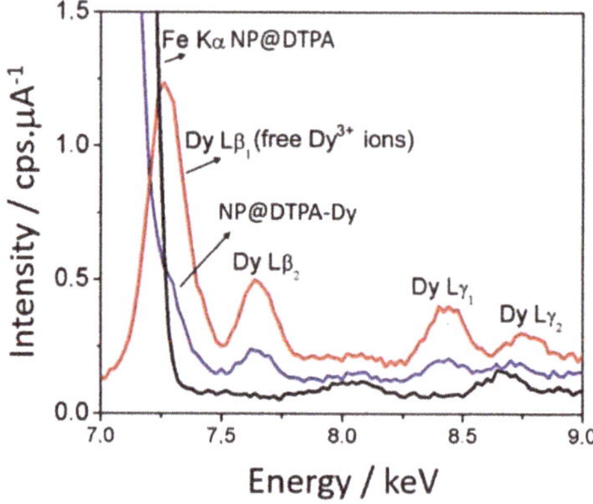

Figure 10. X-ray fluorescence spectra of the NP@DTPA, Dy^{3+} free ions in equilibrium, and NP@DTPA-Dy, after the magnetic confinement (adapted with permission from [35]).

The Langmuir isotherms were elaborated by determining the concentration of the remaining Dy^{3+} ions in solution, after confining the NP@DTPA-Dy particles with an external magnet. From the equilibrium concentrations of the Dy^{3+} ions, C_{eq}, the Langmuir isotherm was elaborated, as shown in Figure 11.

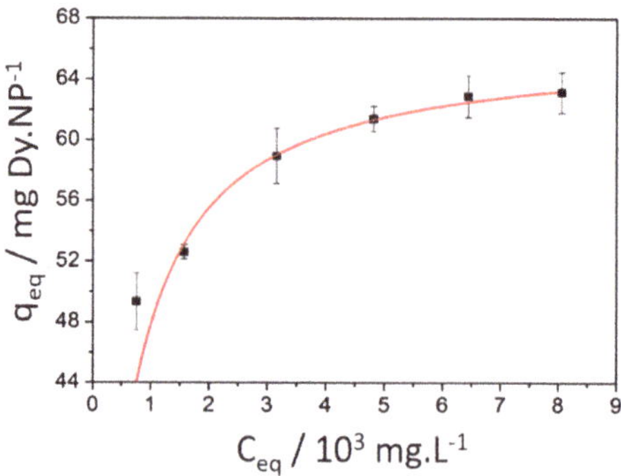

Figure 11. Adsorption isotherm of NP@DTPA in Dy^{3+} solutions at 298 K, pH 6, from the data collected after 500 min, under stirring, for ensuring complete equilibration (adapted with permission from [35]).

Its linearized form is shown in Figure 12.

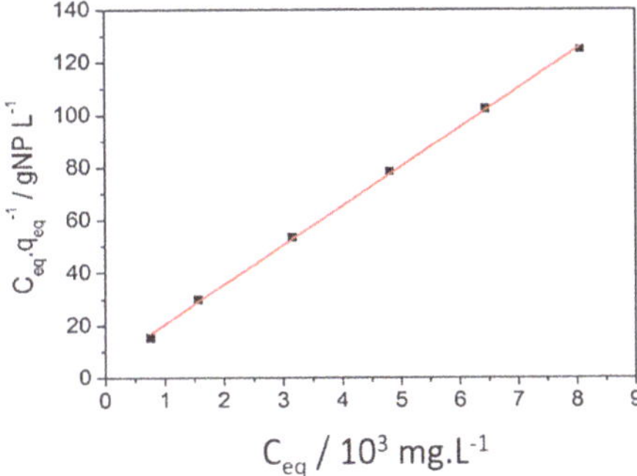

Figure 12. Linear plot of the adsorption isotherm of NP@-DTPA in Dy^{3+} solutions, as expressed by the Langmuir equation.

In this way, the maximum coverage for the Dy^{3+} ions were 63.2 mg(Dy)/g(NP), and the respective adsorption equilibrium constant was 2.57×10^{-3} g(NP) L^{-1}. This value is significantly higher than the equilibrium constant of 4.22×10^{-4} g(NP) L^{-1}, previously reported for the capture of Nd^{3+} ions, respectively, using the same superparamagnetic nanoparticles and experimental conditions. The formation of the surface metal complexes from the lanthanide capture has also been probed by scanning transmission electronic microscopy (STEM) using a high-angle annular dark field arrangement (HAADF) and shown in Figure 13 for the Dy^{3+} complex.

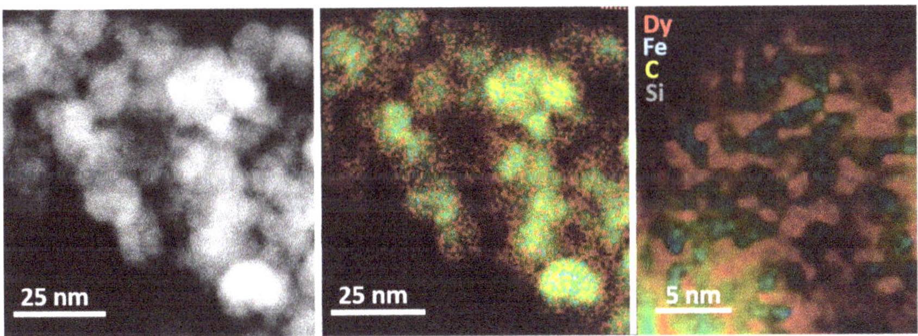

Figure 13. (**Left**) HAADF image of a cluster of NP@DTPA-Dy nanoparticles; (**Middle**) Elemental map image obtained by EDS analyses of the same NP@DTPA-Dy nanoparticles cluster. (**Right**) Enlarged image of the selected area from the middle image. The red, blue, yellow and green colors correspond to the presence of Dy, Fe, C and Si in the nanoparticles, respectively (adapted with permission from [35]).

In Figure 13, one can see the HAADF of the magnetic nanoparticles deposited on the carbon grid. The corresponding elemental map is also shown, revealing the distribution of the Dy (red), Fe (blue), C (yellow), and Si (green) elements at the NP@DTPA-Dy surface. In the more detailed image, one can see the red spots from the dysprosium ions located over the yellow stain of the carbon atoms from the DTPA ligand. As a matter of fact, this image represents the direct observation of the Dy-DTPA complexes at the nanoparticles surface. The surface metal complexes can influence the magnetism of the NP@DTPA-M nanoparticles by changing their electronic content at the surface. Their effect is expected to be more intense in the case of strongly paramagnetic metals, such as the Dy^{3+} ($4f^9$) species, where the intrinsic paramagnetism can couple with the magnetic field of the core. In this case, the magnetization of NP@DTPA-Dy nanoparticles (8.5 mg) containing 6.3% (weight) Dy^{3+} ions were recorded in water/glycerol suspension, using a vibrating sample (VSM) magnetometer. The measurements were carried out in comparison with the NP@DTPA (8.4 mg) sample with no Dy^{3+} ions.

The results are shown in Figure 14. Surprisingly, the binding of paramagnetic Dy^{3+} ions seems to decrease saturation magnetization of the NP@DTPA-Dy nanoparticles, as shown in Figure 14.

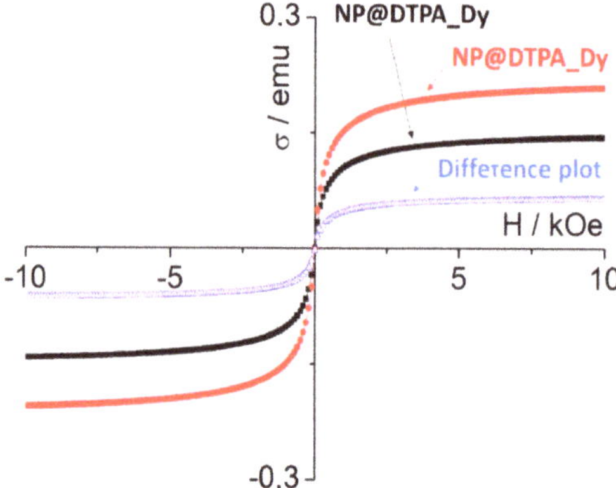

Figure 14. VSM measurements, at room temperature of NP@DTPA and NP@DTPA-Dy suspensions in water/glycerol mixture, showing the decrease of the magnetization promoted by the Dy^{3+} ions, as evaluated by the difference plot (adapted with permission from [35]).

In principle, the paramagnetic lanthanide ions were expected to enhance the saturation magnetization of the nanoparticles. However, the measurements indicated an opposite behavior, raising questions about the occurrence of dipolar interactions with the magnetic core. The HAADF image seems to corroborate the model illustrated in Figure 15, where Dy^{3+} ions are under the influence of opposing magnetic lines of the Fe_3O_4 core, leading to antiferromagnetic dipolar interactions.

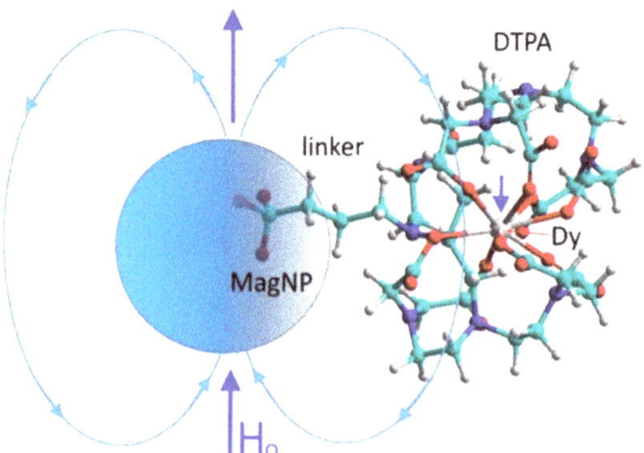

Figure 15. A pictorial representation of the antiferromagnetic dipolar interaction between dysprosium ions and the magnetic nanoparticles (adapted with permission from [35]).

Another way of probing the magnetic behavior of the particles is by monitoring their migration under the influence of an applied magnetic field. This response is known as magnetophoresis and is currently being investigated by many authors, using different techniques. Its understanding is essential for the application of nanoparticles in magnetic separation processes [35–42]. Since magnetic nanoparticles are very small, their magnetophoretic pathway can be perturbed by the thermal energy and the viscous drag forces [42]. For this reason, a strong magnetic force is necessary to overcome the thermal randomization. Because of the high complexity of the analytical solutions dealing with migration in inhomogeneous magnetic fields, the rigorous theoretical formalisms become rather complicated and are beyond the scope of this paper. As a simple approximation, the magnetophoretic velocity v, can be obtained by equilibrating the force involved in the magnetophoretic displacement of the isolated nanoparticles of radius R and magnetic moment m, under a magnetic field H, and gradient ($\partial H/\partial r$), and the opposing viscous drag force, as

$$v = \frac{1}{6\pi\eta R}\chi\mu_o H\frac{\partial H}{\partial r} \tag{12}$$

where μ_o is the magnetic constant (equal 1 in cgs unit), and η is the viscosity.

In this way, the magnetophoretic velocity of the nanoparticles should be directly proportional to the magnetic field and to the magnetic field gradient, as well as to the magnetic moment or susceptibility of the particles, and inversely proportional to their hydrodynamic sizes. A simple arrangement for probing magnetophoresis in the laboratory is by using an analytical balance equipped with a suitable cantilever for allowing external weighing, as shown in Figure 16. The sample probe employed is just a miniature tube, placed above of small $Nd_2Fe_{14}B$ supermagnet (diameter = 1 inch) at a constant distance and position, in order to provide the same magnetic field and gradients. A pictorial representation of the experiment is shown in Figure 17 [35].

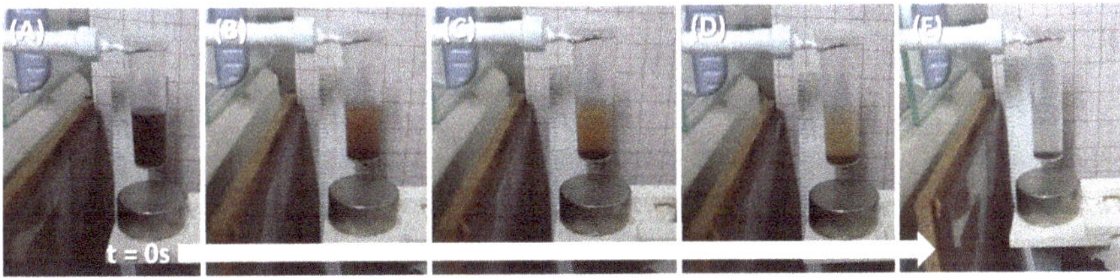

Figure 16. (**A–E**) Monitoring the magnetophoresis of NP@DTPA in the presence of a $Nd_2Fe_{14}B$ magnet, as a function of time by means of an external weight device placed on the analytical balance pan (only the cantilever is shown) (adapted with permission from [43]).

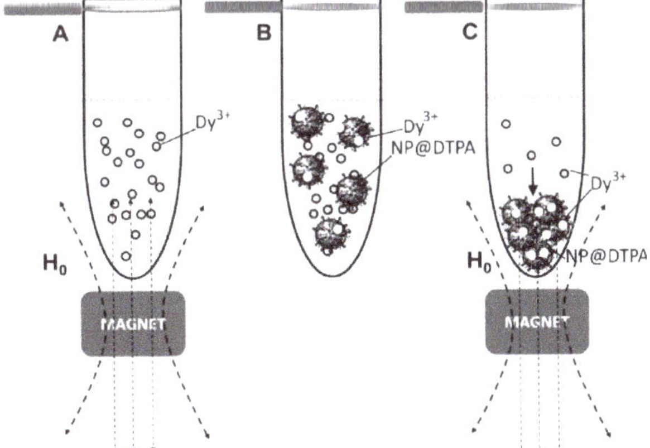

Figure 17. Pictorial representation of the capture of Dy^{3+} ions in the presence of the magnet: (**A**) Initial Dy^{3+} solution (0.1 mol L^{-1}); (**B**) after adding NP@DTPA and stirring without applying the magnet; (**C**) magnetophoretic deposition after the capture of Dy^{3+} ions by the superparamagnetic nanoparticles (adapted with permission from [43]).

By using this arrangement, it is possible to evaluate the average magnetophoretic rates by monitoring the relative weight increase as the magnetic nanoparticles reach the bottom attracted by the magnetic field. The presence of paramagnetic ions in solution has only a negligible effect on the weight increase by the attracted nanoparticles, since their magnetic moments differ by 8 orders of magnitude.

In a typical magnetophoretic experiment, by starting with 1 mL of 0.1 mol L^{-1} Dy^{3+} solution, 8 mg of NP@DTPA was added and the mixture was kept under stirring for 5 h to ensure equilibration. After this, the sample was kept in the external weighing device, and a magnetic field was applied to promote the nanoparticles deposition at the bottom, while measuring the weigh increase as a function of time. A similar experiment was repeated in the absence of Dy^{3+} ions. The comparison, shown in Figure 18, revealed that a systematic decrease of magnetism, by 5.9 ± 1% results after the binding of the Dy^{3+} ions.

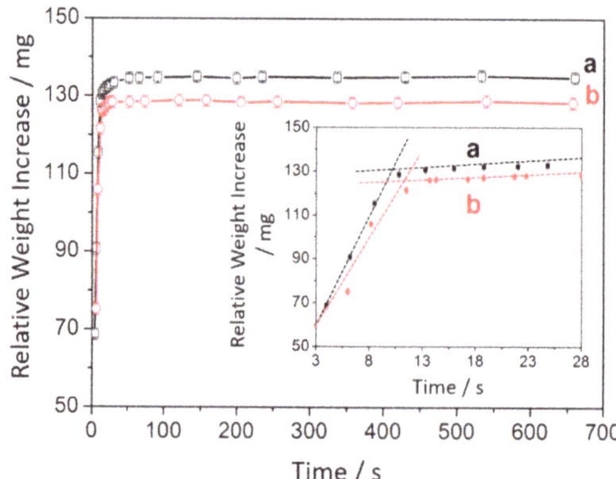

Figure 18. Magnetophoresis of NP@DTPA before (**a**) and after (**b**) the binding of Dy^{3+} ions (magnet distance = 0.40 cm) showing the decrease of magnetization in the presence of the lanthanide ions and the slower magnetophoretic rate (inset) (adapted with permission from [43]).

This result is consistent with the occurrence of an anti-ferromagnetic coupling between the paramagnetic Dy^{3+} ions and the superparamagnetic core, as already observed by VSM, Figures 14 and 15. The magnetophoretic velocity is also slightly reduced after the binding of Dy^{3+} ions, with the curve slopes decreasing from 8.98 to 8.52. This is consistent with the decrease of the magnetization of the nanoparticles and with the increase of the hydrodynamic radii due to the presence of the Dy^{3+} ions. The observed decrease of magnetization is correlated with the amount (6% *w*/*w*) of captured Dy^{3+} ions, allowing to probe the metal ion extraction, using a simple analytical balance.

6. Lanthanide Separation

The equilibrium constants for the complexation of the lanthanide ions have been obtained using the same batch of NP@DTPA nanoparticles, for which all the chemical and physical constants have been previously determined. They have been evaluated from the corresponding Langmuir isotherms, generated under the same conditions, at room temperature, for the La^{3+}, Pr^{3+}, Nd^{3+}, Sm^{3+}, Gd^{3+} and Dy^{3+} ions, leading to the following equilibrium constants: 2.9×10^{-4}, 4.0×10^{-4}, 4.2×10^{-4}, 6.7×10^{-4}, 1.2×10^{-3}, 2.3×10^{-3} g(NP) L^{-1}, respectively. As shown in Figure 19, the trends in the equilibrium constants are dictated by the lanthanide contraction series, where a smaller ionic radius leads to the formation of a stronger complex.

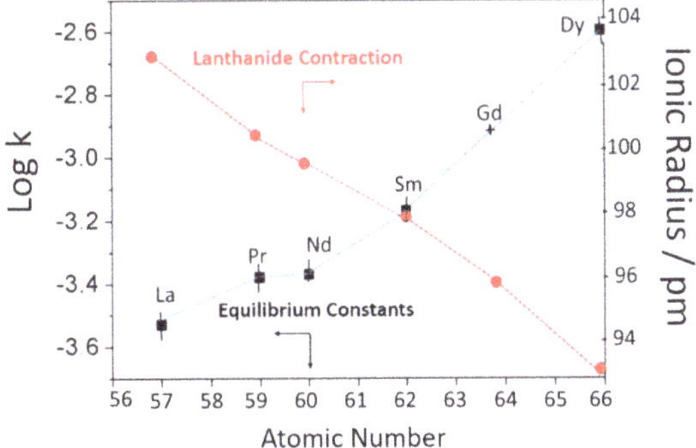

Figure 19. Affinity constants (K) of the lanthanide elements as a function of the atomic number, superimposed to the corresponding radial contraction plot (adapted with permission from [44]).

Lanthanide elements separation using NP@DTPA can be performed using the same classical chelation chemistry employed in hydrometallurgy for rare earths. For instance, using an ionic exchange resin column, the lanthanide ions are loaded at the top and gradually eluted with complexing agents under controlled pH. Complexation can remove the lanthanide ion from the column, but a rather complex equilibrium network takes place, requiring long columns and a very careful procedure. In magnetic nanohydrometallurgy, the lanthanide ions are selectively sequestered by the complexing nanoparticles, according to their specific affinity constants, and can be readily confined with a magnet. The confined nanoparticles are enriched by the heavy elements, considering the specific affinity trends shown in Figure 18. Then, in a second cycle, such confined nanoparticles are suspended in an aqueous solution maintained at pH 2 or less, in order to release the complexed metal ions by the protonation of the carboxylate groups as illustrated in Figure 20.

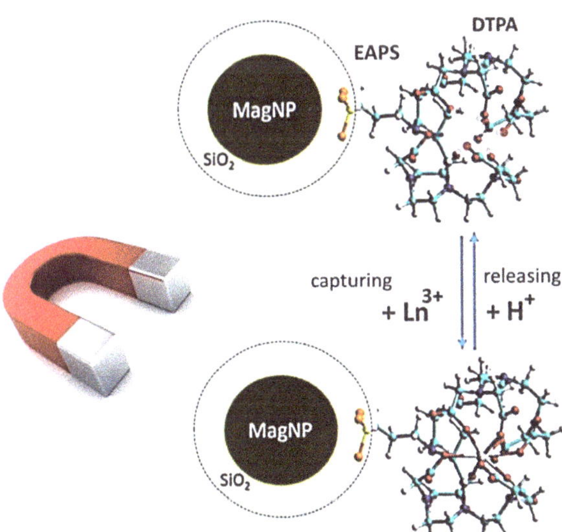

Figure 20. Magnetic capture and releasing of lanthanide ions by the superparamagnetic nanoparticles functionalized with DTPA (adapted with permission from [44]).

Then the pH is adjusted to 6 in order to restore the complexing capability of the particles, but now in the presence of the lanthanide ions in solution, enriched with heavy elements. Complexation does enhance the heavy elements removal from the solution. The magnetic confinement and release cycle can be repeated again and again in order to sequentially separate the heavy lanthanides from the light ones. The chemistry resembles the classical extraction of lanthanides using complexing agents and solvents. However, it is much simpler, involving just a one pot procedure, without using solvents or columns.

6.1. Binary Mixtures

Lanthanide separation studies were initially applied to binary mixtures, such as Nd^{3+} and La^{3+} ions [31,33,43,44] as a proof of concept. The experiment started with an equimolar mixture of the two elements, and the first cycle of complexation with the complexing nanoparticles was performed. In this particular case, the analysis of the elements captured by the nanoparticles led to a distribution of 66% Nd^{3+} and 33% La^{3+}. This result is in perfect agreement with their corresponding affinity constants. In a second step, after confining the nanoparticles with a magnet, a subsequent treatment with diluted acids (pH 2) released the lanthanide ions into a new solution, keeping the 66:33 proportion. Then, the pH was raised up to 6, and a new complexation equilibrium took place. The relative amounts of the two elements captured by the nanoparticles were 94% Nd^{3+} and 6% La^{3+}. In the third step, starting from the 94:6 ratio, the amount of captured Nd^{3+} ions were >99%, since no evidence of La^{3+} could be detected by the X-ray fluorescence technique (Figure 21).

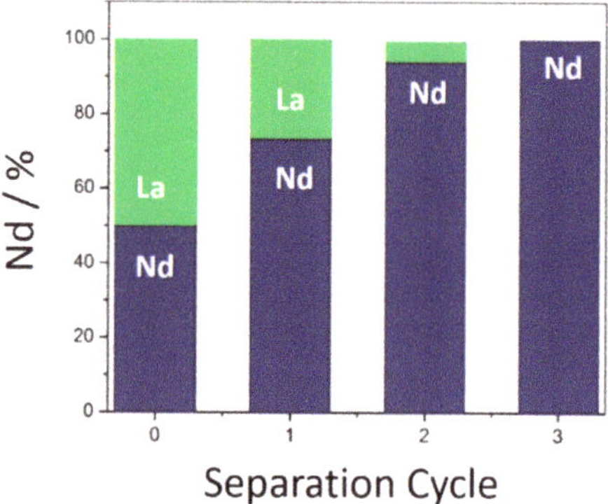

Figure 21. Fractionation of La^{3+} and Nd^{3+} mixtures, starting from an equimolar solution (stage 0) and subsequent MNHM processing by using superparamagnetic nanoparticles functionalized with DTPA, as described in the text (adapted with permission from [29]).

The enrichment plots illustrated in Figure 21 indicate that three partition steps can be enough to obtain Nd^{3+} in pure form, by starting from an equimolar mixture with La^{3+} ions. A similar test was applied for La^{3+} and Sm^{3+}, by starting with an equimolar mixture of the lanthanide. The distribution of the lanthanide ions in the first step was about 81% Sm^{3+} and 19% La^{3+}, reflecting the expected trends from the DTPA-Ln^{3+} stability constants. A subsequent treatment of the collected nanoparticles with diluted acid (pH 2) led to a complete release of the lanthanide ions. After this step, the second cycle was performed, now starting from the 81/19 (Sm^{3+}/La^{3+}) ratio. The analysis of the collected superparamagnetic nanoparticles from the second cycle revealed undetectable amounts of La^{3+} ions, using the X-ray fluorescence instrument. This result is illustrated in Figure 22. Binary mixtures of La^{3+}/Sm^{3+} and La^{3+}/Gd^{3+} ions have also been easily separated after two cycles.

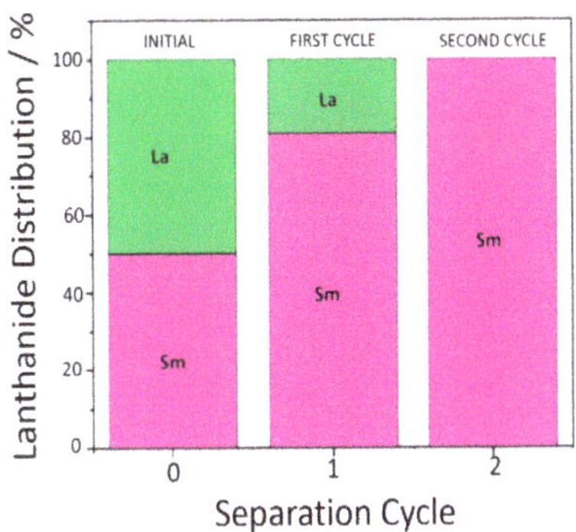

Figure 22. Fractionation of a binary mixture, starting from an equimolar solution (stage 0) of La^{3+} and Sm^{3+} ions, and treating with superparamagnetic nanoparticles functionalized with DTPA, as described in the text (adapted with permission from [44]).

6.2. Ternary Mixtures

Starting with a ternary mixture comprising equimolar amounts of La^{3+}, Nd^{3+} and Gd^{3+} ions (10 mL, 0.10 mol L^{-1}), 10 mg of NP@DTPA was added (pH 6) and kept under gently stirring at room temperature for 6 h. The starting proportion was 33:33:33 (%), respectively. After collecting the iron oxide nanoparticles, the following proportion 13:47:40 (%) of the three elements was obtained, respectively. After releasing the lanthanides at pH 2, the capture process was repeated for the same lanthanide proportion in solution, at pH 6.0. The analytical data collected after the second cycle revealed 75% Gd^{3+} over 25% Nd^{3+}, with no evidence of La^{3+} ions. In the third sequential cycle, the contents of Gd^{3+} ions were practically 100%, as illustrated in Figure 23.

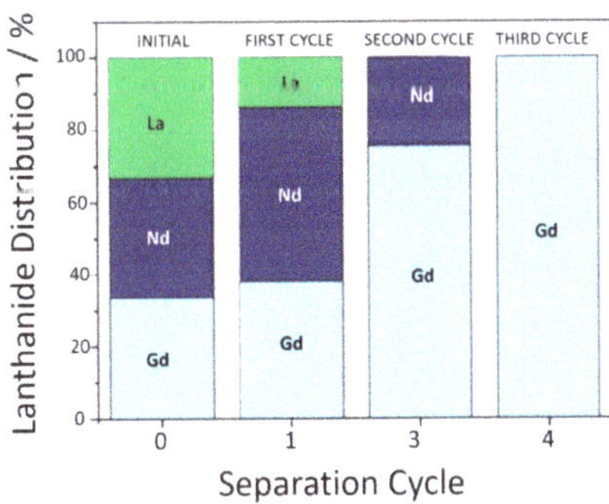

Figure 23. Fractionation of a ternary mixture, starting from equimolar amounts (stage 0) of Gd^{3+}, Nd^{3+} and La^{3+} ions, by using superparamagnetic nanoparticles functionalized with DTPA (adapted with permission from [44]).

6.3. Lanthanide Separation from Monazite

After performing the separations of binary and ternary mixtures, the MNHM process was applied to a lanthanide concentrate from monazite ore. The Ce and Th species present in the mineral were previously removed by precipitation, yielding a lanthanide concentrate containing approximately 45% La^{3+}, 36% Nd^{3+}, 9% Pr^{3+}, 6% Sm^{3+}, and 3% Gd^{3+} ions. The original sample was supplied by the Solvay Laboratory in Brazil, about 30 years ago, when our group was involved in a lanthanide separation project. Because of the ageing, this concentrate was also replicated in our laboratory, using analytical grade compounds. By starting from this mixture, at pH 6, the first cycle of capture exhibited a content of 36% Nd^{3+}, 6% Pr^{3+}, 30% Sm^{3+}, and 28% Gd^{3+} ions. In the second cycle, the following distribution was obtained: 8% Nd^{3+}, 50% Sm^{3+} and 42% Gd^{3+} ions. In the third cycle, only 46% Sm and 54% Gd^{3+} was observed. After the fourth cycle, the proportion became 26% Sm^{3+} and 74% Gd^{3+}. In the fifth cycle, Gd^{3+} ions represented practically 100% of the sample as shown in Figure 24.

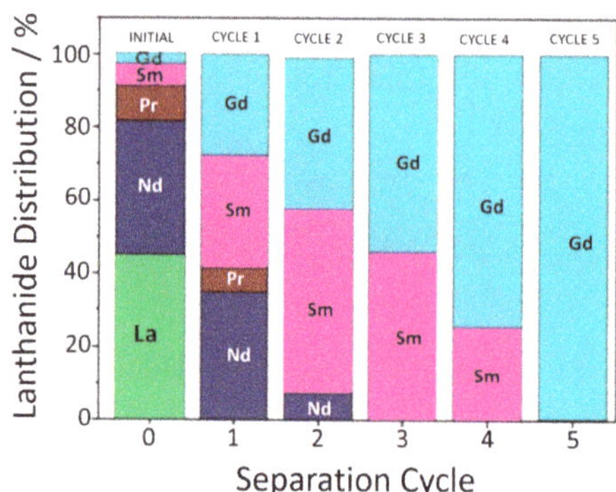

Figure 24. Sequential processing of a monazite concentrate by magnetic nanohydrometallurgy using NP@DTPA nanoparticles in aqueous solution (adapted with permission from [44]).

7. Conclusions

Magnetic nanohydrometallurgy employing iron oxide nanoparticles functionalized with ethylenediaminepropylsilane and the diethylenediaminepentaacetic acid agents can effectively capture lanthanide ions from aqueous solution, allowing their magnetic extraction at pH 6 and release below pH 2, respectively. Preferential binding of heavy lanthanides has been observed and is dictated by their affinity constants reflecting the lanthanide contraction series. For a binary mixture, a high separation efficiency (>99%) can be achieved after two or three successive complexation and release stages. As the number of components increases, more repetitive extraction cycles become necessary. Therefore, MNHM allows the separation of lanthanide elements using engineered nanoparticles, by performing under green and sustainable conditions, i.e., at room temperature, in aqueous solution, and mild pHs, dispensing the use of classical physical separation processes and organic solvents. It can be easily automated, allowing the use of small reactors. The technology can enable urban mining applications, for instance in the neodymium and dysprosium extraction from the magnet components of the electronic wastes.

Author Contributions: Methodology, investigation and writing, F.M.d.M.; conceptualization, investigation and methodology, S.N.A.; conceptualization, supervision and writing, H.E.T. All authors have read and agreed to the published version of the manuscript.

Funding: Fundação de Amparo à Pesquisa do Estado de São Paulo (FAPESP 2018/21489-1) and Conselho Nacional de Desenvolvimento Científico e Tecnológico (CNPq 405301/2013-8).

Acknowledgments: The authors are in debit to Alceu Totti Silveira for his help in X-ray fluorescence measurements and Ulisses Condomitti for this important contributions to the development of the MNHM technology.

Conflicts of Interest: The authors declare no conflict of interest.

References

1. Edelmann, F. Lanthanides and actinides: Annual survey of their organometallic chemistry covering the year. *Coord. Chem. Rev.* **2009**, *253*, 2515–2587. [CrossRef]
2. Wang, Q.; Nono, K.N.; Syrjanpaa, M.; Charbonniere, L.J.; Hovinen, J.; Harma, H. Stable and Highly Fluorescent Europium(III) Chelates for Time-Resolved Immunoassays. *Inorg. Chem.* **2013**, *52*, 8461–8466. [CrossRef] [PubMed]

3. Pihlasalo, S.; Kulmala, A.; Rozwandowicz-Jansen, A.; Hanninen, P.; Harma, H. Sensitive Luminometric Method for Protein Quantification in Bacterial Cell Lysate Based on Particle Adsorption and Dissociation of Chelated Europium. *Anal. Chem.* **2012**, *84*, 1386–1393. [CrossRef] [PubMed]
4. Teo, R.D.; Termini, J.; Gray, H.B. Lanthanides: Applications in Cancer Diagnosis and Therapy. *J. Med. Chem.* **2016**, *59*, 6012–6024. [CrossRef] [PubMed]
5. Shokrollahi, H. Contrast agents for MRI. *Mater. Sci. Eng. C. Mater. Biol. Appl.* **2013**, *33*, 4485–4497. [CrossRef] [PubMed]
6. Coey, J.M.D. Hard Magnetic Materials: A Perspective. *IEEE Trans. Magn.* **2011**, *47*, 4671–4681. [CrossRef]
7. Silva Junior, A.F.; de Campos, M.F. Relevancew of Rare earth for the energy sector. *Holos* **2016**, *32*, 350–363.
8. Silva, G.A.; Petter, C.O.; Albuquerque, N.R. Factors and competitiveness analysis in rare earth mining, new methodology: Case study from Brazil. *Heliyon* **2018**, *4*, e00570. [CrossRef]
9. de Sousa Filho, P.C.; Galaco, A.R.B.S.; Serra, O.A. Rare Earths: Periodic table, discovery, exploration in Bazil and applications. *Quim. Nova* **2019**, *42*, 1208–1224.
10. De Jong, N.; Draye, M.; Favre-Réguillon, A.; LeBuzit, G.; Cote, G.; Foos, J. Lanthanum(III) and gadolinium(III) separation by cloud point extractionNo Title. *J. Colloid Interface Sci.* **2005**, *291*, 303–306. [CrossRef]
11. Hidayah, N.N.; Abidin, S.Z. The evolution of mineral processing in extraction of rare earth elements using solid-liquid extraction over liquid-liquid extraction: A review. *Miner. Eng.* **2017**, *112*, 103–113. [CrossRef]
12. Boubals, N.; Wagner, C.; Dumas, T.; Chaneac, L.; Manie, G.; Kaufholz, P.; Marie, C.; Panak, P.J.; Modolo, G.; Geist, A.; et al. Complexation of Actinide(III) and Lanthanide(III) with H(4)TPAEN for a Separation of Americium from Curium and Lanthanides. *Inorg. Chem.* **2017**, *56*, 7861–7869. [CrossRef] [PubMed]
13. Gras, M.; Papaiconomou, N.; Chainet, E.; Tedjar, F.; Billard, I. Separation of cerium(III) from lanthanum(III), neodymium(III) and praseodymium(III) by oxidation and liquid-liquid extraction using ionic liquids. *Sep. Purif. Technol.* **2017**, *178*, 169–177. [CrossRef]
14. Yamin, D.; Xiangguang, G.; Yanliang, W.; Zeyuan, Z.; Chao, H.; Xiaoqi, S. A separation processing for industrial rare earth feed solution by phosphonium ionic liquid type saponification strategy. *J. Rare Earths* **2017**, *35*, 290–299.
15. Sun, T.; Xu, C.; Chen, J.; Duan, W. Formation of W/O Microemulsions in the Extraction of the Lanthanide Series by Purified Cyanex 301. *Solvent Extr. Exch.* **2017**, *35*, 199–209. [CrossRef]
16. Larsson, K.; Cullen, T.D.; Mezyk, S.P.; McDowell, R.G.; Martin, L.R. Complications in complexation kinetics for lanthanides with DTPA using dye probe molecules in aqueous solution. *Rsc Adv.* **2017**, *7*, 26507–26512. [CrossRef]
17. Condomitti, U.; Zuin, A.; Novak, M.A.; Araki, K.; Toma, H.E. Magnetic coupled electrochemistry: Exploring the use of superparamagnetic nanoparticles for capturing, transporting and concentrating trace amounts of analytes. *Electrochem. Commun.* **2011**, *13*, 72–74. [CrossRef]
18. Condomitti, U.; Silveira, A.T.; Condomitti, G.W.; Toma, S.H.; Araki, K.; Toma, H.E. Silver recovery using electrochemically active magnetite coated carbon particles. *Hydrometallurgy* **2014**, *147–148*. [CrossRef]
19. Condomitti, U.; Almeida, S.N.; Silveira, A.T.; De Melo, F.M.; Toma, H.E. Green processing of strategic elements based on magnetic nanohydrometallurgy. *J. Braz. Chem. Soc.* **2018**, *29*, 948–959. [CrossRef]
20. Toma, H.E. Magnetic nanohydrometallurgy: A nanotechnological approach to elemental sustainability. *Green Chem.* **2015**, *17*, 2027–2041. [CrossRef]
21. Condomitti, U.; Zuin, A.; Silveira, A.T.; Araki, K.; Toma, H.E. Magnetic nanohydrometallurgy: A promising nanotechnological approach for metal production and recovery using functionalized superparamagnetic nanoparticles. *Hydrometallurgy* **2012**, *125*, 148–151. [CrossRef]
22. Mathew, D.S.; Juang, R.S. An overview of the structure and magnetism of spinel ferrite nanoparticles and their synthesis in microemulsions. *Chem. Eng. J.* **2007**, *129*, 51–65. [CrossRef]
23. da Silva, D.G.; Toma, S.H.; de Melo, F.M.; Carvalho, L.V.C.; Magalhaes, A.; Sabadini, E.; dos Santos, A.D.; Araki, K.; Toma, H.E. Direct synthesis of magnetite nanoparticles from iron(II) carboxymethylcellulose and their performance as NMR contrast agents. *J. Magn. Magn. Mater.* **2016**, *397*, 28–32. [CrossRef]
24. Yamaura, M.; Toma, H.E.; Camilo, R.L.; Sampaio, L.C.; Macedo, M.A.; Nakamura, M. Preparation and characterization of (3-aminopropyl) triethoxysilane-coated magnetite nanoparticles. *J. Magn. Magn. Mater.* **2004**, *279*, 210–217. [CrossRef]
25. Park, J.; An, K.; Hwang, Y.; Park, J.-G.; Noh, H.-J.; Kim, J.-Y.; Park, J.-H.; Hwang, N.-M.; Hyeon, T. Ultra-large-scale syntheses of monodisperse nanocrystals. *Nat. Mater.* **2004**, *3*, 891–895. [CrossRef] [PubMed]

26. de Melo, F.M.; Grasseschi, D.; Brandão, B.B.N.S.; Fu, Y.; Toma, H.E. Superparamagnetic Maghemite-Based CdTe Quantum Dots as Efficient Hybrid Nanoprobes for Water-Bath Magnetic Particle Inspection. *Acs Appl. Nano Mater.* **2018**, *1*, 2858–2868. [CrossRef]
27. Melo, F.M.; Silveira, A.T.; Quartarolli, L.F.; Kaid, F.F.; Cornejo, D.R.; Toma, H.E. Magnetic behavior of superparamagnetic nanoparticles containing chelated transition metal ions. *J. Magn. Magn. Mater.* **2019**, *487*, 165324. [CrossRef]
28. Stober, W.; Fink, A.; Bohn, E. Controlled grow of monodisperse silica spheresin micron size range. *J. Colloid Interface Sci.* **1968**, *26*, 62. [CrossRef]
29. Almeida, S.D.N.; Toma, H.E. Neodymium(III) and lanthanum(III) separation by magnetic nanohydrometallurgy using DTPA functionalized magnetite nanoparticles. *Hydrometallurgy* **2016**, *161*, 22–28. [CrossRef]
30. Bhattacharjee, S. DLS and zeta potential—What they are and what they are not? *J. Control. Release* **2016**, *235*, 337–351. [CrossRef]
31. Jesionowski, T.; Ciesielczyk, F.; Krysztafkiewicz, A. Influence of selected alkoxysilanes on dispersive properties and surface chemistry of spherical silica precipitated in emulsion media. *Mater. Chem. Phys.* **2010**, *119*, 65–74. [CrossRef]
32. Zhang, H.; McDowell, R.G.; Martin, L.R.; Qiang, Y. Selective Extraction of Heavy and Light Lanthanides from Aqueous Solution by Advanced Magnetic Nanosorbents. *Acs Appl. Mat. Interfaces* **2016**, *8*, 9523–9531. [CrossRef] [PubMed]
33. Cai, Y.; Yuan, F.; Wang, X.; Sun, Z.; Chen, Y.; Liu, Z.; Wang, X.; Yang, S.; Wang, S. Synthesis of core-shell structured Fe_3O_4@carboxymethyl cellulose magnetic composite for highly efficient removal of Eu(III). *Cellulose* **2017**, *24*, 175–190. [CrossRef]
34. Verwey, E.J.; Overbeek, J.T.G. *The Theory of the Stability of Lyophofic Colloids*; Dover: New York, NY, USA, 2000.
35. Melo, F.M.D.; Almeida, S.D.N.; Uezu, N.S.; Ramirez, C.A.O.; Santos, A.D.D.; Toma, H.E. Extraction of Dysprosium Ions with DTPA Functionalized Superparamagnetic Nanoparticles Probed by Energy Annular Dark Field Imaging. *J. Nanosci. Nanotechnol.* **2017**, *17*, 1–5.
36. Leong, S.S.; Ahmad, Z.; Lim, J. Magnetophoresis of superparamagnetic nanoparticles at low field gradient: Hydrodynamic effect. *Soft Matter* **2015**, *11*, 6968–6980. [CrossRef] [PubMed]
37. Benelmekki, M.; Martinez, L.M.; Andreu, J.S.; Camacho, J.; Faraudo, J. Magnetophoresis of colloidal particles in a dispersion of superparamagnetic nanoparticles: Theory and experiments. *Soft Matter* **2012**, *8*, 6039. [CrossRef]
38. Andreu, J.S.; Camacho, J.; Faraudo, J.; Benelmekki, M.; Rebollo, C.; Martínez, L.M. Simple analytical model for the magnetophoretic separation of superparamagnetic dispersions in a uniform magnetic gradient. *Phys. Rev. E-Stat. Nonlinearsoft Matter Phys.* **2011**, *84*, 1–8. [CrossRef]
39. Faraudo, J.; Andreu, J.S.; Camacho, J. Understanding diluted dispersions of superparamagnetic particles under strong magnetic fields: A review of concepts, theory and simulations. *Soft Matter* **2013**, *9*, 6654. [CrossRef]
40. Friedman, G.; Yellen, B. Magnetic separation, manipulation and assembly of solid phase in fluids. *Curr. Opin. Colloid Interface Sci.* **2005**, *10*, 158–166. [CrossRef]
41. Gómez-Pastora, J.; Bringas, E.; Ortiz, I. Recent progress and future challenges on the use of high performance magnetic nano-adsorbents in environmental applications. *Chem. Eng. J.* **2014**, *256*, 187–204. [CrossRef]
42. Lim, J.; Lanni, C.; Evarts, E.R.; Lanni, F.; Tilton, R.D.; Majetich, S.A. Magnetophoresis of nanoparticles. *Acs Nano* **2011**, *5*, 217–226. [CrossRef] [PubMed]
43. De Melo, F.M.; Almeida, S.D.N.; dos Santos, A.D.; Toma, H.E. Magnetophoresis of Superparamagnetic Nanoparticles Applied to the Extraction of Lanthanide Ions in the Presence of Magnetic Field. *Nanoworld J.* **2017**, *3*, 38–43. [CrossRef]
44. Almeida, S.D.N.; Toma, H.E. Lanthanide ion processing from monazite based on magnetic nanohydrometallurgy. *Hydrometallurgy* **2019**, *189*, 105138. [CrossRef]

Review

A Comprehensive Review of Rare Earth Elements Recovery from Coal-Related Materials

Wencai Zhang [1,*], Aaron Noble [1], Xinbo Yang [2] and Rick Honaker [2]

1 Department of Mining and Minerals Engineering, Virginia Polytechnic Institute and State University, Blacksburg, VA 24061, USA; aaron.noble@vt.edu

2 Department of Mining Engineering, University of Kentucky, Lexington, KY 40506, USA; xinbo.yang@uky.edu (X.Y.); rick.honaker@uky.edu (R.H.)

* Correspondence: wencaizhang@vt.edu

Received: 22 April 2020; Accepted: 14 May 2020; Published: 17 May 2020

Abstract: Many studies have been published in recent years focusing on the recovery of rare earth elements (REEs) from coal-related materials, including coal, coal refuse, coal mine drainage, and coal combustion byproducts particularly fly ash. The scientific basis and technology development have been supported by coal geologists and extractive metallurgists, and through these efforts, the concept has progressed from feasibility assessment to pilot-scale production over the last five years. Physical beneficiation, acid leaching, ion-exchange leaching, bio-leaching, thermal treatment, alkali treatment, solvent extraction, and other recovery technologies have been evaluated with varying degrees of success depending on the feedstock properties. In general, physical beneficiation can be a suitable low-cost option for preliminary upgrading; however, most studies showed exceedingly low recovery values unless ultrafine grinding was first performed. This finding is largely attributed to the combination of small RE-bearing mineral particle size and complex REE mineralogy in coal-based resources. Alternatively, direct chemical extraction by acid was able to produce moderate recovery values, and the inclusion of leaching additives, alkaline pretreatment, and/or thermal pretreatment considerably improved the process performance. The studies reviewed in this article revealed two major pilot plants where these processes have been successfully deployed along with suitable solution purification technologies to continuously produce high-grade mixed rare earth products (as high as +95%) from coal-based resources. This article presents a systematic review of the recovery methods, testing outcomes, and separation mechanisms that are involved in REE extraction from coal-related materials. The most recent findings regarding the modes of occurrence of REEs in coal-related materials are also included.

Keywords: rare earth elements; recovery; coal; acid mine drainage; coal combustion byproducts

1. Introduction

Rare earth elements (REEs) including the 15 lanthanides plus yttrium and scandium have been identified as critical commodities by several international agencies and national governments due to their crucial roles in clean energy, high tech, and national defense industries [1–3]. REEs scarcely form natural ore depositsthat are economically recoverable. As a result, only a few commercial deposits are currently being extracted worldwide. This combination of restricted supply and critical need has prompted many public and private entities to evaluate alternative REE sources. Coal-related materials, including coal refuse, coal fly ash, and coal mine drainage have been identified as a potentially promising resource. The average REE contents in lignite and bituminous coals as well as lignite and bituminous coal ashes worldwide have been estimated to be around 69, 72, 378, and 469 ppm, respectively [4]. Many coal deposits with elevated REE contents have been found and reported in the literature [5–8]. REE concentration in coal mine drainage varies from site-to-site, ranging from several ppb to ppm levels [9,10].

Recovering REEs from coal-related materials has several advantages over commercial rare earth ores: (1) coal-related materials contain more heavy and critical REEs (HREEs and CREEs) relative to the light REEs (LREEs) [5,10]; (2) the mining costs are negligible since REEs can be produced as byproducts from the coal production and utilization processes [11]; (3) coal-based materials, particularly coal mine drainage, tend to have lower concentrations of radionuclide (e.g., U and Th) when compared to traditional ore deposits [12]; and (4) REE recovery from coal materials has the potential to mitigate or eliminate legacy environmental issues [13]. The concept has been discussed in several review articles [5,14–16]; however, these articles have primarily addressed geologic considerations, such as depositional settings and resource characteristics.

To this end, the U.S. Department of Energy initiated research and development efforts in 2014 to evaluate the technical and economicviability of extracting REEs from coal-related materials with a particular focus on technology maturation and process development [17]. During the last five years, the extraction of REEs from coal-related materials has progressed through feasibility assessment, field sampling and characterization, bench-scale REE separation, and pilot-scale REE production stages. During this period, many innovative findings concerning the recovery of REEs from coal-related materials have been published, which significantly contributed to the scientific knowledge in this area [11,12,18–26]. A review article focused on promising methods for REE recovery from coal and coal byproducts was published by the authors of [27] in 2015. However, experimental results and findings were limited at that time. Considerable developments in this area since 2015 led to a need for another review in order to cover the latest findings.

This review article summarizes technical information regarding process flowsheets, metallurgical performance, and economical metrics of the various methods that have been developed for REE extraction from coal-related materials. This review will provide comparative data to prompt further study and analysis in this area. The review is divided into three sections based on the types of the coal-related materials that were investigated as reported in the literature, i.e., coal and coal refuse, coal combustion ash, and acid mine drainage. Recent findings regarding the modes of occurrence of REEs in coal-related materials and how these findings inform process design are also covered.

2. REEs in Coal and Coal Refuse

The geological aspects such as accumulation mechanisms and modes of occurrence of REEs in coal and coal refuse have been extensively investigated by coal geologists, and these topics have been systematically reviewed in several prior publications [5,14,15,28,29]. Therefore, this article only focuses on the recovery aspect of REEs from coal and coal refuse.

2.1. Physical Beneficiation of REEs from Coal and Coal Refuse

Several studies have been performed to concentrate rare earth (RE)-bearing mineral particles from coal and coal refuse using physical beneficiation techniques such as gravity, magnetic, and flotation separations [11,20,30–34]. Table 1 summarizes some of the beneficiation performances. It is worth noting that decarbonization is normally conducted prior to REE beneficiation, which enables the production of clean coal. A systematic study was performed by Honaker et al. [35] to evaluate the viability of using physical separation methods (riffle table, multi-gravity, and wet high intensity magnetic separations as well as froth flotation) to concentrate rare earth minerals from coal and coal refuse, and representative test results are shown in Table 1. With respect to gravity-based separations, the REEs were only enriched by a factor of 1.1 using a riffle table, and the separation performance was still unsatisfactory when using a multi-gravity separator. This finding has been corroborated by a separate study performed by another group of researchers [24]. Physical separations (size, density, and magnetic) performed on a clean coal sample collected from Kentucky, USA showed that a maximum enrichment ratio of 1.21 was obtained. As such, a high degree of enrichment of rare earth minerals from coal and coal refuse is difficult using physical separation methods.

Table 1. A summary of physical beneficiation of rare earth minerals from coal and coal refuse.

Separation Method	Sample	Sources	Separation Method	REE (ppm)	*ER*	*Re* (%)	Reference
Gravity Separation	Coarse refuse (28 × 100 mesh fraction)	Fire Clay	Riffle table	252 [w]	1.1	16.8	[35]
	Coarse refuse (28 × 100 mesh fraction)	Eagle Seam	Riffle table	213 [w]	1.1	16.1	
	Coarse refuse (28 × 100 mesh fraction)	Fire Clay Rider	Riffle table	234 [w]	1.1	24.75	
	Coarse refuse (<100 mesh fraction)	Eagle Seam	Multi-gravity separation	257 [a]	1.2	90	
	Coarse refuse (<100 mesh fraction)	Fire Clay	Multi-gravity separation	290 [a]	1.2	85	
	Coarse refuse (<100 mesh fraction)	Fire Clay Rider	Multi-gravity separation	254 [a]	1.1	87	
Flotation	Decarbonized thickener underflow	Fire Clay	Multi-stage flotation using a conventional cell with sodium oleate as the collector	2300 [a]	5.3	<5	[33]
	Decarbonized thickener underflow	Fire Clay	Multi-stage flotation using a column with sodium oleate as the collector	4700 [a]	10.9	<5	
	Decarbonized thickener underflow	Fire Clay	Single-stage conventional cell flotation using oleic acid as the collector	386 [w]	1.4	23	[35]
	Decarbonized thickener underflow	Eagle Seam	Single-stage flotation using a conventional cell with oleic acid as the collector	367 [w]	1.8	31	
	Decarbonized thickener underflow	Fire Clay Rider	Single-stage conventional cell flotation using oleic acid as the collector	377 [w]	1.3	13	
HHS	Decarbonized thickener underflow	Fire Clay	Potassium octylhydroxamate and sorbitan monooleate were used as the hydrophobizing agent	17,428 [a]	53	5.9	[11]
Magnetic Separation	Decarbonized middling	Fire Clay	Three-stage wet high intensity magnetic separation (1.4 T)	7000 [w]	14	<5	[20]

Note: [w] and [a] represent dry whole sample basis and dry ash basis, respectively; *ER* and *Re* represent enrichment ratio and recovery, respectively.

Froth flotation is normally utilized to treat fine particles based on their varying degrees of surface hydrophobicity. Rare earth mineral particles present in decarbonized thickener underflows of coal preparation plants can be selectivily recovered by froth flotation. As shown in Table 1, concentrates containing 2300 and 4700 ppm of REEs were obtained by using multiple treatment stages involving conventional flotation cells and column flotation, respectively. A limited concentration was also obtained when using a single-stage of flotation. For example, REEs in the decarbonized thickener underflow derived from treating Eagle seam coal was concentated by nearly two times with the flotation product containing 367 ppm of REEs on a dry whole sample basis (see Table 1). Oleic acid was used as the collector for the flotation test results shown in Table 1, which has been widely used to recover rare earth minerals (e.g., monazite and xenotime) from heavy mineral sands [36,37].

As an alternative to flotation, the hydrophobic–hydrophilic separation (HHS) process is a novel ultrafine particle concentrator that provides improved recovery of micron-size material while also providing a dewatered product [38]. The process uses hydrocarbon oils to agglomerate hydrophobic particles and recover the particles through a phase separation. As shown in Table 1, a concentrate containing 17,428 ppm of REEs was obtained from the Fire Clay decarbonized thickener underflow using HHS. Octylhydroxamate was used as the collector, which is another commonly used reagent for rare earth mineral beneficiation [39,40]. The enrichment ratio of the HHS test reached as high as 53:1. In addition to the aforementioned gravity separation and flotation studies, a concentrate containing 7,000 ppm of REEs was also obtained from the decarbonzied Fire Clay middlings using high-intensity magnetic separation [20].

Overall, rare earth minerals in the decarbonized materials can be concentrated by using flotation, HHS, and magnetic separations. However, the recovery values obtained using these technologies are often too low to be economically viable. One explanation for this consistent finding is that RE-bearing particles in coal and coal refuse are extremely fine and often interlocked within the host-particles. SEM characterizations showed that RE-enriched particles occurring in coal refuse and middlings normally have a particle size of <10 μm [20,33]. In this case, extensive grinding is required to liberate the encapsulated rare earth particles, thereby making the operating process cost prohibitive [23]. However, physical separations without significant size reduction can be employed to generate a higher-grade feedstock for downstream recovery and purification processes.

2.2. Chemical Extraction

Given the low recovery and subsequent high production costs associated with physical beneficiation, direct chemical extraction of REEs from coal refuse has been evaluated by a number of researchers [11,20,21,25,41–43]. Moreover, thermal and alkaline treatments prior to acid leaching have also been used to improve the REE leachability [21,44–47]. These technologies have included acid leaching, salt/ion exchange leaching, and leaching with pretreatment, which are reviewed in the following sections.

2.2.1. Salt and Acid Leaching

In the commerical production of REEs from ion-adsorbed clays, ammonium sulfate is commonly used as a salt lixiviant due to the relatively low hydration energy of ammonium ions. Rozelle et al. [48] collected two high-ash content samples from the overlying strata of the Upper Kittanning bed. The samples were crushed and screened to obtain the 595 μm × 150 μm fraction for salt leaching tests. It was found that around 80% of the total REEs were extracted from the solid, which provided promise as an economically viable option. However, in subsequent investigations performed by other researchers, only a small fraction (e.g., 10%) of the total REEs were determined to be ion-exchangeable in the components of the other coal deposits [41,49]. Given these disparate findings, successful extraction of REEs from coal and coal refuse using salt leaching largely depends on the nature of coal deposit.

REEs can be efficiently extracted using acid solutions of relatively low concentration from some coal sources. Laudal et al. successfully extracted nearly 90% of total REEs from a lignite using 0.5 M H_2SO_4 [25]. The high recovery was explained by the fact that the REEs in the lignite are primarily complexed with organic acids, and the complexation can be destroyed under mild acidic conditions. For bituminous coals, sytematic acid leaching studies have been condcuted on decarbonized middlings, fine refuse, and coarse refuse produced from the treatment of coals originating from different seams [11,20,43,50,51]. It was found that REEs in the mineral matter of coal middlings obtained by grinding and flotation were more leachable than those in coal refuse. For example, nearly 83% of REEs were leached from the decarbonized Fire Clay middlings using a nitric acid solution of pH 0 at 75 °C, whereas less than 30% of REEs were extracted under the same conditions from the decarbonized thickener underflow [11]. However, liberation of the mineral matter from coal middlings consumes a significant amount of energy, which requires consideration for determining economic viability [23].

Improvements in the REE leaching efficiency from coal refuse have been achieved by optimizing the operation parameters such as particle size, temperature, and leaching duration. REE recovery values were relatively low (<30%) for most of the investigated sources [11]. A summary of salt and acid leaching of REEs from coal and coal refuse is shown in Table 2. Overall, direct leaching with salt or acid failed to provide satisfactory recovery from high-rank coal and coal refuse, whereas, direct leaching is a promising choice for recovering REEs from low-rank coal. Therefore, thermal and/or alkaline pretreatment has been utilized in subsequent studies to improve the leaching performance.

Table 2. A summary of salt and acid leaching of rare earth elements (REEs) from coal and coal refuse.

Sample	Coal Seam	Extraction Condition	Leaching Recovery	Reference
Decarbonized thickener underflow	West Kentucky No. 13	0.1 M $(NH_4)_2SO_4$, pH 5	Around 10% of total REEs, 7% of LREEs, and 18% of HREEs	[41]
		0.1 M $(NH_4)_2SO_4$, pH 3	Around 12% of total REEs, 10% of LREEs, and 21% of HREEs	
Roof material, 595 μm × 150 μm	Upper Kittanning	1 M $(NH_4)_2SO_4$, 1/2 solid/solution mass ratio, room temperature	Nearly 90% of the total REEs were extracted after 1 h of reaction	[48]
Lignite	Fort Union	0.5 M H_2SO_4, 40 °C, 48 h	Nearly 90% of the total REEs	[25]
Decarbonized middlings	Fire Clay	Nitric acid solution of pH 0 at 75 °C	83% of total REEs, 86% of LREEs, and 69% of HREEs	[11]
	West Kentucky No. 13	Nitric acid solution of pH 0 at 75 °C	15% of La, 21% of Ce, 31% of Nd, 45% of Y	
	Lower Kittanning	Nitric acid solution of pH 0 at 75 °C	41% of total REEs	
Decarbonized thickener underflow	Fire Clay	Nitric acid solution of pH 0 at 75 °C	31% of La, 26% of Ce, 40% of Nd, 36% of Y	
	West Kentucky No. 13	Nitric acid solution of pH 0 at 75 °C	6% of La, 5% of Ce, 16% of Nd, 34% of Y	
	Lower Kittanning	Nitric acid solution of pH 0 at 75 °C	2% of La, 5% of Ce, 8% of Nd, 25% of Y	

2.2.2. Alkali and Thermal Pretreatment

Alkali and thermal pretreatments of coal and coal refuse have been applied to enhance the acid leaching recoveries of REEs from coal and coal refuse [21,41,44,47,52,53]. A summary of the relevant studies reported in the literature is provided in Table 3. Yang et al. used an 8 M NaOH solution to treat decarbonized fine refuse at 75 °C for two hours prior to acid leaching [41]. It was found that the recovery of REEs was significantly increased from 22% to 75% due to the positive impact on the leachability of the light REEs. Under the same experimental conditions, a small incremental increase in recovery was achieved for the HREEs (38% to 48%). Kuppusamy et al. conducted a study of simultaneous production of clean coal and REEs by alkali-acid leaching of a coal fine refuse material [52]. The ash content of the material was reduced from 46.21% to 14.17% after treatment with a NaOH solution (30 wt.%) at 190 °C for 30 min followed by an HCl solution (7.5 wt.%) at 50 °C for 30 min. Simultaneously, 97% of the LREEs and 76% of the HREEs occurring in the material were extracted.

The enhanced leaching efficiency of REEs resulting from alkali treatment was explained by two mechanisms: (1) Difficult-to-leach rare earth minerals such as monazite were converted to more leachable forms [27,41] and (2) Crystal structures of the dominant minerals such as clays were destroyed resulting in liberation of the encapsulated rare earth minerals [52]. The reactions are as follows:

$$REEPO_{4(s)} + 3NaOH_{(aq)} \rightarrow REE(OH)_{3(s)} + Na_3PO_{4(aq)} \tag{1}$$

$$Al_2O_3{\cdot}2SiO_2{\cdot}2H_2O_{(s)} + 6NaOH_{(aq)} \rightarrow 2Na_2SiO_{3(aq)} + 2NaAlO_{2(aq)} + 5H_2O_{(aq)} \tag{2}$$

A schematic diagram incorporating the use of the alkali-acid leaching process to extract REEs from coal and coal refuse is shown in Figure 1. A negative aspect of the alkali leaching step is the considerable amount of contaminants dissolved into solution along with the REEs, which complicates the downstream purification process. Another negative aspect is the chemical cost, which will likely be prohibitively high given the relatively low content of REEs in coal-based feedstock. Additional studies need to be performed to investigate the possibility of selectively increasing the REE leachability using diluted alkaline solutions and/or weak alkalis.

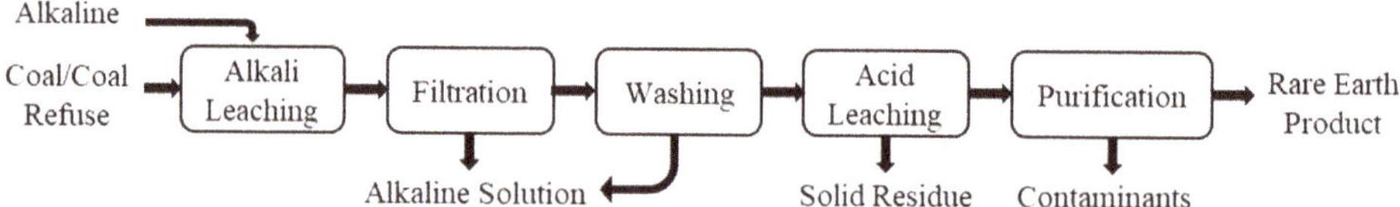

Figure 1. Schematic diagram of the alkaline-acid leaching process to extract REEs from coal and coal refuse [52].

Another scenario to enhance the REE leachability from coal and coal refuse is thermal activation, also known as calcination or roasting [21,41,44,46,47,54]. As reported by Zhang et al., after thermal activation of a coal gangue material at 700 °C for 30 min, 88.6% of the total REEs were leached using 25% HCl at room temperature [46]. The impact of thermal activation on REE leachability has been systematically studied [21,47,54]. As shown in Table 3, for both the clean coal and coal refuse, calcination under 600 °C in static atmosphere without adding any additives significantly improved the REE leaching recovery. For example, the total REE recovery from Pocahontas No. 3 coarse refuse was increased from 14% to 81% by thermal activation using 1.2 M HCl as a lixiviant. In addition, moderate recovery values were also obtained when using mildly acidic conditions. Reducing the acidity by twenty times, i.e., 1.2 M HCl to 0.06 M HCl, resulted in a relatively small drop in REE recovery for thermally activated Pocahontas No. 3 coarse refuse to around 60% [21]. As such, acid consumption is significantly reduced, which is typically the highest cost component of a rare earth extraction process.

Another advantage of thermal activation is that fewer contaminants are dissolved relative to the REEs. During the calcination process, pyrite began to decompose and was converted to iron oxide (primarily hematite) in the temperature range of 400–500 °C. The crystallinity of hematite increased with the elevation in calcination temperature [21,55,56]. When calcined under high temperature, inter-layered structures of clay minerals, especially kaolinite, were destroyed due to dehydration, resulting in disintegration into thin slices and an increase in surface area [21,57,58]. The overall impact of the mineralogical changes was that Al leachability was noticeably increased, while Fe leachability was slightly increased (see Figure 2). However, as shown in Figure 2 the leaching duration can be shortened to less than 15 min, within which the REE leaching reaction is nearly completed while leaching of Al and Fe is minimal. Therefore, a pregnant leaching solution (PLS) containing relatively higher concentrations of REEs can be produced using the thermal activation-mild acid leaching process.

The aforementioned process has been installed and is currently under testing in a rare earth pilot plant funded by the U.S. Department of Energy [59]. A schematic diagram of the process used in the plant is shown in Figure 3. During the testing process, it was found that organic matter associated with the coal refuse helped maintain a constant temperature in the roaster, thus, significantly reducing the energy costs associated with the roaster. In addition to REEs, recent studies [54,60] showed that leaching recovery of other critical metals such as lithium from coal refuse was also positively impacted by calcination. Therefore, given the aforementioned benefits, pre-leach calcination combined with mild acid leaching is one of the most promising approaches for recovering REEs from coal-related materials.

Several mechanisms have been proposed for the positive impacts of thermal activation on REE extraction from coal, i.e.,: (1) Surface area increase resulting from clay dehydration, which liberates some rare earth minerals; (2) Decomposition of the difficult-to-leach rare earth minerals; and (3) Release of a portion of the REEs that were originally associated with the organic matter, which was removed after calcination [21,42,53]. Figure 4 shows the disintegrated kaolinite particles and a REE-enriched particle present in a thermally activated coal refuse sample. Sequential chemical extraction tests have been performed on the calcination products of clean coals of West Kentucky No. 13, Fire Clay, and Illinois No. 6 [47]. It was found that a significant fraction of REEs (50% for the West Kentucky No. 13 material) existed as metal oxides, which originated from the removal of the organic matter present in the clean coals and the decomposition of the difficult-to-leach rare earth minerals.

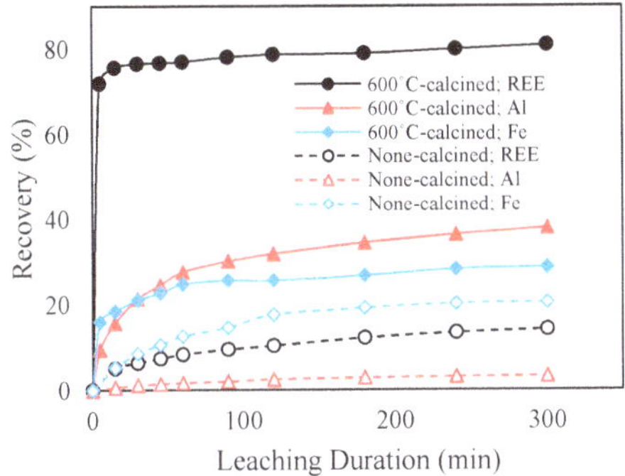

Figure 2. Effects of calcination on the leaching kinetics of REEs, Al, and Fe from Pocahontas No. 3 coarse refuse. (Data were extracted from [21]).

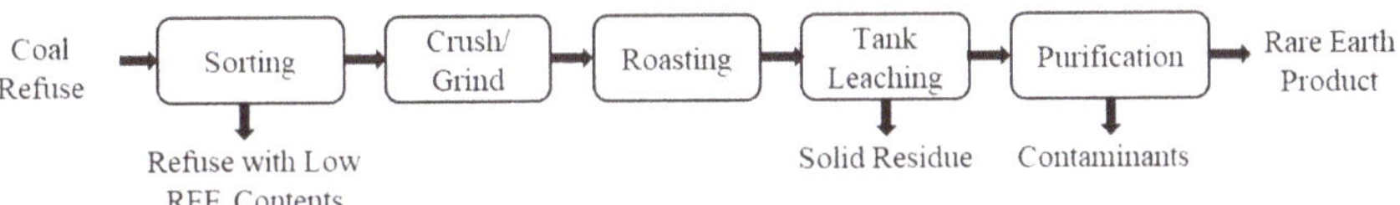

Figure 3. A simplified block diagram of the REE recovery process used in a REE pilot plant located in west Kentucky.

Despite the aforementioned advantages, a few technical and fundamental problems must be resolved to fully validate the approach. As shown in Table 3, the improvements in the HREE recoveries are much smaller than those of the LREEs and elevating the calcination temperature did not enhance the recovery due to the sintering of aluminum silicates [21]. In addition, mineralogical changes of the REEs during calcination is still unclear. Previous studies only listed some possible mechanisms, whereas the direct evidence in supporting the conclusions has not been obtained to date.

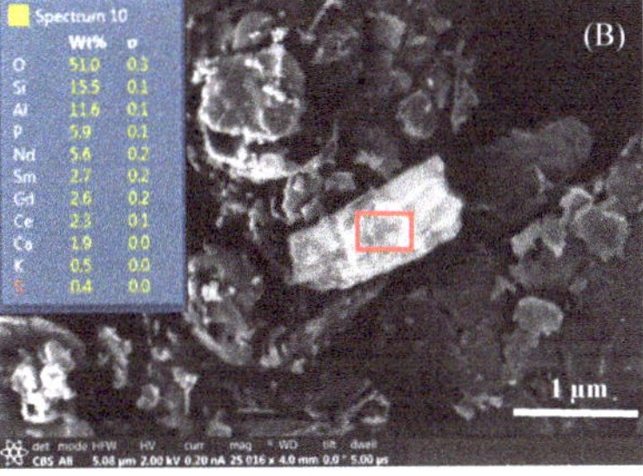

Figure 4. SEM-EDX images of a coal coarse refuse sample after thermal activation: (**A**) Disintegrated kaolinite particles; (**B**) A REE-enriched particle. [21].

Table 3. Summary of the thermal and alkaline pretreatment effects on the leaching recoveries of REEs from coal and coal refuse.

Sample	Coal Seam	Pre-Leach Treatment	Leach Conditions	Recovery			Reference
				TREE	LREE	HREE	
Coarse refuse (2.2 SG float, crushed to below 177 μm)	Pocahontas No. 3	None	1.2 M HCl; 75 °C, 1% (*w/v*) solid concentration, 5 h	14%	12%	23%	[21]
Coarse refuse (2.2 SG float, crushed to below 177 μm)	Pocahontas No. 3	Calcination at 600 °C for 2 h without adding any additives	1.2 M HCl; 75 °C, 1% (*w/v*) solid concentration, 5 h	81%	89%	27%	
Middlings (crushed to below 177 μm)	Pocahontas No. 3	None	1.2 M HCl; 75 °C, 1% (*w/v*) solid concentration, 5 h	28%	31%	19%	
Middlings (crushed to below 177 μm)	Pocahontas No. 3	Calcination at 600 °C for 2 h without adding any additives	1.2 M HCl; 75 °C, 1% (*w/v*) solid concentration, 5 h	76%	80%	57%	
Plant feed (2.2 SG sink, crushed to below 177 μm)	West Kentucky No. 13	None	1.2 M HCl; 75 °C, 1% (*w/v*) solid concentration, 5 h	24%	21%	36%	[54]
Plant feed (2.2 SG sink, crushed to below 177 μm)	West Kentucky No. 13	Calcination at 600 °C for 2 h without adding any additives	1.2 M HCl; 75 °C, 1% (*w/v*) solid concentration, 5 h	79%	87%	41%	
Plant feed (2.2 SG sink, crushed to below 177 μm)	Fire Clay	None	1.2 M HCl; 75 °C, 1% (*w/v*) solid concentration, 5 h	43%	43%	38%	
Plant feed (2.2 SG sink, crushed to below 177 μm)	Fire Clay	Calcination at 600 °C for 2 h without adding any additives	1.2 M HCl; 75 °C, 1% (*w/v*) solid concentration, 5 h	62%	68%	33%	
Plant feed (2.2 SG sink, crushed to below 177 μm)	Illinois No. 6	None	1.2 M HCl; 75 °C, 1% (*w/v*) solid concentration, 5 h	32%	31%	37%	
Plant feed (1.4 SG float, crushed to below 177 μm)	Illinois No. 6	Calcination at 600 °C for 2 h without adding any additives	1.2 M HCl; 75 °C, 1% (*w/v*) solid concentration, 5 h	65%	73%	41%	
Plant feed (1.4 SG float, crushed to below 177 μm)	West Kentucky No. 13	None	1.2 M HCl; 75 °C, 1% (*w/v*) solid concentration, 5 h	25%	30%	15%	[53]
Plant feed (1.4 SG float, crushed to below 177 μm)	West Kentucky No. 13	Calcination at 600 °C for 2 h without adding any additives	1.2 M HCl; 75 °C, 1% (*w/v*) solid concentration, 5 h	86%	88%	82%	
Plant feed (1.4 SG float, crushed to below 177 μm)	Fire Clay	None	1.2 M HCl; 75 °C, 1% (*w/v*) solid concentration, 5 h	41%	47%	20%	
Plant feed (1.4 SG float, crushed to below 177 μm)	Fire Clay	Calcination at 600 °C for 2 h without adding any additives	1.2 M HCl; 75 °C, 1% (*w/v*) solid concentration, 5 h	84%	87%	75%	

Table 3. *Cont.*

Sample	Coal Seam	Pre-Leach Treatment	Leach Conditions	Recovery			Reference
				TREE	LREE	HREE	
Plant feed (1.4 SG float, crushed to below 177 µm)	Illinois No. 6	None	1.2 M HCl; 75 °C, 1% (*w/v*) solid concentration, 5 h	34%	43%	10%	
Plant feed (1.4 SG float, crushed to below 177 µm)	Illinois No. 6	Calcination at 600 °C for 2 h without adding any additives	1.2 M HCl; 75 °C, 1% (*w/v*) solid concentration, 5 h	75%	74%	75%	
Middlings (crushed to below 177 µm)	West Kentucky No. 13	Calcination at 750 °C for 2 h without adding any additives	1.2 M H_2SO_4; 75 °C, 1% (*w/v*) solid concentration, 5 h	41%	81%	40%	[41] Unpublished data
Middlings (crushed to below 177 µm)	West Kentucky No. 13	None	1.2 M H_2SO_4; 75 °C, 1% (*w/v*) solid concentration, 5 h	29%	23%	47%	
Fine refuse	West Kentucky No. 13	Pre-leach using 8 M NaOH solution at a solid/liquid ratio of 1/10 (*w/v*) and 75 °C for 2 h	1.2 M H_2SO_4; 75 °C, 1% (*w/v*) solid concentration, 5 h	75%	82%	48%	
Fine refuse	West Kentucky No. 13	None	1.2 M H_2SO_4; 75 °C, 1% (*w/v*) solid concentration, 5 h	23%	21%	38%	
Flotation Tailings (<500 µm)	East Kootenay	Pre-leach using 30 wt.% NaOH solution at 20% solid concentration and 190 °C for 30 min	7.5 wt.% HCl, 50 °C, 30 min	>85%	97%	76%	[52]
Coal refuse (ground to D_{50} = 3.78 µm)	Junggar coalfield	Calcination at 600 °C for 30 min without adding any additives	25% HCl at 25 °C	88.6%	NA	NA	[46]
Coarse refuse	NA	Calcination at 600 °C for 2 h without adding any additives	6 M HCl, 1/5 solid/liquid ratio, 85–90 °C, 4 h	NA	NA	NA	[44]

3. REE Recovery from Coal Combustion Byproducts

3.1. Modes of Occurrence of REEs in Coal Combustion Ash

Coal combustion fly ash is composed of both amorphous and crystalline phases. The amorphous phases account for 60–90% of bulk fly ash composition, while crystalline material accounts for the remainder [61]. The crystalline phases mainly include quartz, mullite, hematite, magnetite, ferrite spinels, anhydrite, melilite, merwinite, periclase, tricalcium aluminate, and lime [62–64]. Until recently, researchers have found limited success in characterizing the amorphous phases of fly ash due to its disordered nature and heterogeneity [61]. This characteristic along with the low concentration and dispersed nature of REEs have caused difficulties in characterizing the modes of occurrence of REEs in fly ash using traditional approaches. However, many recent studies have addressed this challenge using advanced characterization tools, such as X-ray Absorption Near Edge Structure (XANES), micro-X-Ray Absorption Near Edge Structure (μ-XANES), laser ablation inductively coupled plasma mass spectroscopy (LA-ICP-MS), multimodal image analysis, and sensitive high resolution ion microprobe–reverse geometry SHRIMP-RG [65–70]. In addition, systematic SEM-EDX, TEM-EDX, and sequential chemical extraction (SCE) studies have been performed on coal combustion ashes, which also provided valuable information regarding the REE occurrence modes and potential processing routes [71–78].

Sequential chemical extraction (SCE) tests performed on several class F-type fly ash [79] samples (SiO_2% + Al_2O_3% + Fe_2O_3% > 70%) showed that the majority of REEs were associated with silicates and aluminosilicates (quartz, glass, mullite, zircon, etc.), indicating that REEs are dispersed in the glassy phases and/or associated with the Al–Si–oxide phases [76–78]. Chemical composition analysis of the different size fractions of a class F-type fly ash showed that a strong positive correlation existed between the REEs and the Al plus Si contents [77], which corroborates the above conclusion. In class F fly ashes, REEs associating with carbonates and metal oxides, such as $CaCO_3$ and CaO, accounted for less than 10% of the total REEs, whereas, 50–60% of the total REEs present in class C-type fly ashes produces from Powder River basin coal occurred as carbonates and metal oxides [66,76–78]. Liu et al. combined acid leaching results and solution chemistry modelling findings to predict the percentage of the total REEs that occurred as monazite and hematite in a fly ash sample [66]. It was found that 10–20% of REEs were leached in the pH range of 0–1.5, which corresponds to the range where monazite and hematite dissolve based on solution chemistry modelling. Therefore, 10–20% of the total REEs were reported to exist as monazite and hematite forms. However, many studies have shown that monazite is thermally and chemically stable, and acid cracking or roasting is required to efficiently dissolve monazite [80,81]. Therefore, the solution chemistry modelling findings indicating that monazite and hematite dissolve in solutions having a pH in the range of 0–1.5 is questionable. Furthermore, the conclusion that 10–20% of the total REEs occur as monazite and hematite needs to be re-assessed in further investigations.

The association of REEs with silicates and aluminosilicates has been further proven by SEM-EDX, TEM-EDX, and LA-ICP-MS analyses. Thompson et al. found that, during laser ablation of a REE-enriched fly ash grain, the ion intensities of Al and Si were consistent, whereas the intensities of REEs changed in different ablation periods [70]. This finding indicated that REEs tend to be localized in small grains within fly ash. Using SEM-EDX, some monazite grains within Al–Si cenospheres and aluminosilicate glass particles were found, and particle size of the grains was less than 10 μm [70]. Moreover, several other SEM-EDX studies also showed the dispersion of REE enriched grains within aluminosilicates [66,67,71–73,75,78]. Associations of REEs with the other phases present in fly ash such as iron oxide, zircon, Ca/Fe-rich aluminosilicates, and lime have also been reported [65,66,82]. In addition to associations with the major phases, discrete REE enriched grains such as apatite (Ca, LREE, and P) and monazite (LREE and P) were also found in fly ash [66,70].

Overall, many RE-bearing minerals such as monazite, xenotime, rhabdophane, zircon, ilmenite, lime, and calcite have been found in fly ash, and those minerals have complex association characteristics with the major phases [66,67,70,73,75]. Taggart et al. tested both the bulk and micro speciation of yttrium (Y) in fly ashes using bulk and micro XANES [69]. Bulk XANES analysis indicated that Y coordination states in the fly ashes resembled a combination of Y-oxide, Y-carbonate, and Y-doped glass. However, using micro XANES, some "hotspots" of Y were observed including different mineral forms (e.g., Y-phosphate), which were not observed in the bulk measurements. This result indicated the heterogeneity of REEs in fly ash, and microscale analysis may be unable to represent the REEs in bulk fly ash. Hower et al. reported that the distribution of REEs in a coal ash sample seemed to be in the form of nanoscale crystalline minerals with additional distributions corresponding to overlapping ultra-fine minerals as well as atomic dispersion within the fly ash glass [71]. Therefore, the heterogeneity of REEs in fly ash occurs in nano- and atomic-scales.

3.2. *Physical Beneficiation of REEs from Coal Combustion Ash*

Coal fly ash particles can be separated into different fractions based on their contrasts in physical characteristics such as density, particle size, magnetism, and surface hydrophobicity [19,24,83–87]. Rather than processing the bulk ash material, REE extraction from certain fractions that are relatively more enriched in REEs will make the overall recovery more economically viable. REEs in fly ash are more concentrated in the finer fractions relative to the entire bulk material [19,24,74,83–86]. Size fractionation analyses of REEs in fly ashes collected from power plants of various countries (China, USA, United Kingdom, Poland) showed that REE contents gradually increased with a decrease in particle size [19,74,83,85]. For example, the minus 500 mesh fraction of a fly ash sample that was collected from the Jungar power plant [74] contained 648 ppm of total REEs, which is more than two times higher than the plus 120 mesh fraction (277 ppm). Several explanations have been suggested for this phenomenon: (1) The finer fractions in fly ash usually contain more glass phase relative to the coarser fractions, while REEs are preferentially associated with the glass phase in fly ash [85]; (2) The organic-bound REEs partially volatize and deposit on the fine particles of fly ash [84,88]; and (3) REEs associated with organic matter may form extremely small particles when the organic matter is combusted and such small particles tend to enrich in the finer fractions of fly ash [84].

Fly ash particles can also be partitioned into different fractions using magnetic separation. Dai et al. analyzed the magnetic, non-magnetic, and glassy fractions of a fly ash and discovered that the magnetic fraction contained less REEs relative to the bulk fly ash (202 ppm versus 261 ppm) [84]. Blissett et al. obtained a magnetic fraction containing 270 ppm of rare earth oxide from a fly ash with 505 ppm of rare earth oxide [83]. Lin et al. separated a fly ash into five fractions by using different magnetic field intensities [24]. It was found that REEs were more enriched in the weak- and non-magnetic fractions. For example, the non-magnetic fraction of a fly ash sample contained more than 600 ppm of REEs, whereas the strongest magnetic fraction only contained around 200 ppm of REEs. Therefore, based on these studies, it can be concluded that REEs are preferentially enriched in the non-magnetic fraction of fly ash.

Fractionation of REEs was also observed in gravity and flotation separations [24,27,83]. In one of our prior studies [19], a float-sink test was performed on a fly ash sample at a density cut point of 1.8 specific gravity (SG). It was found that the 1.8 SG sink fraction contained more REEs than the 1.8 SG float fraction (521 ppm versus 376 ppm). Flotation tests using different collectors and collector dosages also produced a series of products with different REE contents varying from 400 to 650 ppm. Lin et al. performed density fractionation tests on two fly ash samples and it was found that maximum REE contents occurred in the medium density fractions (2.71–2.95 SG and 2.45–2.71 SG) [24].

Given these results, physical separation can be used to pre-concentrate REEs from coal combustion ash, thereby providing a higher-grade feed material to the downstream extraction processes (e.g., acid leaching), resulting in a reduction in the overall recovery cost ($/kg of rare earth oxide). In the patents [89,90], magnetic separation was used to produce a feed material for the subsequent acid leaching process. Two physical beneficiation processes are shown in Figure 5. In both circuits, magnetic separation is performed prior to size fractionation, whereas in another study by Pan et al. [87], the sequence is reversed. Therefore, no fixed strategy has been proposed for the physical beneficiation process. The selection of pre-concentration strategy for a specific ash sample should be based on a comprehensive laboratory evaluation of the REE partitioning characteristics.

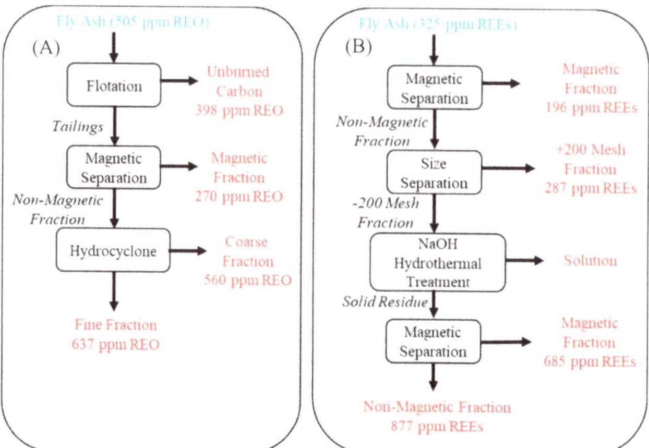

Figure 5. Flowsheet reported in the literature for REE pre-concentration using physical methods: (**A**) reported by [83] and (**B**) reported by [76].

3.3. *Chemical Extraction of REEs from Coal Combustion Ash*

Many studies have been published in recent years focused on the chemical extraction of REEs from coal combustion ashes [26,53,76,88,89,91–101]. As shown in Table 4, acid leaching has been extensively exploited to extract REEs from coal combustion ash, and often the ash materials must be chemically and/or thermally treated prior to acid leaching to achieve better extraction performance. Chemical and physical characteristics of coal combustion ash change significantly relative to the raw materials after pretreatment [26,96,98]. Therefore, the leaching mechanisms are distinct.

3.3.1. Acid Leaching of REEs from Coal Combustion Ashes

Satisfactory recoveries of REEs from some coal combustion ash produced from conventional boilers has been achieved by using acid leaching alone. For example, Taggart et al. extracted more than 70% of the total REEs from fly ashes of the Powder River basin using 15 M HNO_3 at 85–90 °C [102]. Nearly 100% recovery was obtained from the samples using 12 M HCl at 85 °C, and a considerable amount of REEs (71%) were extracted from a fly ash sample of the same source even under much weaker acidity (1 M HCl). Cao et al. conducted a parametric study to optimize the leaching recovery of REEs from a fly ash sample that was collected from a power plant located in Guizhou, China [91]. It was found that 71.9% of La, 66.0% of Ce, and 61.9% of Nd were leached using 3 M HCl at 60 °C. However, due to the fact that most of the REEs are encapsulated in the amorphous structures of fly ash generated by combusting pulverized coal under high temperature (~1400 °C), relatively low recoveries were usually achieved using acid leaching alone. As shown in Table 4, 35–43% and 40–57% of total REEs were extracted from fly ashes of Illinois and Appalachian basins, respectively, using strong acidity at high temperature (12 M HCl at 85 °C; [96]).

Several studies prepared coal ash samples by combusting coals in muffle furnaces, which were used to simulate fly ash and bottom ash produced from pulverized coal-fired boilers [97,103]. Relatively high leaching recoveries were achieved from artificially prepared coal ashes. However, the temperatures used were much lower than the typical temperatures used in conventional pulverized coal boilers. For example, Kumari et al. burned a coal at 450 °C for 8 h and treated the material as coal bottom ash [97]. Acid leaching optimization showed that 90% of Ce and Nd as well as 35% of Y were extracted using 4 M HCl at 90 °C. It has been realized that mineral matter in coal may oxidize, decompose, fuse, disintegrate, or agglomerate under temperatures as high as 1400 °C, and rapid cooling in the post-combustion zone in boilers results in the formation of spherical, amorphous particles [104,105]. However, when combusting in a muffle furnace at a temperature lower than 900 °C, no glassy phases were detected [21,47,53,57]. Therefore, REEs in the laboratory prepared ash material are more readily leached compared with the REEs in ashes produced in pulverized coal-fired boilers.

Leachabilities of REEs from fly ash produced in fluidized bed combustion (FBC) systems have also been evaluated [53,101]. REEs present in FBC ash are more leachable than ashes produced from pulverized coal-fired boilers. Tuan et al. extracted 62.1% Y, 55.5% Nd, and 65.2% Dy from a FBC bottom ash using 2 M HCl at 80 °C [101]. Honaker et al. achieved around 80% of total REE recovery from an FBC bottom ash using 1.2 M HCl at 75 °C [53]. Relatively low burning temperatures (750–900 °C) are typically utilized in FBC units [106]. Therefore, glassy phases are less likely formed during the combustion and cooling processes in FBC, which contribute to the higher REE leaching efficiency. As shown in Figure 6, the FBC ash samples maintained good crystallization and no noticeable glassy phases were found. Furthermore, similar to the thermal activation of REEs present in coal and coal refuse (Section 3.3.2), combustion in a FBC system may enhance the REE leaching efficiency due to dehydration of the clays, decomposition of the hard-to-leach rare earth minerals, and removal of the organic matters [21,47,53].

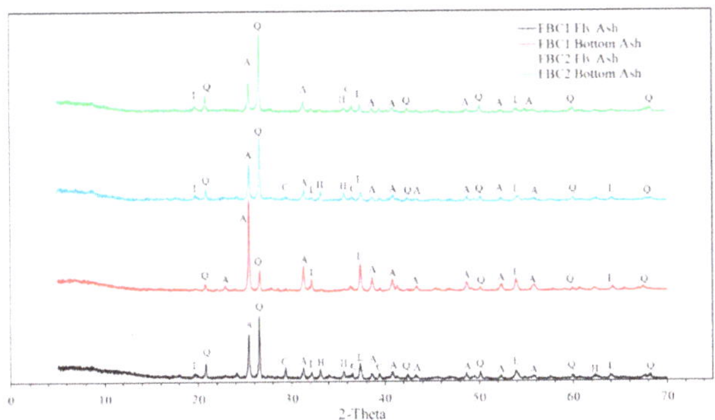

Figure 6. XRD patterns of the fluidized bed combustion (FBC) ash samples: A, anhydrite; C, calcite; H, hematite; I, illite; L, Lime; and Q, Quartz [53].

Parametric studies have been performed to optimize the leaching recovery of REEs from coal combustion ash. Kumari et al. performed leaching tests on a coal ash using three different types of acids, i.e., hydrochloric acid, nitric acid, and sulfuric acid, under the same conditions [97]. It was found that the leaching recoveries follow the order of HCl > HNO_3 > H_2SO_4. However, to the authors' knowledge, sulfuric acid is less volatile and more cost-effective compared with the other two acids. Tuan et al. did not observe a significant difference among the three mineral acids in terms of REE extraction from FBC bottom ashes [101]. Despite the inconsistent findings, hydrochloric acid is the most commonly used lixiviant per literature (see Table 4). Sulfuric acid is expected to provide the worst efficiency since the large amount of calcium present in coal combustion ash will complex with sulfate to form gypsum, and some REEs will be incorporated into the gypsum structure and lost to the precipitate [107]. Leaching recovery of REEs is sensitive to liquid/solid ratio, acid concentration, temperature, and leaching duration [91,97,100]. Leaching recoveries of La, Ce, and Nd from a coal fly ash were nearly doubled when increasing the liquid/solid ratio from 5/1 to 20/1 or prolonging the reaction time from 30 to 180 min [91].

Two flowsheets that were reported in the literature for recovering REEs from coal combustion ash using acid leaching are shown in Figure 7. Solvent extraction was used in both circuits. In the first circuit (Figure 7A), tris-2-ethylhexyl amine (TEHA) was used to complex hydrogen ions in the solvent extraction step, and acid regenerated in the stripping step, which reduced the chemical cost. REEs were recovered from the raffinate of solvent extraction using precipitation. In the second circuit (Figure 7B), fly ash and bed ash produced from FBC combustors are leached using diluted acid (e.g., 1.2 M hydrochloric acid). REEs in the pregnant leach solution are extracted using a three-stage counter-current solvent extraction (SX) system. Finally, a concentrated solution of REEs containing minimal contaminants is produced from the SX-stripping stage. High-purity REE concentrates are produced by selective precipitation using oxalic acid as the precipitant.

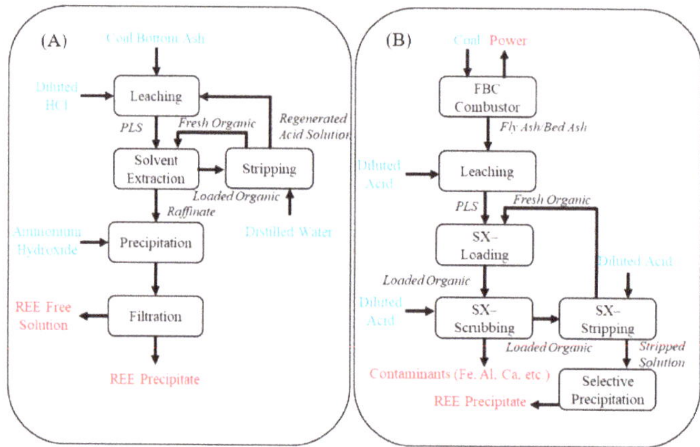

Figure 7. Flowsheets reported in the literature for recovering REEs from coal combustion ash using acid leaching: (**A**) reported by [97] and (**B**) reported by [53].

Overall, REE recovery from coal combustion ashes using direct acid leaching is inefficient. Harsh leaching conditions with concentrated acid solutions at higher temperatures are normally required to achieve satisfactory recovery. In addition to the direct acid leaching, REE extraction from fly ash using bioleaching was also reported [108]. Three microbial strains, *Candida bombicola*, *Phanerochaete chrysosporium*, and *Cryptococcus curvatus*, were tested by Park and Liang in terms of REE extraction from fly ash. *Candida bombicola* provided the optimal results with 63% Sc, 62.2% Y, 67.7% Yb, 64.4% Er, 60% Dy, and 51.9% Gd being extracted at 28 °C for 6 h [108].

3.3.2. Chemical/Thermal Pretreatment

Coal combustion ash was chemically and/or thermally pretreated prior to acid leaching to achieve high REE recoveries [18,26,76,98,100,109]. Lin et al. performed hydrothermal treatment on a coal fly ash and found that 21.3% of the material was dissolved by using 5 M NaOH with a solid/liquid ratio of 1/20 at 100 °C for 120 min [76]. REE content in the solid material was increased from 366 to 803 ppm after hydrothermal treatment, which indicates that REEs present in coal combustion ash remain with the solid residue after the hydrothermal treatment. Wang et al. used an 8 M HCl solution to leach a hydrothermally treated fly ash and achieved a total REE recovery of 88.15% [18]. As shown in Figure 8A, Si, Ga, and Al were also recovered from the fly ash alongside REEs. Ma et al. proposed an alternative NaOH–HCl leaching process, which extracted 55% of REEs, 63% of Si, 72% of Ga, and 78% of Al from the fly ash (see Figure 8B) [98]. Unfortunately, none of the studies produced high-grade final rare earth products.

In the hydrothermal treatment process, NaOH reacts with the major components of fly ash according to the following reactions [110,111]:

$$SiO_2 + 2NaOH \rightarrow Na_2SiO_3 + H_2O \quad (3)$$

$$3Al_2O_3{\cdot}2SiO_2 + 10NaOH \rightarrow 6NaAlO_2 + 2Na_2SiO_3 + 5H_2O \quad (4)$$

$$Al_2O_3 + 2NaOH \rightarrow 2NaAlO_2 + H_2O \quad (5)$$

The above reactions destroy the amorphous glassy structure of fly ash and liberate the RE-bearing particles, which are dissolved in the acid leaching step. In addition, hydrothermal treatment using NaOH is also able to convert hard-to-dissolve rare earth minerals into soluble forms (Equation (1)). All of the above reactions contribute to the extraction of REEs from coal combustion ash.

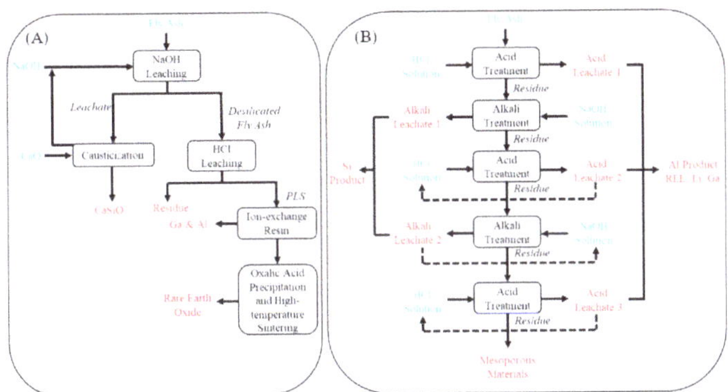

Figure 8. Flowsheets reported in the literature for recovering REEs from coal combustion ash using NaOH hydrothermal treatment followed by acid leaching: (**A**) reported by [18] and (**B**) reported by [98].

Several studies have reported that REE recovery from fly ash was improved by alkali roasting treatment prior to acid leaching [26,100,112]. Taggart et al. compared the performance of various roasting additives, including Na_2O_2, NaOH, CaO, Na_2CO_3, $CaSO_4$, and $(NH_4)_2SO_4$, by performing roasting tests at 450 °C on fly ash samples originating from power stations in the Appalachian, Illinois, and Powder River basins followed by leaching with 1 M HNO_3 [26]. It was found that NaOH roasting often recovered more than 90% of the total REEs, which is more efficient than the other additives. Tang et al. optimized Na_2CO_3 roasting on a coal fly ash collected from Guizhou China [100]. Mass ratio of 1/1 (fly ash/ Na_2CO_3) and roasting temperature of 860 °C provided the maximum total REE recovery (~90%) when leaching with 2 M HCl. Liu et al. proposed a flowsheet to achieve the simultaneous extraction of REEs, Ga, and Nb from a coal fly ash of the Songzao coalfield, which is famous for its significant enrichment in trace elements [84,112,113]. As shown in Figure 9, water leaching was used to extract Ga and Al from the roasting product, and REEs were recovered from the water leaching solid residue by acid leaching. Ion adsorption resin was used to separate Ga and Al. Laboratory test results showed that 68.62% Al, 76.11% Ti, and 80.07% REEs were extracted from the fly ash using the alkali roasting–water leaching–acid leaching method. The reactions between sodium carbonate and fly ash during roasting are as follows:

$$Al_2O_3{\cdot}SiO_2 + 2Na_2CO_3 \rightarrow Na_2SiO_3 + 2NaAlO_2 + 2CO_{2(g)} \quad (6)$$

$$Fe_2O_3 + Na_2CO_3 \rightarrow 2NaFeO_2 + CO_{2(g)} \quad (7)$$

$$TiO_2 + Na_2CO_3 \rightarrow Na_2TiO_3 + CO_{2(g)} \quad (8)$$

$$Ga_2O_3 + Na_2CO_3 \rightarrow 2NaGaO_2 + CO_{2(g)} \quad (9)$$

$$Nb_2O_5 + 3Na_2CO_3 \rightarrow 2Na_3NbO_4 + 3CO_{2(g)} \quad (10)$$

$$REE_2O_3 + Na_2CO_3 \rightarrow 2NaREEO_2 + CO_{2(g)} \quad (11)$$

$$2REEPO_4 + 3Na_2CO_3 \rightarrow REE_2O_3 + 2Na_3PO_4 + 3CO_{2(g)} \quad (12)$$

Table 4. Summary of REE extraction from coal combustion ash using chemical methods.

Sample	Source	Pretreatment	Lixiviant	Leaching Condition	REE Recovery	Reference
Bottom ash	Laboratory prepared	None	4 M HCl	50 g/L pulp density, 90 °C, 120 min	Around 90% for Ce and Nd, and 35% for Y	[97]
Fly ash	Guizhou, China	None	2 M HCl	Liquid/solid ratio 10/1, 120 min	Around 20% for La, 40% for Ce, 5% for Pr, 20% for Nd, and 10% for Y	[100]
Fly ash	Guizhou, China	Na_2CO_3, 1/1 solid/solid ratio, 860 °C	3 M HCl	Liquid/solid ratio 20/1 (*v/w*), 400 rpm stirring speed	72.78% for total REEs	
Fly ash	Upper, Middle, and Lower Kittanning seams, United States	None	1.2 M HCl	1% solid concentration, 75 °C, 5 h	Around 60% for total REEs	[53]
Bottom ash	Illinois No.6 seam, United States	None	1.2 M HCl	1% solid concentration, 75 °C, 5 h	Around 80% for total REEs	
Bottom ash	South Korea	None	2 M HCl	100 g/L pulp density, 80 °C, 12 h	62.1% Y, 55.5% Nd, 65.2% Dy	[101]
Fly ash	Guizhou, China	None	3 M HCl	Liquid/solid ratio 10/1 (*v/w*), 60 °C, 120 min	71.9% La, 66.0% Ce, 61.9% Nd	[91]
Fly ash/Bottom ash	Sichuan, China	None	4% HF	50 g/L, 23–25 °C, 24 h	>90% for total REEs	[18]
Fly ash	Sichuan, China	None	8 M HCl	Liquid/solid ratio 40/1(*v/w*), 80 °C, 6 h	32.36% for total REEs	
Flay ash	Sichuan, China	40% NaOH, 10/1 (*v/w*) solid/liquid ratio, 150 °C, 2 h	8 M HCl	Liquid/solid ratio 30/1 (*v/w*), 60 °C, 2 h	88.15% for total REEs	
Fly ash	Shanxi, China	Acid-alkali based alternate extraction (230 g/L HCl, 200 g/L NaOH, liquid to solid ratio 5/1 (*v/w*), 90 °C)			65% for total REEs	[98]
Fly ash	Powder River Basin, United States	None	12 M HCl	Liquid/solid ratio 100/1, 85 °C, 4 h	Neary 100% for total REEs	[96]
Fly ash	Illinois Basin, United States	None	12 M HCl	Liquid/solid ratio 100/1, 85 °C, 4 h	35–43% recovery for total REEs	
Fly ash	Appalachian Basin, United States	None	12 M HCl	Liquid/solid ratio 100/1, 85 °C, 4 h	40–57% for total REEs	
Fly ash	Appalachian Basin, United States	6.25 M NaOH, liquid/solid mass ratio 10/1, 85 °C, 4 h	20% HCl	NA	48.8–85.9% for total REEs	

Table 4. *Cont.*

Sample	Source	Pretreatment	Lixiviant	Leaching Condition	REE Recovery	Reference
Coal ash	Not Available	None	Super critical CO_2		No experimental tests were performed.	[92]
Fly ash	Ohio, United States	5 M NaOH, solid/liquid ratio 1:20, 100 °C, 2 h	None	None	REE was enriched from 325 to 877 ppm	[76]
Fly ash	Appalachian, Illinois, and Powder River Basins, United States	NaOH roasting (1:1 additive-ash ratio, 450 °C, 30 min)	2 M HNO_3	Room temperature	100% total REE recovery for the Powder River Basin, >70% for the other sources	[26]
Coal ash	Laboratory prepared	None	6 M HNO_3	Liquid/ratio of 33/1 (*v/w*), 60 min	90.5% for total REEs and 90.9% for LREEs	[103]
Fly ash	Powder River Basin, United States	None	15 M HNO_3	10–50 g/L, 85–90 °C, 4 h	69.9% for total REEs	[102]
Coal ash	NA	Physical beneficiation	HNO_3	Approximately 90 °C	NA	[90]
Fly ash	Chongqing, China	Na_2CO_3 roasting (1.5:1 additive-ash ratio, 860 °C, 30 min)	6 M HCl	Liquid/solid ratio 20/1 (*v/w*), 60 °C, 4 h	Around 80% of total REEs were extracted	[112]
Fly ash	Japan	None	9.5% H_2SO_4	Liquid/solid ratio 100/1 (*v/w*), 80 °C, 2 h	Around 10–45% of La was extracted	[88]

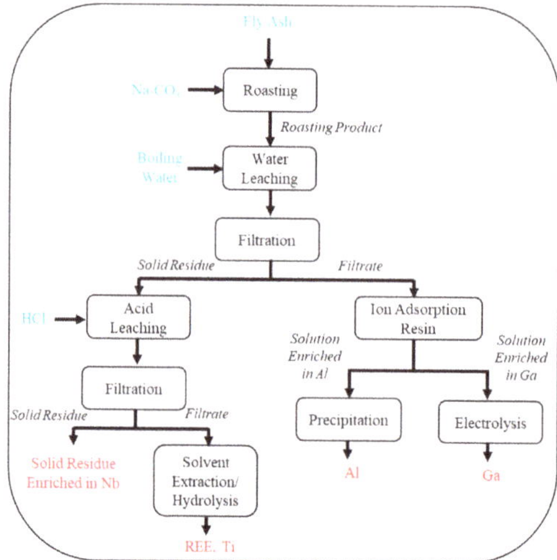

Figure 9. A flowsheet for recovering REEs, Ga, Nb, and Al from a fly ash using alkali roasting [112].

4. REE Recovery from Acid Mine Drainage and Sludge

4.1. REEs in Acid Mine Drainage

The occurrence of REEs in acid coal mine drainage (AMD) is mainly due to the dissolution of RE-bearing species under acidic conditions created by the natural oxidization of sulfide minerals, primarily pyrite. When exposed to the natural environment, pyrite is oxidized as described by the following reactions [114–116]:

$$FeS_{2(s)} + 14Fe^{3+}_{(aq)} + 18H_2O_{(l)} \rightarrow 15Fe^{2+}_{(aq)} + 2SO^{2-}_{4(aq)} + 16H^+_{(aq)} \tag{13}$$

$$FeS_{2(s)} + 7/2O_{2(aq)} + H_2O_{(l)} \rightarrow Fe^{2+}_{(aq)} + 2SO^{2-}_{4(aq)} + 2H^+_{(aq)} \tag{14}$$

$$Fe^{2+}_{(aq)} + 1/4O_{2(aq)} + H^+_{(aq)} \rightarrow Fe^{3+}_{(aq)} + 1/2H_2O_{(l)} \tag{15}$$

As shown in the above equations, both ferric ions and oxygen serve as oxidants for pyrite. Moreover, it has been well realized that Fe^{3+}/S^0-oxidizing microorganisms can significantly accelerate the reaction [116,117].

Many articles have been published focusing on the characterization of REEs in AMD [9,118–123]. Based on the data drawn from 233 samples collected by the United States Geological Survey (USGS) in 1999 and 2011, Ziemkiewicz et al. found that the total REE concentration increases exponentially with a decrease in pH [124]. Other studies also observed the same phenomenon [9,121–123]. For example, Stewart et al. reported that the total REE concentration and solution pH in 18 AMD samples collected from the Appalachian Basin ranged from 0.29 to 1134 μg/L and 2.8–6.6, respectively, with the higher concentration occurring in low pH solutions [9]. Total REE concentration in AMD is normally less than coal and coal refuse, whereas, when reported based on total dissolved solids in AMD, the concentration is similar to or even higher than coal and coal refuse. For example, Honaker et al. obtained the dissolved solid from an AMD sample containing 6.7 mg/L total REEs by completely evaporating the liquid phase [115]. REE content in the dissolved solid was measured to be 380 ppm, which is much higher than the average content of World coals (68 ppm, [4]). Moreover, AMD samples with more than 10 mg/L REEs have been reported in the literature [125]. Therefore, AMD can be used as an alternative resource of REEs. Extremely high concentrations of REEs have been detected in some sludges generated during the passive treatment process [9,22,124]. For example, 3037 mg/kg of total

REEs were estimated from sludge samples from the Saxman Run treatment plant [124]. The same group later conducted a broad survey of 141 treatment sites in the Northern and Central Appalachian coal basins and found that more that 20 of the 623 AMD sludge samples had concentrations exceeding 2000 mg/kg on a dry whole mass basis [22]. Northern Appalachian samples tended to have higher REE concentrations than those from Central Appalachia with statistically significant deviations for all REEs except Ce, Pr, and Nd.

AMD and AMD sludge typically contain more valuable REEs such as Y relative to La and Ce. For example, Y in the coarse refuse collected from the West Kentucky No. 13 and Illinois No. 6 seams represented less than 10% of the total REEs; whereas the REEs in AMD generated from the refuse piles was more than 25% Yttrium. In addition, in the basin comparison study by Vass et al., the ratio of critical REEs (defined as Y, Nd, Eu, Tb, and Dy) to total REEs was nearly 50% for the Northern Appalachian samples [22]. The enrichment of HREEs in AMD, especially the elements located in the middle of lanthanide series (Sm-Dy, middle REEs), has been well recognized by geologists, which may be explained by several mechanisms: (1) the abundance and distribution of mineral phases containing REEs, (2) the stability of RE-bearing mineral phases with respect to the aqueous fluids, (3) the chemistry of the aqueous fluids, and (4) the immobilizing capacity of minerals, precipitates, and colloidal materials to REEs [120,122,123,126,127]. Overall, AMD can be used as a potential resource of REEs due to the relatively high concentration (reported on dissolved solid basis) and the preferential enrichment in HREEs.

4.2. REE Recovery from Acid Mine Drainage

Many studies have been performed to recover valuable components such as Fe, Al, Cu, Zn, Ni, and sulfuric acid from acid mine drainage (AMD). The recovery methods can be classified as precipitation [128–132], adsorption [133,134], diffusion dialysis [135], and ion-exchange [129]. Seo et al. used an oxidation-sequential precipitation method to recover Fe, Al, and Mn from a coal mine drainage [131]. The laboratory test results showed that 99.2–99.3% of Fe, 70.4–82.2% of Al, and 37.8–87.5% of Mn were recovered at pH 4.5, 5.5, and 8.5, respectively. Furthermore, Cu and Zn can be selectively concentrated by collecting the precipitates formed in the pH ranges of 4.49–6.11 and 5.50–7.23, respectively [128]. Instead of artificially adding alkalis, dissolved metals in AMD can also be precipitated and recovered using OH^- produced from electrochemical reactions [136]. Chockalingam and Subramanian found that rice husk is able to uptake 99% Fe^{3+}, 98% Fe^{2+}, 98% Zn^{2+}, and 95% Cu^{2+} from an acid mine water, with a concomitant increase in the pH value by two absolute units [133]. Crane and Sapsford reported that nanoscale zerovalent iron (nZVI) selectively adsorbed Cu, Cd, and Al with more than 99.9% recovery in 1 h [137]. In addition, an acidic pH buffer enabled the formation of copper–bearing nanoparticles from AMD in presence of nZVI. Magnetite nanoparticles have also been successfully prepared from AMD [138–140].

Due to the much higher economic values of REEs relative to the major metals such as iron and aluminum present in AMD, several studies regarding REE recovery from AMD have been reported recently [9,10,12,22,124,141–144]. In one of our prior studies [10], staged precipitation tests were performed on a coal mine drainage (6.14 ppm of REEs), and a REE pre-concentrate containing 1.1% of REEs was produced in the pH range of 4.85–6.11. In addition, the pre-concentrate also contained 17.1% Al, 1.7% Zn, 1.4% Cu, 1.14% Mn, 0.5% Ni, and 0.2% Co, indicating that multiple valuable components can be pre-concentrated simultaneously using staged precipitation. By using selective re-dissolution and oxalic precipitation, a product containing 94% rare earth oxides was finally obtained from the pre-concentrate. A process flowsheet has been developed in a recent study by the authors [145]. The staged precipitation results also explained the observations in the passive treatment systems of coal mine drainage, i.e., >90% REEs were sequestered in the treatment solids when pH was raised above 6.0, and REEs were preferentially retained in the basaluminite ($Al_4(SO_4)(OH)_{10}{\cdot}5H_2O$) [141,142]. Ramasamy et al. (2018) synthesized N- and O- ligand doped mesoporous silica-chitosan hybrid beads for extracting REEs from AMD and the test results showed that more than 90% of REEs were recovered in 5 min [143].

According to Vass et al., REEs in AMD treatment sludge can also be economically recovered by solubilization followed by REE extraction from the solution [22]. Further work by the same group has led to the design and construction of an acid leaching/solvent extraction mini-pilot plant for the recovery of REEs from AMD sludge. Leaching data show that high recoveries of REEs (>80%) can be achieved at a pH value of 1.0 using sulfuric acid. Moreover, the addition of a leaching modifier increased the leaching recovery from 65% to >95% at a pH of 2.0. Together with the downstream solvent extraction operation, the continuous pilot process was able to produce high-grade mixed rare earth oxide products exceeding 80% purity. Techno-economic analysis shows favorable economic outcomes; however, the authors note that the results are very sensitive to consumable costs [146].

Based on the above discussion, a flowsheet for REE recovery and comprehensive utilization of AMD is shown in Figure 10. Cost for the staged precipitation step is minimal given the fact that treatment of AMD is mandated by regulatory agencies. In addition, since only a small quantity of REE-enriched precipitate is obtained from the staged precipitation step, chemical consumptions associated with the downstream processes are low. Therefore, AMD can be considered as a promising source of REEs.

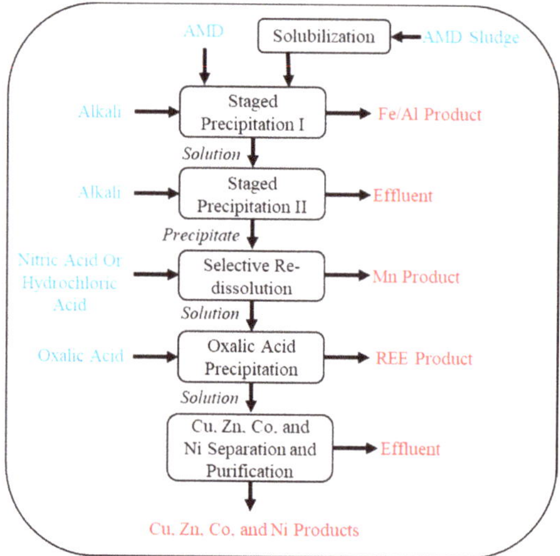

Figure 10. A flowsheet for multiple metals recovery from acid mine drainage (AMD) and acid mine drainage sludge.

5. Summary and Conclusions

Many studies have been conducted for recovering REEs from coal-related materials, primarily coal refuse, coal combustion ash, and acid coal mine drainage. High-purity rare earth concentrates have been successfully produced from coal refuse and acid coal mine drainage. A rare earth pilot plant was constructed and tested, enabling a continuous production of REEs from coal refuse. Reasonable recovery values also have been achieved from coal combustion ash. Altogether, these findings show that significant accomplishments have been made over the past several years in this area.

A summary of the advantages and disadvantages of the methods that have been used for recovering REEs from coal-related materials is shown in Table 5. To achieve optimum recovery performance with minimum cost, process flowsheets integrating various technologies, such as physical beneficiation, acid leaching, pre-leach roasting, and hydrothermal pretreatment, need to be designed and tested. More fundamental studies are required to understand the positive impacts caused by pre-leach calcination on REE recovery from coal refuse. Moreover, this review indicates that not enough attention has been paid to the extraction behavior of other critical elements as well as major elements. This deficiency limits the development of multi-element recovery strategies from coal-related materials. In addition, downstream recovery and purification plans are also ambiguous since the extraction behavior of major elements such as Al, Fe, and Ca were rarely reported.

Table 5. A summary of the advantages and disadvantages of the methods that have been used for REE recovery from coal-related materials.

Material	Method	Advantage	Disadvantage
Coal preparation byproducts	Physical beneficiation	Can be used as a pre-concentration step to generate a higher-grade feed for downstream recovery processes.	Low recovery; Ultrafine grinding is required if a considerable enrichment ratio is expected to be achieved.
	Direct acid leaching	Provide relatively higher recovery compared with physical beneficiation.	Leaching performance depends on the nature of tested samples; Acid consumption is usually high.
	Pre-leach calcination followed by acid leaching	Higher REE recovery; Quick leaching kinetics; Mild leaching conditions; Low chemical consumption.	Recovery of contaminant ions, such as Al^{3+}, is also increased.
Coal combustion ashes	Physical beneficiation	Can be used as a pre-concentration step to generate a higher-grade feed for downstream recovery processes.	Unable to provide a considerable enrichment.
	Direct acid leaching	Able to transfer a portion of REEs from solid into solution, which can be further recovered and purified using other approaches.	Low recovery; High chemical cost; Harsh leaching conditions.
	Alkaline/hydrothermal treatment followed by leaching	High recovery; Quick leaching kinetics.	High alkali consumption; Low selectivity and a lot of contaminants are extracted along with REEs
Acid coal mine drainage	Staged precipitation followed by redissolution and selective precipitation	High recovery; Simple process flowsheet; Low chemical consumption; Can be integrated with existing AMD treatment systems.	AMD containing several ppm of REEs with a large volume may be difficult to find.

Author Contributions: Conceptualization, W.Z.; methodology, A.N. and X.Y.; resources, R.H.; writing—original draft preparation, W.Z. and X.Y.; writing—review and editing, R.H., A.N., and X.Y. All authors have read and agreed to the published version of the manuscript.

Funding: This research received no external funding.

Conflicts of Interest: The authors declare no conflict of interest.

References

1. Chu, S. *Critical Materials Strategy*; DIANE Publishing: Darby, PA, USA, 2011.
2. Trump, D. A federal strategy to ensure secure and reliable supplies of critical minerals. *Donald Trump Washington DC Accessed April* **2018**, 3, 2018.
3. Blengini, G.A.; Nuss, P.; Dewulf, J.; Nita, V.; Talens Peiró, L.; Vidal-Legaz, B.; Latunussa, C.; Mancini, L.; Blagoeva, D.; Pennington, D.; et al. EU methodology for critical raw materials assessment: Policy needs and proposed solutions for incremental improvements. *Resour. Policy* **2017**, *53*, 12–19. [CrossRef]

4. Ketris, M.P.; Yudovich, Y.E. Estimations of clarkes for carbonaceous biolithes: World averages for trace element contents in black shales and coals. *Int. J. Coal Geol.* **2009**, *78*, 135–148. [CrossRef]
5. Seredin, V.V.; Dai, S. Coal deposits as potential alternative sources for lanthanides and yttrium. *Int. J. Coal Geol.* **2012**, *94*, 67–93. [CrossRef]
6. Dai, S.; Xie, P.; Jia, S.; Ward, C.R.; Hower, J.C.; Yan, X.; French, D. Enrichment of U-Re-V-Cr-Se and rare earth elements in the Late Permian coals of the Moxinpo Coalfield, Chongqing, China: Genetic implications from geochemical and mineralogical data. *Ore Geol. Rev.* **2017**, *80*, 1–17. [CrossRef]
7. Seredin, V.V. Rare earth element-bearing coals from the Russian Far East deposits. *Int. J. Coal Geol.* **1996**, *30*, 101–129. [CrossRef]
8. Hower, J.C.; Ruppert, L.F.; Eble, C.F. Lanthanide, yttrium, and zirconium anomalies in the Fire Clay coal bed, Eastern Kentucky. *Int. J. Coal Geol.* **1999**, *39*, 141–153. [CrossRef]
9. Stewart, B.W.; Capo, R.C.; Hedin, B.C.; Hedin, R.S. Rare earth element resources in coal mine drainage and treatment precipitates in the Appalachian Basin, USA. *Int. J. Coal Geol.* **2017**, *169*, 28–39. [CrossRef]
10. Zhang, W.; Honaker, R.Q. Rare earth elements recovery using staged precipitation from a leachate generated from coarse coal refuse. *Int. J. Coal Geol.* **2018**, *195*, 189–199. [CrossRef]
11. Honaker, R.Q.; Groppo, J.; Yoon, R.-H.; Luttrell, G.H.; Noble, A.; Herbst, J. Process evaluation and flowsheet development for the recovery of rare earth elements from coal and associated byproducts. *Miner. Metall. Process.* **2017**, *34*, 107–115. [CrossRef]
12. Vass, C.R.; Noble, A.; Ziemkiewicz, P.F. The occurrence and concentration of rare earth elements in acid mine drainage and treatment by-products: Part 1—Initial survey of the Northern Appalachian Coal Basin. *Min. Metall. Explor.* **2019**, 917–929. [CrossRef]
13. Sarswat, P.K.; Leake, M.; Allen, L.; Free, M.L.; Hu, X.; Kim, D.; Noble, A.; Luttrell, G.H. Efficient recovery of rare earth elements from coal based resources: A bioleaching approach. *Mater. Today Chem.* **2020**, *16*. [CrossRef]
14. Seredin, V.V.; Dai, S.; Sun, Y.; Chekryzhov, I.Y. Coal deposits as promising sources of rare metals for alternative power and energy-efficient technologies. *Appl. Geochem.* **2013**, *31*, 1–11. [CrossRef]
15. Dai, S.; Yan, X.; Ward, C.R.; Hower, J.C.; Zhao, L.; Wang, X.; Zhao, L.; Ren, D.; Finkelman, R.B. Valuable elements in Chinese coals: A review. *Int. Geol. Rev.* **2018**, *60*, 590–620. [CrossRef]
16. Dai, S.; Xie, P.; Ward, C.R.; Yan, X.; Guo, W.; French, D.; Graham, I.T. Anomalies of rare metals in Lopingian super-high-organic-sulfur coals from the Yishan Coalfield, Guangxi, China. *Ore Geol. Rev.* **2017**, *88*, 235–250. [CrossRef]
17. Alvin, M.A. The rarity of rare earth elements (PowerPoint slides). Presentated at National Assocation Regulatorgy Utility Commissioners (NARUC) Winter Policy Summit, Washington, DC, USA, 12 February 2018.
18. Wang, Z.; Dai, S.; Zou, J.; French, D.; Graham, I.T. Rare earth elements and yttrium in coal ash from the Luzhou power plant in Sichuan, Southwest China: Concentration, characterization and optimized extraction. *Int. J. Coal Geol.* **2019**, *203*, 1–14. [CrossRef]
19. Zhang, W.; Groppo, J.; Honaker, R. Ash beneficiation for REE recovery. In Proceedings of the 2015 World Coal Ash Conference, Nashville, TN, USA, 5–7 May 2015.
20. Zhang, W.; Yang, X.; Honaker, R.Q. Association characteristic study and preliminary recovery investigation of rare earth elements from Fire Clay seam coal middlings. *Fuel* **2018**, *215*, 551–560. [CrossRef]
21. Zhang, W.; Honaker, R. Calcination pretreatment effects on acid leaching characteristics of rare earth elements from middlings and coarse refuse material associated with a bituminous coal source. *Fuel* **2019**, *249*, 130–145. [CrossRef]
22. Vass, C.R.; Noble, A.; Ziemkiewicz, P.F. The occurrence and concentration of rare earth elements in acid mine drainage and treatment byproducts. Part 2: Regional survey of Northern and Central Appalachian Coal Basins. *Min. Metall. Explor.* **2019**, *36*, 917–929. [CrossRef]
23. Huang, Q.; Noble, A.; Herbst, J.; Honaker, R. Liberation and release of rare earth minerals from Middle Kittanning, Fire Clay, and West Kentucky No. 13 coal sources. *Powder Technol.* **2018**, *332*, 242–252. [CrossRef]
24. Lin, R.; Howard, B.H.; Roth, E.A.; Bank, T.L.; Granite, E.J.; Soong, Y. Enrichment of rare earth elements from coal and coal by-products by physical separations. *Fuel* **2017**, *200*, 506–520. [CrossRef]
25. Laudal, D.A.; Benson, S.A.; Addleman, R.S.; Palo, D. Leaching behavior of rare earth elements in Fort Union lignite coals of North America. *Int. J. Coal Geol.* **2018**, *191*, 112–124. [CrossRef]

26. Taggart, R.K.; Hower, J.C.; Hsu-Kim, H. Effects of roasting additives and leaching parameters on the extraction of rare earth elements from coal fly ash. *Int. J. Coal Geol.* **2018**, *196*, 106–114. [CrossRef]
27. Zhang, W.; Rezaee, M.; Bhagavatula, A.; Li, Y.; Groppo, J.; Honaker, R. A review of the occurrence and promising recovery methods of rare earth elements from coal and coal by-products. *Int. J. Coal Prep. Util.* **2015**, *35*, 281–294. [CrossRef]
28. Dai, S.; Ren, D.; Chou, C.L.; Finkelman, R.B.; Seredin, V.V.; Zhou, Y. Geochemistry of trace elements in Chinese coals: A review of abundances, genetic types, impacts on human health, and industrial utilization. *Int. J. Coal Geol.* **2012**, *94*, 3–21. [CrossRef]
29. Dai, S.; Finkelman, R.B. Coal as a promising source of critical elements: Progress and future prospects. *Int. J. Coal Geol.* **2018**, *186*, 155–164. [CrossRef]
30. Gupta, T.; Ghosh, T.; Akdogan, G.; Bandopadhyay, S. Maximizing REE enrichment by froth flotation using Box-Behnken design in Alaskan coal. In Proceedings of the 2017 SME Annual Conference & Expo, Denver, CO, USA, 19–22 February 2017; pp. 408–412.
31. Gupta, T.; Ghosh, T.; Akdogan, G.; Srivastava, V.K. Characterizing rare earth elements in Alaskan coal and ash. *Miner. Metall. Process.* **2017**, *34*, 138–145. [CrossRef]
32. Gao, W.; Zhang, X.; Zheng, X.; Lin, X.; Cao, H.; Zhang, Y.; Sun, Z. Lithium carbonate recovery from cathode scrap of spent lithium-ion battery: A closed-loop process. *Environ. Sci. Technol.* **2017**, *51*, 1662–1669. [CrossRef]
33. Zhang, W.; Honaker, R.; Groppo, J. Concentration of rare earth minerals from coal by froth flotation. *Miner. Metall. Process.* **2017**, *34*, 132–137. [CrossRef]
34. Honaker, R.; Groppo, J.; Bhagavatula, A.; Rezaee, M.; Zhang, W. Recovery of rare earth minerals and elements from coal and coal byproducts. In Proceedings of the International Coal Preparation Conference, Lousiville, KY, USA, 25–27 April 2016; pp. 25–27.
35. Honaker, R.; Hower, J.; Eble, C.; Weisenfluh, J.; Groppo, J.; Rezaee, M.; Bhagavatula, A.; Luttrell, G.H.; Bratton, R.C.; Kiser, M.; et al. Laboratory and bench-scale testing for rare earth elements. *Cell* **2014**, *724*, 554–3652.
36. Cheng, T.-W.; Holtham, P.N.; Tran, T. Froth flotation of monazite and xenotime. *Miner. Eng.* **1993**, *6*, 341–351. [CrossRef]
37. Pavez, O.; Peres, A.E.C. Effect of sodium metasilicate and sodium sulphide on the floatability of monazite-zircon-rutile with oleate and hydroxamates. *Miner. Eng.* **1993**, *6*, 69–78. [CrossRef]
38. Gupta, N.; Li, B.; Luttrell, G.; Yoon, R.H.; Bratton, R.; Reyher, J. Hydrophobic-hydrophilic separation (HHS) process for the recovery and dewatering of ultrafine coal. In Proceedings of the 2016 SME Annual Conference and Expo, Phoenix, AZ, USA, 21–24 February 2016; pp. 706–709.
39. Pradip, P.; Fuerstenau, D.W. Design and development of novel flotation reagents for the beneficiation of Mountain Pass rare-earth ore. *Miner. Metall. Process.* **2013**, *30*, 1–9. [CrossRef]
40. Pradip; Fuerstenau, D.W. The adsorption of hydroxamate on semi-soluble minerals. Part I: Adsorption on barite, Calcite and Bastnaesite. *Colloids Surf.* **1983**, *8*, 103–119. [CrossRef]
41. Yang, X.; Werner, J.; Honaker, R.Q. Leaching of rare earth elements from an Illinois basin coal source. *J. Rare Earths* **2019**, *37*, 312–321. [CrossRef]
42. Wei, G.; Bo, F.; Jinxiu, P.; Wenpu, Z.; Xianwen, Z. Depressant behavior of tragacanth gum and its role in the flotation separation of chalcopyrite from talc. *J. Mater. Res. Technol.* **2019**, *8*, 697–702. [CrossRef]
43. Yang, X. Leaching Characteristics of Rare Earth Elements from Bituminous Coal-Based Sources. Ph.D. Thesis, University of Kentucky, Lexington, KY, USA, 2019.
44. Bo, C.; Ya, L.A.I.; Guo, X.I.A.O.; Chang, X.U. Technique for extraction and concentration of rare earth elements in gangue. *Glob. Geol.* **2010**, *28*, 257–260.
45. Kuppusamy, V.K.; Holuszko, M. Rare earth elements in flotation products of coals from East Kootenay coalfields, British Columbia. *J. Rare Earths* 2019. [CrossRef]
46. Zhang, P.; Han, Z.; Jia, J.; Wei, C.; Liu, Q.; Wang, X.; Zhou, J.; Li, F.; Miao, S. Occurrence and Distribution of Gallium, Scandium, and Rare Earth Elements in Coal Gangue Collected from Junggar Basin, China. *Int. J. Coal Prep. Util.* **2017**, *39*, 389–402. [CrossRef]
47. Zhang, W.; Honaker, R. Enhanced leachability of rare earth elements from calcined products of bituminous coals. *Miner. Eng.* **2019**, *142*, 105935. [CrossRef]

48. Rozelle, P.L.; Khadilkar, A.B.; Pulati, N.; Soundarrajan, N.; Klima, M.S.; Mosser, M.M.; Miller, C.E.; Pisupati, S.V. A study on removal of rare earth elements from U.S. coal byproducts by ion exchange. *Metall. Mater. Trans. E* **2016**, *3*, 6–17. [CrossRef]
49. Finkelman, R.B.; Palmer, C.A.; Wang, P. Quantification of the modes of occurrence of 42 elements in coal. *Int. J. Coal Geol.* **2018**, *185*, 138–160. [CrossRef]
50. Honaker, R.; Yang, X.; Chandra, A.; Zhang, W.; Werner, J. *Hydrometallurgical Extraction of Rare Earth Elements from Coal*; Springer International Publishing: Berlin, Germany, 2018. ISBN 978-3-319-95021-1.
51. Zhang, W.; Noble, A. Mineralogy characterization and recovery of rare earth elements from the roof and floor materials of the Guxu coalfield. *Fuel* **2020**, *270*, 117533. [CrossRef]
52. Kuppusamy, V.K.; Kumar, A.; Holuszko, M. Simultaneous extraction of clean coal and rare earth elements from coal tailings using alkali-acid leaching process. *J. Energy Resour. Technol. Trans. ASME* **2019**, *141*, 1–7. [CrossRef]
53. Honaker, R.Q.; Zhang, W.; Werner, J. Acid leaching of rare earth elements from coal and coal ash: Implications for using fluidized bed combustion to assist in the recovery of critical materials. *Energy Fuels* **2019**, *33*, 5971–5980. [CrossRef]
54. Zhang, W.; Honaker, R. Characterization and recovery of rare earth elements and other critical metals (Co, Cr, Li, Mn, Sr, and V) from the calcination products of a coal refuse sample. *Fuel* **2020**, *267*, 117236. [CrossRef]
55. Hu, G.; Dam-Johansen, K.; Wedel, S.; Hansen, J.P. Decomposition and oxidation of pyrite. *Prog. Energy Combust. Sci.* **2006**, *32*, 295–314. [CrossRef]
56. Music, S.; Popović, S.; Ristić, M. Thermal decomposition of pyrite. *J. Radioanal. Nucl. Chem. Artic.* **1992**, *162*, 217–226. [CrossRef]
57. Cao, Z.; Cao, Y.; Dong, H.; Zhang, J.; Sun, C. Effect of calcination condition on the microstructure and pozzolanic activity of calcined coal gangue. *Int. J. Miner. Process.* **2016**, *146*, 23–28. [CrossRef]
58. De la Villa, R.V.; García, R.; Martínez-Ramírez, S.; Frías, M. Effects of calcination temperature and the addition of ZnO on coal waste activation: A mineralogical and morphological evolution. *Appl. Clay Sci.* **2017**, *150*, 1–9. [CrossRef]
59. Honaker, R.Q.; Zhang, W.; Werner, J.; Noble, A.; Luttrell, G.H.; Yoon, R.-H. Enhancement of a process flowsheet for recovering and concentrating critical materials from bituminous coal sources. *Min. Metall. Explor.* **2019**, accepted. [CrossRef]
60. Zhang, W.; Noble, A.; Yang, X.; Honaker, R. Lithium leaching recovery and mechanisms from density fractions of an Illinois Basin bituminous coal. *Fuel* **2020**, *268*, 117319. [CrossRef]
61. Chancey, R.T.; Stutzman, P.; Juenger, M.C.G.; Fowler, D.W. Comprehensive phase characterization of crystalline and amorphous phases of a Class F fly ash. *Cem. Concr. Res.* **2010**, *40*, 146–156. [CrossRef]
62. Goodarzi, F. Characteristics and composition of fly ash from Canadian coal-fired power plants. *Fuel* **2006**, *85*, 1418–1427. [CrossRef]
63. Kukier, U.; Ishak, C.F.; Sumner, M.E.; Miller, W.P. Composition and element solubility of magnetic and non-magnetic fly ash fractions. *Environ. Pollut.* **2003**, *123*, 255–266. [CrossRef]
64. McCarthy, G.J.; Solem, J.K.; Manz, O.E.; Hassett, D.J. Use of a database of chemical, mineralogical and physical properties of North American fly ash to study the nature of fly ashand its utilization as a mineral admixture in concrete. *MRS Online Proc. Libr. Arch.* **1989**, *178*, 3. [CrossRef]
65. Kolker, A.; Scott, C.; Hower, J.C.; Vazquez, J.A.; Lopano, C.L.; Dai, S. Distribution of rare earth elements in coal combustion fly ash, determined by SHRIMP-RG ion microprobe. *Int. J. Coal Geol.* **2017**, *184*, 1–10. [CrossRef]
66. Liu, P.; Huang, R.; Tang, Y. Comprehensive understandings of rare earth element (REE) speciation in coal fly ashes and implication for REE extractability. *Environ. Sci. Technol.* **2019**, *53*, 5369–5377. [CrossRef]
67. Montross, S.N.; Verba, C.A.; Chan, H.L.; Lopano, C. Advanced characterization of rare earth element minerals in coal utilization byproducts using multimodal image analysis. *Int. J. Coal Geol.* **2018**, *195*, 362–372. [CrossRef]
68. Stuckman, M.Y.; Lopano, C.L.; Granite, E.J. Distribution and speciation of rare earth elements in coal combustion by-products via synchrotron microscopy and spectroscopy. *Int. J. Coal Geol.* **2018**, *195*, 125–138. [CrossRef]

69. Taggart, R.K.; Rivera, N.A.; Levard, C.; Ambrosi, J.P.; Borschneck, D.; Hower, J.C.; Hsu-Kim, H. Differences in bulk and microscale yttrium speciation in coal combustion fly ash. *Environ. Sci. Process. Impacts* **2018**, *20*, 1390–1403. [CrossRef] [PubMed]
70. Thompson, R.L.; Bank, T.; Montross, S.; Roth, E.; Howard, B.; Verba, C.; Granite, E. Analysis of rare earth elements in coal fly ash using laser ablation inductively coupled plasma mass spectrometry and scanning electron microscopy. *Spectrochim. Acta-Part B At. Spectrosc.* **2018**, *143*, 1–11. [CrossRef]
71. Hower, J.C.; Cantando, E.; Eble, C.F.; Copley, G.C. Characterization of stoker ash from the combustion of high-lanthanide coal at a Kentucky bourbon distillery. *Int. J. Coal Geol.* **2019**, *213*, 103260. [CrossRef]
72. Hower, J.C.; Qian, D.; Briot, N.J.; Santillan-Jimenez, E.; Hood, M.M.; Taggart, R.K.; Hsu-Kim, H. Nano-scale rare earth distribution in fly ash derived from the combustion of the fire clay coal, Kentucky. *Minerals* **2019**, *9*, 206. [CrossRef]
73. Hower, J.C.; Qian, D.; Briot, N.J.; Henke, K.R.; Hood, M.M.; Taggart, R.K.; Hsu-Kim, H. Rare earth element associations in the Kentucky State University stoker ash. *Int. J. Coal Geol.* **2018**, *189*, 75–82. [CrossRef]
74. Hower, J.; Groppo, J.; Henke, K.; Hood, M.; Eble, C.; Honaker, R.; Zhang, W.; Qian, D. Notes on the Potential for the Concentration of Rare Earth Elements and Yttrium in Coal Combustion Fly Ash. *Minerals* **2015**, *5*, 356–366. [CrossRef]
75. Hower, J.C.; Dai, S.; Seredin, V.V.; Zhao, L.; Kostova, I.J.; Silva, L.F.O.; Mardon, S.M.; Gurdal, G. A note on the occurrence of yttrium and rare earth elements in coal combustion byproducts. *Coal Combust. Gasif. Prod.* **2013**, 39–47. [CrossRef]
76. Lin, R.; Stuckman, M.; Howard, B.H.; Bank, T.L.; Roth, E.A.; Macala, M.K.; Lopano, C.; Soong, Y.; Granite, E.J. Application of sequential extraction and hydrothermal treatment for characterization and enrichment of rare earth elements from coal fly ash. *Fuel* **2018**, *232*, 124–133. [CrossRef]
77. Pan, J.; Zhou, C.; Tang, M.; Cao, S.; Liu, C.; Zhang, N.; Wen, M.; Luo, Y.; Hu, T.; Ji, W. Study on the modes of occurrence of rare earth elements in coal fly ash by statistics and a sequential chemical extraction procedure. *Fuel* **2019**, *237*, 555–565. [CrossRef]
78. Pan, J.; Zhou, C.; Liu, C.; Tang, M.; Cao, S.; Hu, T.; Ji, W.; Luo, Y.; Wen, M.; Zhang, N. Modes of occurrence of rare earth elements in coal fly ash: A case study. *Energy Fuels* **2018**, *32*, 9738–9743. [CrossRef]
79. ASTM standard specification for coal fly ash and raw or calcined natural pozzolan for use. *Annu. B. ASTM Stand.* **2010**, 3–6. [CrossRef]
80. Kumari, A.; Panda, R.; Jha, M.K.; Kumar, J.R.; Lee, J.Y. Process development to recover rare earth metals from monazite mineral: A review. *Miner. Eng.* **2015**, *79*, 102–115. [CrossRef]
81. Cetiner, Z.S.; Wood, S.A.; Gammons, C.H. The aqueous geochemistry of the rare earth elements. Part XIV. The solubility of rare earth element phosphates from 23 to 150 °C. *Chem. Geol.* **2005**, *217*, 147–169. [CrossRef]
82. Hower, J.; Groppo, J.; Joshi, P.; Dai, S.; Moecher, D.; Johnston, M. Location of cerium in coal-combustion fly ashes: Implications for recovery of lanthanides. *Coal Combust. Gasif. Prod.* **2003**, *5*, 73–78. [CrossRef]
83. Blissett, R.S.; Smalley, N.; Rowson, N.A. An investigation into six coal fly ashes from the United Kingdom and Poland to evaluate rare earth element content. *Fuel* **2014**, *119*, 236–239. [CrossRef]
84. Dai, S.; Zhao, L.; Peng, S.; Chou, C.L.; Wang, X.; Zhang, Y.; Li, D.; Sun, Y. Abundances and distribution of minerals and elements in high-alumina coal fly ash from the Jungar Power Plant, Inner Mongolia, China. *Int. J. Coal Geol.* **2010**, *81*, 320–332. [CrossRef]
85. Dai, S.; Zhao, L.; Hower, J.C.; Johnston, M.N.; Song, W.; Wang, P.; Zhang, S. Petrology, mineralogy, and chemistry of size-fractioned fly ash from the Jungar power plant, Inner Mongolia, China, with emphasis on the distribution of rare earth elements. *Energy Fuels* **2014**, *28*, 1502–1514. [CrossRef]
86. Lanzerstorfer, C. Pre-processing of coal combustion fly ash by classification for enrichment of rare earth elements. *Energy Rep.* **2018**, *4*, 660–663. [CrossRef]
87. Pan, J.; Nie, T.; Vaziri Hassas, B.; Rezaee, M.; Wen, Z.; Zhou, C. Recovery of rare earth elements from coal fly ash by integrated physical separation and acid leaching. *Chemosphere* **2020**, *248*, 126112. [CrossRef]
88. Kashiwakura, S.; Kumagai, Y.; Kubo, H.; Wagatsuma, K. Dissolution of rare earth elements from coal fly ash particles in a dilute H2SO4 solvent. *Open J. Phys. Chem.* **2013**, *03*, 69–75. [CrossRef]
89. Joshi, P.B.; Preda, D.V.; Skyler, D.A.; Tsinberg, A.; Green, B.D.; Marinelli, W.J. Recovery of Rare Earth Elements and Compounds from Coal Ash. U.S. Patent 8,968,688, 3 March 2015.
90. Joshi, P.B.; Preda, D.V.; Skyler, D.A.; Scherer, A.; Green, B.D.; Marinelli, W.J. Recovery of rare earth Elements and Compounds from Coal Ash. U.S. Patent 9,394,586, 19 July 2016.

91. Cao, S.; Zhou, C.; Pan, J.; Liu, C.; Tang, M.; Ji, W.; Hu, T.; Zhang, N. Study on influence factors of leaching of rare earth elements from coal fly ash. *Energy Fuels* **2018**, *32*, 8000–8005. [CrossRef]
92. Das, S.; Gaustad, G.; Sekar, A.; Williams, E. Techno-economic analysis of supercritical extraction of rare earth elements from coal ash. *J. Clean. Prod.* **2018**, *189*, 539–551. [CrossRef]
93. Fan, M.; Co, E.M.; Zhao, Y.; Long, Z.; Liu, W. Preprint 17–103. Recovery of valuable elements from Chinese coal by-products. In Proceedings of the Preprint 17–103, 1–6. SME Annual Meeting, Dever, CO, USA, 19–22 February 2017.
94. Huang, Z.; Fan, M.; Tian, H. Rare earth elements of fly ash from Wyoming's Powder River Basin coal. *J. Rare Earths* **2019**. [CrossRef]
95. Huang, C.; Wang, Y.; Huang, B.; Dong, Y.; Sun, X. The recovery of rare earth elements from coal combustion products by ionic liquids. *Miner. Eng.* **2019**, *130*, 142–147. [CrossRef]
96. King, J.F.; Taggart, R.K.; Smith, R.C.; Hower, J.C.; Hsu-Kim, H. Aqueous acid and alkaline extraction of rare earth elements from coal combustion ash. *Int. J. Coal Geol.* **2018**, *195*, 75–83. [CrossRef]
97. Kumari, A.; Parween, R.; Chakravarty, S.; Parmar, K.; Pathak, D.D.; Lee, J.; Jha, M.K. Novel approach to recover rare earth metals (REMs) from Indian coal bottom ash. *Hydrometallurgy* **2019**, *187*, 1–7. [CrossRef]
98. Ma, Z.; Zhang, S.; Zhang, H.; Cheng, F. Novel extraction of valuable metals from circulating fluidized bed-derived high-alumina fly ash by acid–alkali–based alternate method. *J. Clean. Prod.* **2019**, *230*, 302–313. [CrossRef]
99. Shimizu, R.; Sawada, K.; Enokida, Y.; Yamamoto, I. Supercritical fluid extraction of rare earth elements from luminescent material in waste fluorescent lamps. *J. Supercrit. Fluids* **2005**, *33*, 235–241. [CrossRef]
100. Tang, M.; Zhou, C.; Pan, J.; Zhang, N.; Liu, C.; Cao, S.; Hu, T.; Ji, W. Study on extraction of rare earth elements from coal fly ash through alkali fusion–Acid leaching. *Miner. Eng.* **2019**, *136*, 36–42. [CrossRef]
101. Tuan, L.Q.; Thenepalli, T.; Chilakala, R.; Vu, H.H.T.; Ahn, J.W.; Kim, J. Leaching characteristics of low concentration rare earth elements in Korean (Samcheok) CFBC bottom ash samples. *Sustainability* **2019**, *11*, 2562. [CrossRef]
102. Taggart, R.K.; Hower, J.C.; Dwyer, G.S.; Hsu-Kim, H. Trends in the rare earth element content of U.S.-based coal combustion fly ashes. *Environ. Sci. Technol.* **2016**, *50*, 5919–5926. [CrossRef] [PubMed]
103. Peiravi, M.; Ackah, L.; Guru, R.; Mohanty, M.; Liu, J.; Xu, B.; Zhu, X.; Chen, L. Chemical extraction of rare earth elements from coal ash. *Miner. Metall. Process.* **2017**, *34*, 170–177. [CrossRef]
104. Kutchko, B.G.; Kim, A.G. Fly ash characterization by SEM-EDS. *Fuel* **2006**, *85*, 2537–2544. [CrossRef]
105. Ward, C.R.; French, D. Determination of glass content and estimation of glass composition in fly ash using quantitative X-ray diffractometry. *Fuel* **2006**, *85*, 2268–2277. [CrossRef]
106. Mastral, A.M.; Callén, M.S.; García, T. Fluidized bed combustion (FBC) of fossil and nonfossil fuels. A comparative study. *Energy Fuels* **2000**, *14*, 275–281. [CrossRef]
107. Dutrizac, J.E. The behaviour of the rare earth elements during gypsum (CaSO4·2H2O) precipitation. *Hydrometallurgy* **2017**, *174*, 38–46. [CrossRef]
108. Park, S.; Liang, Y. Bioleaching of trace elements and rare earth elements from coal fly ash. *Int. J. Coal Sci. Technol.* **2019**, *6*, 74–83. [CrossRef]
109. Matyas, B.; Gerber, P.; Solymos, A.; Kaszanitzky, F.; Panto, G.; Leffler, J. Process for Recovering Rare Metals from the Combustion Residue of Coal by Digestion. U.S. Patent 4,649,031, 10 March 1987.
110. Ding, J.; Ma, S.; Shen, S.; Xie, Z.; Zheng, S.; Zhang, Y. Research and industrialization progress of recovering alumina from fly ash: A concise review. *Waste Manag.* **2017**, *60*, 375–387. [CrossRef]
111. Yao, Z.T.; Xia, M.S.; Sarker, P.K.; Chen, T. A review of the alumina recovery from coal fly ash, with a focus in China. *Fuel* **2014**, *120*, 74–85. [CrossRef]
112. Liu, H.; Tian, H.; Zou, J. Combined extraction of rare metals Ga-Nb-REE from fly ash. *Sci. Technol. Rev.* **2015**, *33*, 39–43. (In Chinese with English Abstract)
113. Dai, S.; Wang, X.; Zhou, Y.; Hower, J.C.; Li, D.; Chen, W.; Zhu, X.; Zou, J. Chemical and mineralogical compositions of silicic, mafic, and alkali tonsteins in the late Permian coals from the Songzao Coalfield, Chongqing, Southwest China. *Chem. Geol.* **2011**, *282*, 29–44. [CrossRef]
114. Fernando, W.A.M.; Ilankoon, I.M.S.K.; Syed, T.H.; Yellishetty, M. Challenges and opportunities in the removal of sulphate ions in contaminated mine water: A review. *Miner. Eng.* **2018**, *117*, 74–90. [CrossRef]
115. Honaker, R.Q.; Zhang, W.; Yang, X.; Rezaee, M. Conception of an integrated flowsheet for rare earth elements recovery from coal coarse refuse. *Miner. Eng.* **2018**, *122*, 233–240. [CrossRef]

116. Mousavi, S.M.; Jafari, A.; Yaghmaei, S.; Vossoughi, M.; Roostaazad, R. Bioleaching of low-grade sphalerite using a column reactor. *Hydrometallurgy* **2006**, *82*, 75–82. [CrossRef]
117. Casas, J.M.; Martinez, J.; Moreno, L.; Vargas, T. Bioleaching model of a copper-sulfide ore bed in heap and dump configurations. *Metall. Mater. Trans. B Process Metall. Mater. Process. Sci.* **1998**, *29*, 899–909. [CrossRef]
118. Merten, D.; Geletneky, J.; Bergmann, H.; Haferburg, G.; Kothe, E.; Büchel, G. Rare earth element patterns: A tool for understanding processes in remediation of acid mine drainage. *Chemie der Erde* **2005**, *65*, 97–114. [CrossRef]
119. Pérez-López, R.; Delgado, J.; Nieto, J.M.; Márquez-García, B. Rare earth element geochemistry of sulphide weathering in the São Domingos mine area (Iberian Pyrite Belt): A proxy for fluid-rock interaction and ancient mining pollution. *Chem. Geol.* **2010**, *276*, 29–40. [CrossRef]
120. Prudêncio, M.I.; Valente, T.; Marques, R.; Sequeira Braga, M.A.; Pamplona, J. Geochemistry of rare earth elements in a passive treatment system built for acid mine drainage remediation. *Chemosphere* **2015**, *138*, 691–700. [CrossRef]
121. Sahoo, P.K.; Tripathy, S.; Equeenuddin, S.M.; Panigrahi, M.K. Geochemical characteristics of coal mine discharge vis-à-vis behavior of rare earth elements at Jaintia Hills coalfield, northeastern India. *J. Geochem. Explor.* **2012**, *112*, 235–243. [CrossRef]
122. Sun, H.; Zhao, F.; Zhang, M.; Li, J. Behavior of rare earth elements in acid coal mine drainage in Shanxi Province, China. *Environ. Earth Sci.* **2012**, *67*, 205–213. [CrossRef]
123. Zhao, F.; Cong, Z.; Sun, H.; Ren, D. The geochemistry of rare earth elements (REE) in acid mine drainage from the Sitai coal mine, Shanxi Province, North China. *Int. J. Coal Geol.* **2007**, *70*, 184–192. [CrossRef]
124. Ziemkiewicz, P.; He, T.; Noble, A.; Liu, X. *Recovery of Rare Earth Elements (REEs) from Coal Mine Drainage*; West Virginia Mine Drainage Task Force Symposium: Morgantown, WV, USA, 2016.
125. Cravotta, C.A. Dissolved metals and associated constituents in abandoned coal-mine discharges, Pennsylvania, USA. Part 1: Constituent quantities and correlations. *Appl. Geochem.* **2008**, *23*, 166–202. [CrossRef]
126. Grawunder, A.; Merten, D.; Büchel, G. Origin of middle rare earth element enrichment in acid mine drainage-impacted areas. *Environ. Sci. Pollut. Res.* **2014**, *21*, 6812–6823. [CrossRef] [PubMed]
127. Ferreira da Silva, E.; Bobos, I.; Xavier Matos, J.; Patinha, C.; Reis, A.P.; Cardoso Fonseca, E. Mineralogy and geochemistry of trace metals and REE in volcanic massive sulfide host rocks, stream sediments, stream waters and acid mine drainage from the Lousal mine area (Iberian Pyrite Belt, Portugal). *Appl. Geochem.* **2009**, *24*, 383–401. [CrossRef]
128. Balintova, M.; Petrilakova, A. Study of pH influence on selective precipitation of heavy metals from acid mine drainage. *Chem. Eng. Trans.* **2011**, *25*, 345–350. [CrossRef]
129. Feng, D.; Aldrich, C.; Tan, H. Treatment of acid mine water by use of heavy metal precipitation and ion exchange. *Miner. Eng.* **2000**, *13*, 623–642. [CrossRef]
130. Park, S.M.; Yoo, J.C.; Ji, S.W.; Yang, J.S.; Baek, K. Selective recovery of Cu, Zn, and Ni from acid mine drainage. *Environ. Geochem. Health* **2013**, *35*, 735–743. [CrossRef]
131. Seo, E.Y.; Cheong, Y.W.; Yim, G.J.; Min, K.W.; Geroni, J.N. Recovery of Fe, Al and Mn in acid coal mine drainage by sequential selective precipitation with control of pH. *Catena* **2017**, *148*, 11–16. [CrossRef]
132. Wei, X.; Viadero, R.C.; Buzby, K.M. Recovery of iron and aluminum from acid mine drainage by selective precipitation. *Environ. Eng. Sci.* **2005**, *22*, 745–755. [CrossRef]
133. Chockalingam, E.; Subramanian, S. Studies on removal of metal ions and sulphate reduction using rice husk and Desulfotomaculum nigrificans with reference to remediation of acid mine drainage. *Chemosphere* **2006**, *62*, 699–708. [CrossRef]
134. Mohan, D.; Chander, S. Removal and recovery of metal ions from acid mine drainage using lignite-A low cost sorbent. *J. Hazard. Mater.* **2006**, *137*, 1545–1553. [CrossRef] [PubMed]
135. Wei, C.; Li, X.; Deng, Z.; Fan, G.; Li, M.; Li, C. Recovery of H2SO4 from an acid leach solution by diffusion dialysis. *J. Hazard. Mater.* **2010**, *176*, 226–230. [CrossRef] [PubMed]
136. Park, S.M.; Shin, S.Y.; Yang, J.S.; Ji, S.W.; Baek, K. Selective recovery of dissolved metals from mine drainage using electrochemical reactions. *Electrochim. Acta* **2015**, *181*, 248–254. [CrossRef]
137. Crane, R.A.; Sapsford, D.J. Selective formation of copper nanoparticles from acid mine drainage using nanoscale zerovalent iron particles. *J. Hazard. Mater.* **2018**, *347*, 252–265. [CrossRef]

138. Kefeni, K.K.; Msagati, T.M.; Mamba, B.B. Synthesis and characterization of magnetic nanoparticles and study their removal capacity of metals from acid mine drainage. *Chem. Eng. J.* **2015**, *276*, 222–231. [CrossRef]
139. Silva, R.D.A.; Castro, C.D.; Vigânico, E.M.; Petter, C.O.; Schneider, I.A.H. Selective precipitation/UV production of magnetite particles obtained from the iron recovered from acid mine drainage. *Miner. Eng.* **2012**, *29*, 22–27. [CrossRef]
140. Wei, X.; Viadero, R.C. Synthesis of magnetite nanoparticles with ferric iron recovered from acid mine drainage: Implications for environmental engineering. *Colloids Surf. A Physicochem. Eng. Asp.* **2007**, *294*, 280–286. [CrossRef]
141. Ayora, C.; Macías, F.; Torres, E.; Lozano, A.; Carrero, S.; Nieto, J.M.; Pérez-López, R.; Fernández-Martínez, A.; Castillo-Michel, H. Recovery of rare earth elements and yttrium from passive-remediation systems of acid mine drainage. *Environ. Sci. Technol.* **2016**, *50*, 8255–8262. [CrossRef]
142. Hedin, B.C.; Capo, R.C.; Stewart, B.W.; Hedin, R.S.; Lopano, C.L.; Stuckman, M.Y. The evaluation of critical rare earth element (REE) enriched treatment solids from coal mine drainage passive treatment systems. *Int. J. Coal Geol.* **2019**, *208*, 54–64. [CrossRef]
143. Ramasamy, D.L.; Puhakka, V.; Iftekhar, S.; Wojtuś, A.; Repo, E.; Ben Hammouda, S.; Iakovleva, E.; Sillanpää, M. N- and O- ligand doped mesoporous silica-chitosan hybrid beads for the efficient, sustainable and selective recovery of rare earth elements (REE) from acid mine drainage (AMD): Understanding the significance of physical modification and conditioning of th. *J. Hazard. Mater.* **2018**, *348*, 84–91. [CrossRef]
144. Wei, X.; Zhang, S.; Shimko, J.; Dengler, R.W. Mine drainage: Treatment technologies and rare earth elements. *Water Environ. Res.* **2019**, 1–8. [CrossRef] [PubMed]
145. Zhang, W.; Honaker, R. Process development for the recovery of rare earth elements and critical metals from an acid mine drainage. *Miner. Eng.* **2020**, 106382. [CrossRef]
146. Ziemkiewicz, P.; Noble, A. *Recovery of Rare Earth Elements (REEs) from Coal Mine Drainage*; NETL REE Review Meeting: Pittsburgh, PA, USA, 2019.

Review

Hydrometallurgical Recovery and Process Optimization of Rare Earth Fluorides from Recycled Magnets

Prince Sarfo *, Thomas Frasz, Avimanyu Das and Courtney Young *

Metallurgical and Materials Engineering, Montana Tech, 1300 West Park Street, Butte, MT 59701, USA; tfrasz@mtech.edu (T.F.); adas@mtech.edu (A.D.)

* Correspondence: psarfo@mtech.edu (P.S.); cyoung@mtech.edu (C.Y.)

Received: 28 February 2020; Accepted: 4 April 2020; Published: 10 April 2020

Abstract: Magnets containing substantial quantities of rare earth elements are currently one of the most sought-after commodities because of their strategic importance. Recycling these rare earth magnets after their life span has been identified to be a unique approach for mitigating environmental issues that originate from mining and also for sustaining natural resources. The approach is hydrometallurgical, with leaching and precipitation followed by separation and recovery of neodymium (Nd), praseodymium (Pr) and dysprosium (Dy) in the form of rare earth fluorides (REF) as the final product. The methodology is specifically comprised of sulfuric acid (H_2SO_4) leaching and ammonium hydroxide (NH_4OH) precipitation followed by reacting the filtrate with ammonium bifluoride ($NH_4F{\cdot}HF$) to yield the REF. Additional filtering also produces ammonium sulfate ($(NH_4)_2SO_4$) as a byproduct fertilizer. Quantitative and qualitative evaluations by means of XRD, ICP and TGA-DSC to determine decomposition of ammonium jarosite, which is an impurity in the recovery process were performed. Additionally, conditional and response variables were used in a surface-response model to optimize REF production from end-of-life magnets. A REF recovery of 56.2% with a REF purity of 62.4% was found to be optimal.

Keywords: rare earth elements; magnets; recycling; recovery; fluorides; modelling

---

1. Introduction

Disquiet around the sustainability of rare-earth elements (REE) provisions has stimulated determination not only to recycle but also to improve the proficiency of the materials they are used to make [1]. With respect to the amount of REEs produced, neodymium (Nd) usage in the production of neodymium-Iron-Boron (NdFeB) magnets from mine output is about 13%. Out of that, 34% of the magnets produced are used in the manufacturing of actuator hard disk drives [1–3]. With their life span centered on their application, rare earth materials in hard disk drives are also applied in parts such as printed circuit boards (PCB), spindles, and so on [4].

In the production of these magnets, specific additions of elements are also used to adjust their properties [5,6]. For example, cobalt (Co) is used to substitute REE and iron (Fe) materials (up to over 5%) to increase the Curie temperature [7,8]. Dysprosium (Dy) addition increases the temperature characteristics such that the compound has a better stability against demagnetization. It also decreases the residual induction of the magnet, which leads to lower magnetic field properties [6,9]. To lower the production costs, Pr is now used as a substitute of Nd (up to 20–25%) in the production of magnets. Additionally, the magnet material is coated with a protective layer, such as copper (Cu), aluminum (Al), or nickel (Ni), as well as polymeric material to hold the magnet on the steel plate [4]. Although this shows that recycling of end-of-life magnets can help reduce the criticality of these REE in the near future, commercial recycling of REE is low, at less than 1%. This is mainly due to inefficient collection, technological difficulties, and high cost of processing [10–15].

Separation stage(s) have always been an important step in recycling. For rare earth magnets, since the element of interest is found with two or more others, they may require different extraction technology. Hydrometallurgical approaches have been shown to be a very efficient way to separate the REEs in which chemical separation by leaching is performed [16,17]. For REE leaching, lixiviants are directly added with or without heat treatment to dissolve the solid materials. Once the materials are in solution, various processes such as precipitation, solvent extraction, and ion exchange can be used to economically produce individual REE in the required form.

As a part of the impurities encountered in REE recovery, some of the leached Fe is also recovered as ammonium jarosite in the final REF product. To remove this impurity, the final product form is subjected to high temperature (250–500 °C) to decompose the ammonium jarosite [18–21]

For solvent extraction and ion exchange, different cationic, anionic and solvating extractants such as di (2-ethyl-hexyl) phosphoric acid (D2EHPA), dialkyl phosphonic acid (Cyanex 272), 2-ethyl-hexyl phosphonic acid mono-2-ethyl-hexyl ester (PC 88A), neodecanoic acid (Versatic 10), tributyl phosphate (TBP), and tricaprylylmethylammonium chloride (Aliquat 336) have been reported for the separation of REEs from solution with D2EHPA being more commonly used with nitrate, sulfate, chloride and perchlorate solutions, PC 88A with chloride solutions, and TBP with nitrate solutions [22–35]. Interestingly, many of the same chemical types used as solvent extractants are also used in solid form as ion-exchange resins from the same type of leaching solutions [23,24,28,32]. For both solvent extraction and ion exchange, the REE-loaded material must then be selectively stripped. The resulting solutions are then predominantly processed to precipitate the individual REEs, often as REOs, but not always [36–39].

This paper presents the recycling Nd magnet scrap using a novel hydrometallurgical process involving H_2SO_4 leach, NH_4OH precipitation, and $NH_4F{\cdot}HF$ reaction. The latter step transforms the precipitate into rare earth fluorides (REF) which should be appropriate feedstock for subsequent pyrometallurgical processing into metal in molten fluoride electrolysis [39]. The application of hydrofluoric acid (HF) was completely avoided and the process was optimized through statistical analysis and modelling.

2. Experimental

2.1. Materials

Samples of scrap magnets were obtained from end-of-life computer hard drives from the IT Department of Montana Tech. Ammonium hydroxide (NH_4OH) dissolved in water at 28–30% concentration was obtained from VWR International LLC. Sulfuric acid (98% H_2SO_4) from Pharmco Products Inc. and ammonium bifluoride (98.8% $NH_4F{\cdot}HF$) manufactured by J.T. Baker were the other reagents used in the work.

2.2. Sample Preparation

2.2.1. Demagnetization

Nd magnets were obtained by disassembling the actuators in various hard drives. After loading in a ThermoScientific Lindberg/Blue M box furnace, the furnace was programmed to heat up to 500 °C at 5 °C per minute under ambient air. This was done in order to cause demagnetization and weaken the adhesive used to hold the magnet on the steel plate. The temperature was held at 500 °C for 60 min. Afterwards, the demagnetized magnets were air-cooled and sorted from the steel plates.

2.2.2. Comminution

Comminution was done to liberate the NdFeB part of the demagnetized sample so that ground mass could be easily leached with the lixiviant in view of the increased surface area of the material. Using a disc pulverizer from Bico Inc. (Burbank, CA, USA), the samples were initially comminuted with a set of 3.1 mm. That product was further comminuted with a set of 0.3 mm. A sieve analysis was performed to determine the size distribution of the comminuted material and is discussed later.

2.3. Hydrometallurgical Processing

REEs were leached from the comminuted sample in a 2M H_2SO_4 acid solution with sample to solution ratio of 1g:10mL [4]. This process was done under a fume hood for 2 h with the acid solution being added in small amounts because of the aggressiveness of the reaction as shown in Equation (1).

$$Nd_2Fe_{14}B + 45H^+ + 3H_2O \rightarrow 2Nd^{3+} + 14Fe^{3+} + BO_3^{3-} + 25.5H_2 \quad (1)$$

Filtration was performed after leaching to separate the REE-acidic pregnant solution from the residue. NH_4OH was then added to the filtrate in a ratio of 1 mL NH_4OH to 20 mL rare earth rich pregnant solution to adjust the pH to 1.2, as shown in Equation (2).

$$NH_4OH + Nd^{3+} + 2SO_4^{2-} + H^+ + 2H_2O \rightarrow (NH_4)Nd(SO_4)_2(H_2O)_3 \quad (2)$$

Upon addition, the solution was stirred at 90 rpm to completely dissolve back into solution anything that formed when the NH_4OH was added. After that, the solution was allowed to sit for 12 h so the REE-rich precipitate could fully form and settle. Finally, filtration was performed so that the REE-rich residue could be collected and allowed to air dry. The dry REE-rich precipitate is then added into a mixture of $NH_4F{\cdot}HF$ and deionized water in a ratio of 1g:1.5g:10g and stirred for 45 min to enhance the formation of REF. The residue obtained after filtration was air dried and analyzed for REF content. The process flow sheet is shown in Figure 1. Various stages are also identified in the figure to facilitate further discussion later in this paper.

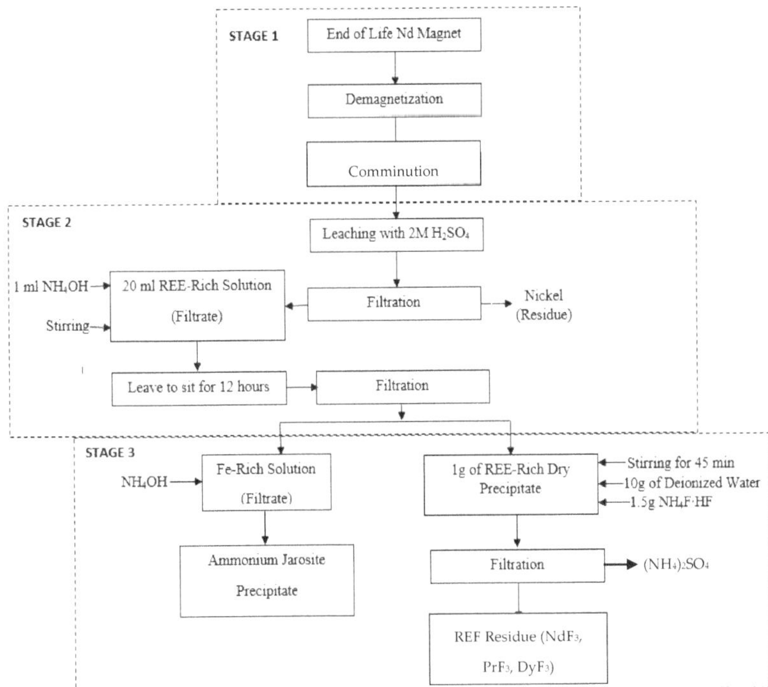

Figure 1. Flow sheet of Nd magnet recycling to produce REF.

Recovery of Fe in the form of ammonium jarosite, as shown in Equation (3), can also be done by the addition of more NH_4OH into the Fe-rich solution (as shown in Stage 3). With Equation (3) representing the reaction, the pH of the solution increased until a pH of 2 where maximum ammonium jarosite precipitated [17].

$$NH_4OH + 3Fe^{3+} + 2SO_4^{2-} + 5H_2O \rightarrow (NH_4)Fe_3(OH)_6(SO_4)_2 + 5H^+ \quad (3)$$

2.4. Modelling of REF Recovery

Stage 1 and Stage 2 of the flow sheet in Figure 1 were identified as less critical compared to Stage 3 of the process. Hence, it was decided to optimize Stage 3 in the present work. To optimize the REF recovery, Response Surface Methodology (RSM) was pragmatic in the analysis of the experiments. RSM is a mathematical and statistical technique that employs empirical models to fit the experimental data with reference to the Design of Experiments (DOE). With the process responses not following a linear model, the Box-Behnken design was employed for designing the experimental matrix to delineate the response surfaces generated by the condition variables [39–44]. This design selects points in the experimental domain for a three-level factorial arrangement in such a way that permits proficient approximation of the first and second order coefficients for the mathematical model [42]. The user identifies a high level and a low level of each condition variable and the mid-point is automatically identified for the point selection. Several experiments (usually 3 or 5) are conducted at the mid-point of all variables to estimate the inherent variability associated with the experimental technique.

In this work, the objectives were to maximize the amount of REF recovered from the precipitate along with their purity. This RSM was used at the point in the experiment where the REE-rich precipitate is added to $NH_4F{\cdot}HF$ to produce the REF in Stage 3. The experiments were performed as per the RSM design of experiments developed using the statistical software Design Expert 9 procured from Stat-Ease Inc., Minneapolis, MN, USA [43]. During the scoping tests, several condition variables were identified that affected the amount of REF recovered from the precipitate as well as its purity. To limit the number of experiments, the most important three condition variables were chosen, namely, the amount of deionized water (mL), amount of $NH_4F{\cdot}HF$ (g) and degree of stirring (min). Of course, the identified responses were the amount of REF recovered from the precipitate and the purity of the REF. The selected points in the experimental domain and the responses obtained are discussed later, along with the response surfaces, model equations, optimization and the interaction of the condition variables.

2.5. Material Characterization

After demagnification, the feed materials as well as intermediate and final products from this study were characterized using Scanning Electron Microscopy with Energy Dispersive X-ray (SEM-EDX), X-ray Diffractometry (XRD), Inductively Coupled Plasma–Optical Emission Spectrometry (ICP-OES), and a thermal analyzer with thermogravimetric (TG) and differential scanning calorimetry (DSC) capabilities.

2.5.1. SEM/EDX

The SEM-EDX analyzer was employed to determine the chemical compositions of all phases in the demagnetized material. The SEM-EDX system uses a TESCAN TIMA with a tungsten filament and an EDAX Z2 analyzer (TESCAN ORSAY HOLDING, a.s., Kohoutovice, Czech Republic). Cross-sectioned sides of a representative sample were hand-separated and cold-mounted in epoxy using molds approximately 25 mm in diameter and 10 mm in thickness. Resulting mounts were ground and polished to a smooth finish and then conductively coated with carbon to obtain SEM images by backscattered electron (BSE) detection. EDX analyses helped determine the chemical compositions of all products.

2.5.2. XRD

X-ray diffraction was carried out with a Rigaku Ultima IV X-ray Diffractometer (XRD) (RIGAKU AMERICAS CORPORATION, Woodland, TX, USA) using Cu-Kα radiation at 40 kV and 40 mA. This was used for quantitative analysis and also to determine the various phases of the precipitates and the products obtain after $NH_4 \cdot HF$ addition.

2.5.3. ICP-OES

A ratio of 3:1 *v*/*v* HCl to HNO_3 was used to digest a representative sample of the final product obtained at the end of each experiment and resulting solutions were analyzed by ICP-OES to determine the elemental content of the sample (Thermo Electron Duo View iCAP 6500, Waltham, MA, USA).

2.5.4. TGA-DSC

Thermal decomposition of the ammonium jarosite, obtained in the final stage, was carried out in a TA Instruments SDT Q650 simultaneous thermal analyzer with thermogravimetric (TG) and differential scanning calorimetry (DSC) capabilities (TA Instruments, New Castle, DE, USA). Analyses were performed using 10 mg sample with a heating rate of 10 °C/min to 400 °C under argon atmosphere. The furnace temperature was regulated precisely to provide a uniform rate of decomposition.

3. Results and Discussions

3.1. Granulometry

Regarding Stage 1 of Figure 1 and using sieves ranging in aperture size from 0.149 to 4.699 mm, the size distribution of the comminuted sample is shown in Figure 2. It can be seen from this figure that nearly 50% of the material passed through the 0.6 mm sieve size. The 80% passing size (d80) of the comminuted mass is about 1.7 mm.

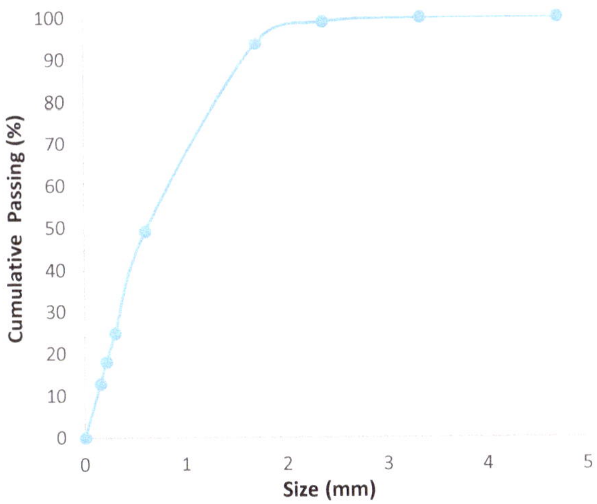

Figure 2. Cumulative particle size distribution of the demagnetized and comminuted sample.

3.2. SEM-EDX Analysis

Image from the Backscatter Electron (BSE) detector on the SEM-EDX demonstrate some dissimilar phases in the demagnetized and comminuted sample (see Stage 1 of Figure 1). "Three Dots" analysis was performed using the EDX. Of the three indicated points shown in Figure 3, each has a distinctive chemistry: point (a) appears to be Ni, which is the outer layer; point (b) is a combination of Pr, Nd and

Fe, where Pr is used to substitute Nd (up to 20–25%) to lower production costs; and point (c) shows a mixture of Nd and Fe. It is noted that boron (B) was not detected due to its low intensity, resulting from it being a light element (i.e., element #5 on the periodic table), as well as from the inherent disability of the BSE detector that was used [45].

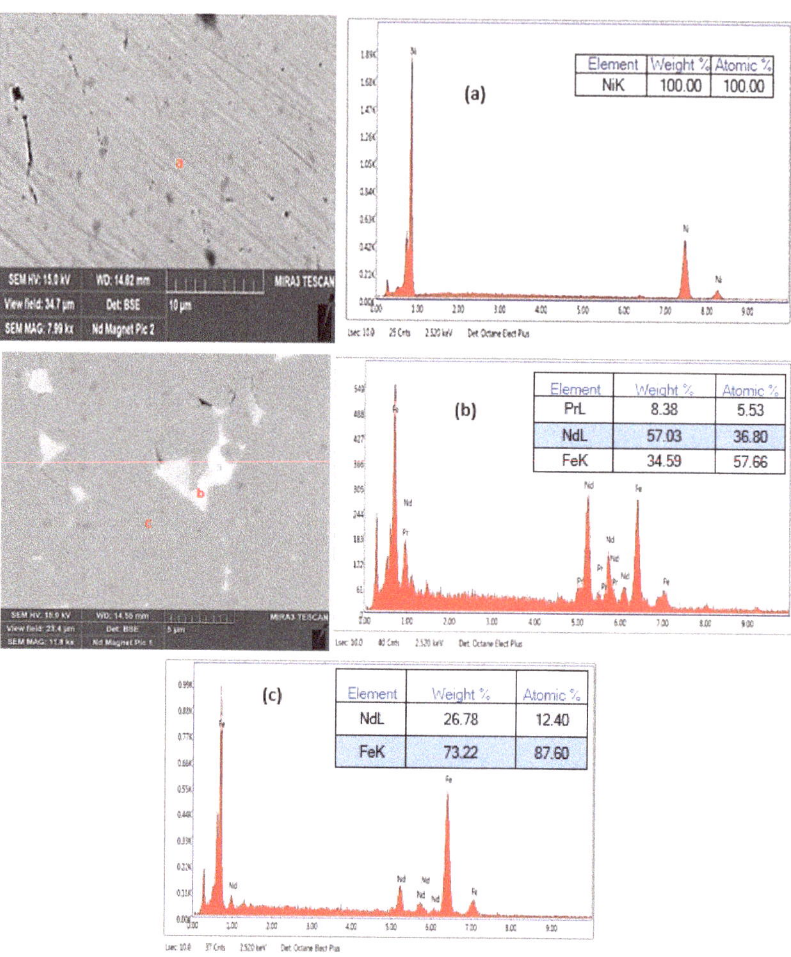

Figure 3. SEM image and EDX composition of the demagnetized and comminuted sample (see text for details). (**a**) Ni; (**b**) Pr, Nd, Fe; (**c**) Nd, Fe.

3.3. XRD Results

Figures 4 and 5 show the phase compositions of the precipitates (Stage 2 feed to Stage 3 rich, dry precipitate, as per Figure 1) and the final product (Stage 3 REF residue, as per Figure 1), respectively. From Figure 4, it can be said that most of the Fe remains in solution and that the dry precipitate is made up of mainly $(NH_4)Nd(SO_4)_2(H_2O)_3$. This compound is an ammonium neodymium sulfate double salt and is noted to be similar to ammonium jarosite [17] with Nd substituting for Fe. Figure 5 also shows the phases of the product made after adding and stirring the dry precipitate in $NH_4F{\cdot}HF$ for 45 min, which is then heated under argon atmosphere to 400 °C and allowed to cool. Equation (4) depicts a possible mechanism for REF formation with the REE being Nd assuming ammonium (NH_4^+) is in excess:

$$2(NH_4)Nd(SO_4)_2(H_2O)_3 + 3NH_4F{\cdot}HF + 3NH_4^+ \rightarrow 2NdF_3 + 4(NH_4)_2SO_4 + 6H_2O + 3H^+ \quad (4)$$

The product is clearly a REF containing REEs of Nd, Dy and Pr. It can therefore be concluded that REF can be produced by this process and the resulting fluorides can be used later as feedstock for pyrometallurgical metal production [39].

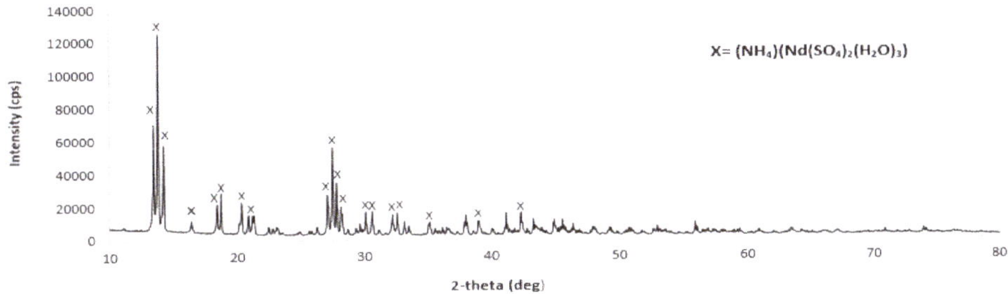

Figure 4. Crystal phases in the REE rich precipitate after adding NH_4OH.

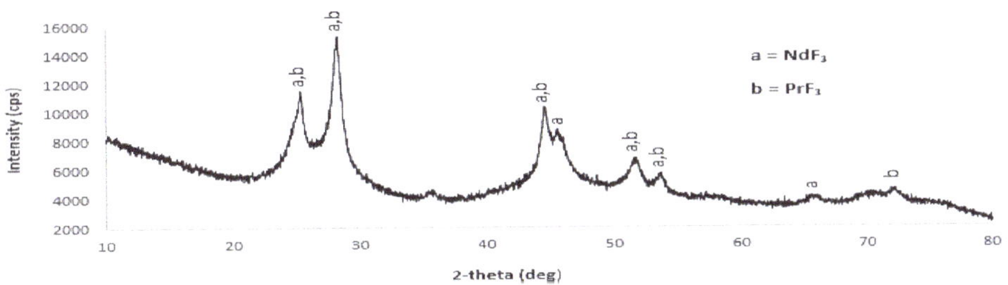

Figure 5. Crystal phases in the REF filtrate after 45 min in $NH_4F{\cdot}HF$ and heated to 400 °C.

3.4. ICP Results

Table 1 shows the chemical composition of REEs in the final REF product indicating Dy and Fe are present, apparently in amounts below the XRD detection limit. Together, the total REE amount sums to 63.25%. Based on stoichiometry and assuming the REEs exist as REF (i.e., NdF_3, PrF_3 and DyF_3), the F-content would be 24.94%. This leaves a balance of 10.59%, which is expected to be ammonium sulfate [$(NH_4)_2SO_4$] and ammonium jarosite [$(NH_4)Fe_3(SO_4)_2(OH)_6$], as per Equation (4). Furthermore, based on the stoichiometry of the jarosite, the Fe content would equate to 3.44% of these other components, suggesting there is 7.15% ammonium sulfate. Both ammonium sulfate and ammonium jarosite can be monohydrated [46] which will thermally decompose at low temperature. In this regard, a TGA-DSC study was undertaken on the final REF residue from Stage 3. Because the product was heated to 400 °C under argon, a subsequent TGA-DSC scan did not exceed 400 °C. It is important to note that this REF residue, as discussed above, is not pure, and therefore contains some ammonium sulfate and ammonium jarosite.

Table 1. The chemical composition of the final product.

Sample Composition	Dy	Fe	Nd	Pr	Others
Weight (%)	2.89	1.22	52.95	7.41	35.53

3.5. TGA-DSC Results

The TGA-DSC graphs of the REF residue are shown in Figure 6. The plot shows that mass loss occurs in five steps which agrees with the literature [46–49]. In this regard, the first two steps equate simply to crystalline water loss of the ammonium sulfate (Equation (5)) and ammonium jarosite (Equation (7)), respectively:

25–50 °C

$$(NH_4)_2SO_4{\cdot}H_2O \rightarrow (NH_4)_2SO_4 + H_2O \tag{5}$$

75–125 °C

$$(NH_4)Fe_3(SO_4)_2(OH)_6{\cdot}H_2O \rightarrow (NH_4)Fe_3(SO_4)_2(OH)_6 + H_2O \tag{6}$$

The third step corresponds to the decomposition of dehydrated ammonium sulfate to ammonium bisulfate:

140–210 °C

$$(NH_4)_2SO_4 \rightarrow NH_4HSO_4 + NH_3 \tag{7}$$

The fourth step is likely caused by complete thermal decomposition of the ammonium bisulfate:

220–250 °C

$$NH_4HSO_4 \rightarrow 1/3NH_3 + 1/3N_2 + SO_2 + 2H_2O \tag{8}$$

Finally, the fifth step appears to be the dehydroxylation of the dehydrated ammonium jarosite, thereby accounting for the continued loss in weight as the temperature increased to 380 °C:

>260 °C

$$(NH_4)Fe_3(SO_4)_2(OH)_6 \rightarrow (NH_4)(FeO)_3(SO_4)_2 + 3H_2O \tag{9}$$

If the temperature had been increased to 600 °C, three additional steps involving the sequential reactions of the dehydroxylated ammonium jarosite would be observed such that all of the Fe ultimately becomes hematite [46–49]:

>385 °C

$$2(NH_4)(FeO)_3(SO_4)_2 \rightarrow 2NH_3 + H_2O + 2Fe_3O_{2.5}(SO_4)_2 \tag{10}$$

>510 °C

$$2Fe_3O_{2.5}(SO_4)_2 \rightarrow 2Fe_2O_3 + Fe_2(SO_4)_3 + SO_3 \tag{11}$$

>540 °C

$$Fe_2(SO_4)_3 \rightarrow Fe_2O_3 + 3SO_3 \tag{12}$$

where $Fe_3O_{2.5}(SO_4)_2$ is essentially equivalent to a solid solution of 2/3 $Fe_2(SO_4)_3$ and 5/6 Fe_2O_3.

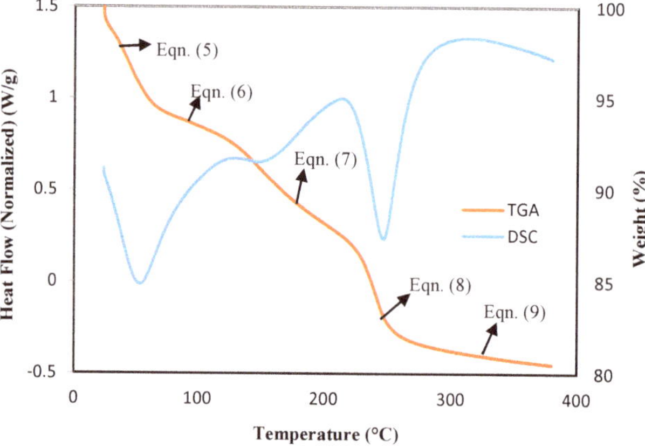

Figure 6. TGA-DSC curve for REF residue from Stage 3.

4. Model Development and Process Optimization Using RSM

Table 2 shows the Design of Experiments (DOE) along with the condition variables and responses for Stage 3 processing. In all, 17 experiments were performed, each using 2 g of REE-rich dry precipitate from Stage 2. The following ranges of the condition variables were employed: deionized water volume was 10–30 mL; 2–4 g of $NH_4F{\cdot}HF$ was added; the degree of stirring was 15–45 min. Responses were REE recovery and REF purity where recovery refers to the amount of REE from Stage 2 being converted to REF residue and purity refers to the quality of the REF residue based on ICP analysis.

Table 2. DOE conditions and their responses.

Standard	Run	Volume of Deionized Water (mL)	Amount of $NH_4{\cdot}HF$ (g)	Degree of Stirring (mins)	REE Recovery (%)	Purity of REF Residue (%)
2	1	30	2	30	46.76	60.59
8	2	30	3	45	55.58	63.26
10	3	20	4	15	53.15	54.01
7	4	10	3	45	63.03	46.12
4	5	30	4	30	49.63	58.38
17	6	20	3	30	47.16	56.93
3	7	10	4	30	64.09	49.66
1	8	10	2	30	50.44	45.52
12	9	20	4	45	55.41	51.84
16	10	20	3	30	49.67	53.08
5	11	10	3	15	57.48	51.19
11	12	20	2	45	54.65	48.59
15	13	20	3	30	50.09	54.94
14	14	20	3	30	52.74	51.39
6	15	30	3	15	52.13	50.85
9	16	20	2	15	53.56	50.65
13	17	20	3	30	51.38	57.42

Furthermore, five experiments were completed with all condition variables being at their midpoints. The results were analyzed to develop a statistically significant model for the responses of the REE recovery and REF purity as shown in Equations (13) and (14):

$$\textit{REE Recovery}\ (\%) = 50.14 - 3.87A + 0.087B + 0.84C - 2.70AB - 0.53AC + 2.68A^2 + 4.15C^2 + 4.04A^2B + 1.41A^2C \quad (13)$$

$$\textit{REF Purity}\ (\%) = 54.75 + 4.20A + 1.65B - 1.06C - 1.59AB + 4.37AC + 0.18\ A^2 - 1.40B^2 - 2.08C^2 - 1.17A^2B + 2.89A^2C \quad (14)$$

where A denotes the volume of deionized water used, B denotes the amount of $NH_4F{\cdot}HF$ added, and C denotes the degree of stirring. In both cases, a cubic model represented the data best and the R^2 values were 0.94 and 0.93 for recovery and purity, respectively.

Using Equations (13) and (14), 3-D surface plots in Figures 7 and 8 were generated to illustrate the effects of the condition variables on the responses. From Figure 7a,b, it can be said that $NH_4F{\cdot}HF$ addition appears to have more prominent effect on the process efficiency. The impact of deionized water addition is more prominent at higher stirring rates and higher $NH_4F{\cdot}HF$ additions. With respect to stirring, it can be observed from Figure 7a that the amount of REF recovery increases with stirring time for both 10 and 30 mL of water used. However, recovery is higher for lower volumes of water than for higher ones, and this same pattern is observed with respect to $NH_4F{\cdot}HF$ addition as illustrated in Figure 7b when using Equation (13).

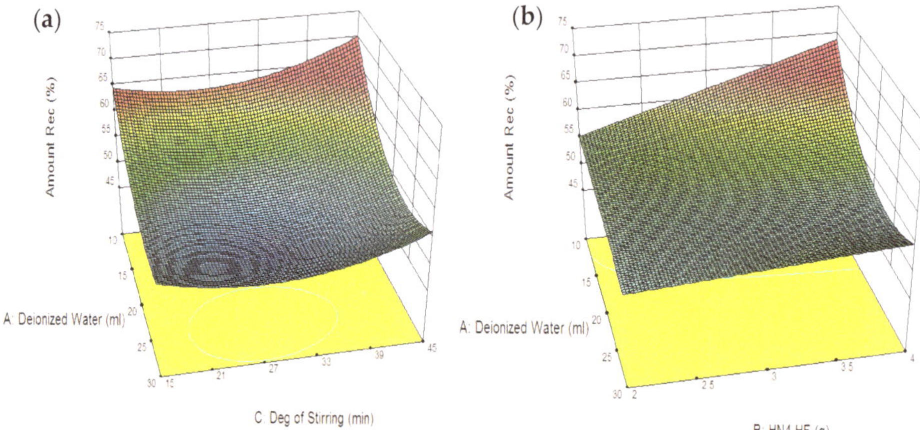

Figure 7. 3D plots of REF recovery with respect to process variables at (**a**) 4 g of $NH_4F{\cdot}HF$ and (**b**) 43 min for degree of stirring.

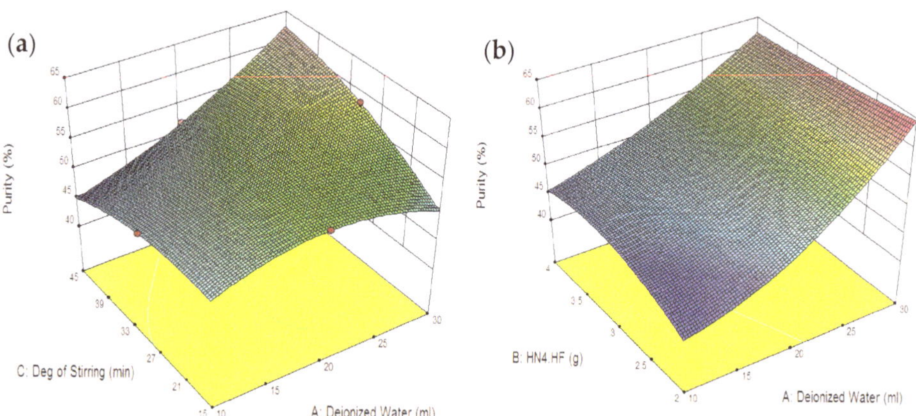

Figure 8. 3D plots of REF purity with respect to process variables at (**a**) 4 g of $NH_4F{\cdot}HF$ and (**b**) 44 min for degree of stirring.

Figure 8a,b were generated from Equation (14) and illustrates the response surfaces for REF purity in relations to the process variables. From these plots, it is observed that the amount of deionized water again did influence purity of REF recovery to a significant extent at all $NH_4F{\cdot}HF$ additions and also at higher stirring degrees. The degree of stirring, as observed in Figure 8a, seems to affect the REF insignificantly at low water additions. However, REF purity increased sharply with stirring time at a high volume of water addition. Figure 8b shows that $NH_4F{\cdot}HF$ addition has less impact on the purity of REF. Sharp increase in the purity of the REF is seen with increase in the volume of water.

It may also be noted that, in Equations (13) and (14), the process variables are all in terms of the coded values and differ in the range +1 to −1, with the mid-point having a value of zero. Thus, for B, the maximum amount of $NH_4F{\cdot}HF$ used (4 g) corresponds to +1 and the lowest (2 g) corresponds to −1, whereas the midpoint of 3 g corresponds to 0. To use these statistically significant equations, the actual variable value needs to be converted to the coded form lying between +1 and −1. Using the above model equations optimization was carried out to find the conditions that maximized both REF recovery and REF purity. The identified conditions and their maximum corresponding responses are as follows:

Amount of Deionized Water Used: 30 mL
$NH_4F{\cdot}HF$ Added: 4 g
Degree of Stirring: 45 min

REF Recovery: 56.23%
REF Purity: 62.42%

5. Conclusions

From this study, it is established REF can be produced from end-of-life Nd magnet from hard disk drive by hydrometallurgical techniques using $NH_4F{\cdot}HF$. The final product (REF) can also be used as part of the salt bath for molten salt electrolysis using fluoride salt [39]. This process is preferred over those reported in the literature because the use of hazardous HF was avoided. Employing statistical design of experiments (DOE), process parameters such as volume of deionized water, amount of $NH_4F{\cdot}HF$ and degree of stirring were defined in detail, and a conducive process regime for the recovery of the REFs was identified. It was established that at higher volumes of water, $NH_4F{\cdot}HF$ addition, and degree of stirring, optimum REF recovery with reasonably high purity can be achieved.

Author Contributions: Conceptualization, P.S., C.Y., and A.D.; methodology, P.S., T.F., and A.D.; software, P.S.; validation, P.S., A.D., and C.Y.; formal analysis, P.S., and T.F.; investigation, T.F.; resources, P.S.; data curation, P.S., and T.F.; writing—original draft preparation, P.S. and T.F.; writing—review and editing, P.S., A.D., and C.Y.; visualization, P.S., and T.F.; supervision, C.Y., and A.D..; project administration, C.Y.; funding acquisition, C.Y. All authors have read and agreed to the published version of the manuscript.

Funding: Research was sponsored by the Army Research Laboratory and was accomplished under Cooperative Agreement Number W911NF-15-2-0020.

Acknowledgments: The views and conclusions contained in this document are those of the authors and should not be interpreted as representing the official policies, either expressed or implied, of the Army Research Laboratory or the U.S. Government. The U.S. Government is authorized to reproduce and distribute reprints for Government purposes notwithstanding any copyright notation herein. The authors acknowledge Grant Wallace and Jan Chorney for their support.

Conflicts of Interest: The authors declare no conflict of interest.

References

1. Yang, Y.; Walton, A.; Sheridan, R.; Güth, K.; Gauß, R.; Gutfleisch, O.; Buchert, M.; Binnemans, K. REE recovery from end-of-life NdFeB permanent magnet scrap: A critical review. *J. Sustain. Metall.* **2017**, *3*, 122–149. [CrossRef]
2. Walters, A.; Lusty, P. *Rare Earth Elements*; British Geological Survey: Nottingham, UK, 2010.
3. Ueberschaar, M.; Rotter, V.S. Enabling the recycling of rare earth elements through product design and trend analyses of hard disk drives. *J. Mater. Cycles Waste Manag.* **2015**, *17*, 266–281. [CrossRef]
4. Önal, M.A.R.; Borra, C.R.; Guo, M.; Blanpain, B.; Van, G.T. Hydrometallurgical recycling of NdFeB magnets: Complete leaching, iron removal and electrolysis. *J. Rare Earths* **2017**, *35*, 574–584. [CrossRef]
5. Rademaker, J.H.; Kleijn, R.; Yang, Y. Recycling as a strategy against rare earth element criticality: A systemic evaluation of the potential yield of NdFeB magnet recycling. *Environ. Sci. Technol.* **2013**, *47*, 10129–10136. [CrossRef]
6. Zepf, V. *Rare Earth Elements: A New Approach to the Nexus of Supply, Demand and Use: Exemplified along the Use of Neodymium in Permanent Magnets*; Springer: Heidelberg, Germany, 2013.
7. Goldman, A. *Handbook of Modern Ferromagnetic Materials*; Kluwer Academic Publishers: Dordrecht, The Netherlands, 1999; p. 649.
8. Rotter, V.S.; Chancerel, P.; Ueberschaar, M. Recycling-oriented product characterization for electric and electronic equipment as a tool to enable recycling of critical metals. In *Rewas*; Springer: Berlin, Germany, 2013.
9. Nguyen, R.T.; Diaz, L.A.; Imholte, D.D.; Lister, T.E. Economic assessment for recycling critical metals from hard disk drives using a comprehensive recovery process. *JOM* **2017**, *69*, 1546–1552. [CrossRef]
10. Meyer, L.; Bras, B. Rare earth metal recycling, sustainable systems and technology. In Proceedings of the IEEE International Symposium, Dana Point, CA, USA, 16–18 May 2011.

11. Tanaka, M.; Oki, T.; Koyama, K.; Narita, H.; Oishi, T. Recycling of rare earths from scrap. In *Handbook on the Physics and Chemistry of Rare Earths*; Bunzli, J.C.G., Pecharsky, V.K., Eds.; Elsevier: Amsterdam, The Netherlands, 2013; Volume 43, pp. 159–211.
12. Anderson, C.D.; Anderson, C.G.; Taylor, P.R. A survey of recycled rare earths metallurgical processing. In Proceedings of the 51st Annual Conference of Metallurgists of CIM, Niagara Falls, ON, Canada, 30 Septembere–3 October 2012; pp. 411–422.
13. Graedel, T.E.; Allwood, J.; Birat, J.P.; Reck, B.K.; Sibley, S.F.; Sonnemann, G.; Buchert, M.; Hagelüken, C. *Recycling Rates of Metals—A Status Report*; United Nations Environment Programme: Nairobi, Kenya, 2011.
14. Reck, B.K.; Graedel, T.E. Challenges in metal recycling. *Science* **2012**, *337*, 690–695. [CrossRef]
15. Eggert, R.; Wadia, C.; Anderson, C.; Bauer, D.; Fields, F.; Meinert, L.; Taylor, P. Rare earths: Market disruption, innovation, and global supply chains. *Annu. Rev. Environ. Resour.* **2016**, *41*, 199–222. [CrossRef]
16. Goonan, T.G. *Rare Earth Elements—End Use and Recyclability*; US Geological Survey: Reston, VA, USA, 2011.
17. Lyman, J.W.; Palmer, G.R. *Recycling of Neodymium Iron Boron Magnet Scrap*; University of Michigan Library: Ann Arbor, MI, USA, 1993.
18. Weber, R.J.; Reisman, D.J. *Rare Earth Elements: A Review of Production, Processing, Recycling, and Associated Environmental Issues*; US EPA Region: Washington, DC, USA, 2012.
19. Majzlan, J.; Stevens, R.; Boerio-Goates, J.; Woodfield, B.F.; Navrotsky, A.; Burns, P.C.; Crawford, M.K.; Amos, T.G. Thermodynamic properties, low-temperature heat-capacity anomalies, and single-crystal X-ray refinement of hydronium jarosite,$(H_3O)Fe_3(SO_4)_2(OH)_6$. *Phys. Chem. Miner.* **2004**, *31*, 518–531. [CrossRef]
20. Alonso, M.; López-Delgado, A.; López, F.A. A kinetic study of the thermal decomposition of ammoniojarosite. *J. Mater. Sci.* **1998**, *33*, 5821–5825. [CrossRef]
21. Dutrizac, J.E. Converting jarosite residues into compact hematite products. *JOM* **1990**, *42*, 36–39. [CrossRef]
22. Thakur, N.V.; Jayawant, D.V.; Iyer, N.S.; Koppiker, K.S. Separation of neodymium from lighter rare earths using alkyl phosphonic acid, PC 88A. *Hydrometallurgy* **1993**, *34*, 99–108. [CrossRef]
23. Preston, J.S. *The Recovery of Rare Earth Oxides from A Phosphoric Acid Byproduct. Part 4. The Preparation of Magnet-Grade Neodymium Oxide from The Light Rare Earth Fraction*; Mintek: Randburg, South Africa, 1996.
24. Lu, D.; Horng, J.S.; Hoh, Y.C. The separation of neodymium by quaternary amine from didymium nitrate solution. *J. Less Common Met.* **1989**, *149*, 219–224. [CrossRef]
25. Morais, C.A.; Ciminelli, V.S.T. Process development for the recovery of high-grade lanthanum by solvent extraction. *Hydrometallurgy* **2004**, *73*, 237–244. [CrossRef]
26. Radhika, S.; Nagaphani, K.B.; Lakshmi, K.M.; Ramachandra, R.B. Solvent extraction and separation of rare earths from phosphoric acid solutions with TOPS 99. *Hydrometallurgy* **2001**, *110*, 50–55. [CrossRef]
27. Panda, N.; Devi, N.; Mishra, S. Solvent extraction of neodymium (III) from acidic nitrate medium using Cyanex 921 in kerosene. *J. Rare Earths* **2012**, *30*, 794–797. [CrossRef]
28. El-Hefny, N.E. Kinetics and mechanism of extraction and stripping of neodymium using a Lewis cell. *Chem. Eng. Process. Process. Intensif.* **2007**, *46*, 623–629. [CrossRef]
29. Lee, M.-S.; Lee, J.-Y.; Kim, J.-S.; Lee, G.-S. Solvent extraction of neodymium ions from hydrochloric acid solution using PC88A and saponified PC88A. *Sep. Purif. Technol.* **2006**, *46*, 72. [CrossRef]
30. Banda, R.; Jeon, H.; Lee, M. Solvent extraction separation of Pr and Nd from chloride solution containing La using Cyanex 272 and its mixture with other extractants. *Sep. Purif. Technol.* **2012**, *98*, 481–487. [CrossRef]
31. Guo, C.; Zhi, Y. Effective solvent extraction of La, Ce and Pr from hydrochloric acid with a novel extractant N, N-dihexyloxamic acid. *J. Chem. Technol. Biotechnol.* **2016**, *92*, 1596–1600. [CrossRef]
32. Chen, J.; Guo, L.; Deng, Y.; Li, D. Application of [A336][P507]/[P204] on High Selective Extraction and Separation of Rare Earths (III) from Mechanism to Techniques. In Proceedings of the 52nd Conference of Metallurgists (COM), Montreal, QC, Canada, 27–31 October 2013; pp. 367–374.
33. Rout, A.; Kotlarska, J.; Dehaen, W.; Binnemans, K. Liquid–liquid extraction of neodymium(iii) by dialkylphosphate ionic liquids from acidic medium: The importance of the ionic liquid cation. Electronic supplementary information (ESI) available: NMR spectra of the ionic liquids. *Phys. Chem. Chem. Phys.* **2013**, *15*, 16533–16541. [CrossRef]
34. Esmaeil, J.; Malek, S. The production of rare earth elements group via tributyl phosphate extraction and precipitation stripping using oxalic acid. *Arab. J. Chem.* **2016**, *9*, S1532–S1539.
35. Belova, V.V.; Voshkin, A.A.; Kholkin, A.I.; Payrtman, A.K. Solvent extraction of some lanthanides from chloride and nitrate solutions by binary extractants. *Hydrometallurgy* **2009**, *97*, 198–203. [CrossRef]

36. London, I.M.; Goode, J.R.; Moldoveanu, G.; Rayat, M.S. Rare Earth Elements Symposium. In Proceedings of the 52nd Conference of Metallurgists (COM), Montreal, QC, Canada, 27–31 October 2013.
37. Yan, C.; Liao, C.; Jia, J.; Wang, M.; Li, B. Comparison of economical and technical indices on rare earth separation processes of bastnasite by solvent extraction. *J. Rare Earths* **1999**, *17*, 58–63.
38. Krishnamurthy, N.; Gupta, C.K. *Extractive Metallurgy of Rare Earths*; CRC Press: Boca Raton, FL, USA, 2015; p. 869.
39. Sarfo, P. Recovery of Rare Eearth Elements by Advanced Processing Technologies. Ph.D. Thesis, Montana Tech, Butte, MT, USA, October 2019.
40. Sarfo, P.; Wyss, G.; Ma, G.; Das, A.; Young, C. Carbothermal reduction of copper smelter slag for recycling into pig iron and glass. *Miner. Eng.* **2017**, *107*, 8–19. [CrossRef]
41. Sarfo, P.; Das, A.; Wyss, G.; Young, C. Recovery of metal values from copper slag and reuse of residual secondary slag. *Waste Manag.* **2017**, *70*, 272–281. [CrossRef]
42. Box, G.E.P.; Behnken, D.W. Some new three level designs for the study of quantitative variables. *Technometrics* **1960**, *2*, 455–475. [CrossRef]
43. Bezerra, M.A.; Santelli, R.E.; Oliveira, E.P.; Villar, L.S.; Escaleira, L.A. Response surface methodology (RSM) as a tool for optimization in analytical chemistry. *Talanta* **2008**, *76*, 965–977. [CrossRef]
44. Stat-Ease, Inc. Stat-Ease® DOE Software. 2003. Available online: https://www.hearne.software/Resources/Design-Ease/softoverview.aspx (accessed on 10 October 2018).
45. Ingemarsson, L.; Halvarsson, M. *Sem/edx Analysis of Boron*; High Temperature Corrosion Centre-Chalmer University of Technology: Gothenburg, Sweden, 2011.
46. Das, G.K.; Anand, S.; Acharya, S.; Das, R.P. Preparation and decomposition of ammoniojarosite at elevated temperatures in H_2O–$(NH_4)_2SO_4$–H_2SO_4 media. *Hydrometallurgy* **1995**, *38*, 263–276. [CrossRef]
47. Frost, R.L.; Wills, R.-A.; Kloprogge, J.T.; Martens, W. Thermal decomposition of ammonium jarosite $(NH_4)Fe_3(SO_4)_2(OH)_6$. *J. Therm. Anal. Calorim.* **2006**, *84*, 489–496. [CrossRef]
48. Kunda, W.; Veltman, H. Decomposition of jarosite. *Metall. Trans. B* **1979**, *10*, 439–446. [CrossRef]
49. Spratt, H.; Rintoul, L.; Avdeev, M.; Martens, W. The thermal decomposition of hydronium jarosite and ammoniojarosite. *J. Therm. Anal. Calorim.* **2014**, *15*, 101–109. [CrossRef]

MDPI
St. Alban-Anlage 66
4052 Basel
Switzerland
Tel. +41 61 683 77 34
Fax +41 61 302 89 18
www.mdpi.com

Minerals Editorial Office
E-mail: minerals@mdpi.com
www.mdpi.com/journal/minerals

www.ingramcontent.com/pod-product-compliance
Lightning Source LLC
LaVergne TN
LVHW071000170726
843515LV00006B/1029

* 9 7 8 3 0 3 6 5 2 5 6 6 2 *